BRITISH ZOOLOGY.

CLASS III. REPTILES

IV. FISH.

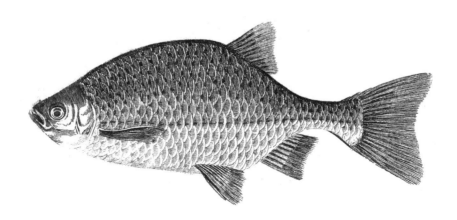

LONDON.
Printed for Benj. White,
MDCCLXXVI.

PLATES

TO

BRITISH ZOOLOGY.

VOL. III. OCTAVO.

VOL. III.   a   XVII.

PLATES.

PLATES.

PLATES.

C L A S S III.

R E P T I L E S.

All the works of the LORD are good, and he will give every needful thing in due feafon.

So that a man cannot fay this is worfe than that, for in time they fhall all be well approved.

ECCLESIASTICUS XXXIX. 33, 34,

VOL. III.　　　　　　　B

R E P T I L E S.

WE are now to confider the clafs of Reptiles, which are, for the moft part, objects of deteftation; but however the opinion of the world may be, if a writer undertakes a general hiftory of animals, he muft include them: they form at left one link in the chain of beings, and may therefore be viewed with a degree of pleafure by a philofophic eye.

But notwithftanding the prejudice againft this clafs is almoft univerfal, is it founded on reafon? In fome it may be owned that the outward form is difagreeable, while the noxious qualities of others are juftly productive of terror: but are we on that account to reject them? The more fatal they are, the more deeply we fhould enquire into their effects, that we may be capable of relieving thofe who are fufferers, and fecure others from the fame misfortune. But if we duly weigh their noxious qualities, we fhall, with our moral poet, find

" *All partial evil univerfal good.*"

B 2   The

The teeth of wild beafts, and of ferpents, are not only created as inftruments of vengeance, but are falutary in leffening the numbers of thofe animals which are highly ufeful in the degree, and only hurtful in their excefs; but if their bad qualities are ferviceable, we are more indebted to their good ones than we chufe to acknowlege.

But many of the animals that form this clafs are of immediate benefit to mankind. The Turtle, or Sea-Tortoife, fupplies the torrid zone with a wholefome and delicious food, as the epicures of our own country can atteft. Frogs are a food in feveral parts, as Lizards and Serpents are in others.

The medicinal virtues of the Viper are partly exploded by the moderns, but time, the overthrower of fyftems, as well as empires, may reftore it to the rank it held with the antients. The *Lacerta Scincus* is, however, yet efteemed in the *Eaft* for its falubrious qualities, and even Toads have contributed to the eafe of patients in the moft inveterate of all difeafes.

Had I followed *Linnæus*, and included the Cartilaginous Fifh in this clafs, there would have been ample room for panegyric, for it is very doubtful whether any are pernicious; but the ufe of many, either as food or for mechanical purpofes, were never queftioned.

But if the external figure of the reptile tribe is difgufting, they have one general beauty, an apt configuration of parts for their way of life, nor are

are they deftitute of their peculiar graces: the fine
difpofition of plates in the fhell of the Tortoife,
with the elegant fymmetry of their colors, muft
ftrike even common obfervers, while the eye of
the defpifed Toad has a luftre denied to more
pleafing forms. The frolicfome agility of Lizards
enlivens the dried banks in hot climates; and the
great affection which fome of them fhew to mankind, fhould farther engage our regard and
attention.

The wreathing of the fnake, with the vivid die
of its fkin, are certainly graceful, tho' from the
dread of fome particular fpecies which are venemous, we have acquired an antipathy for the whole.
The antients, who confidered the Serpent as an
emblem of health, could affociate pleafing ideas
with this animal. We therefore find it an orna-
ment at every entertainment, and in every fcene of
mirth, both in painting and in fculpture. *Virgil*
adopted this notion, and has accordingly defcribed
it with every beauty both of form and color,

Adytis cum lubricus anguis ab imis
Septem ingens gyros, feptena volumina traxit;
Amplexus placidé tumulum, lapfufque per aras:
Cærulæa cui terga notæ, maculofus et auro
Squamam incendebat fulgor; ceu nubibus arcus
Mille trahit varios adverfo fole colores.

V. 84.

B 3   From

From the deep tomb, with many a ſhining fold,
An azure ſerpent roſe, in ſcales that flam'd with gold :
Like heaven's bright bow his varying beauties ſhone
That draws a thouſand colors from the ſun :
Pleas'd round the altars and the tomb to wind,
His glittering length of volumes trails behind.

PITT.

But if after all ſome lively writer ſhould purſue the Naturaliſts with more wit than argument, and more humor than good-nature, it ſhould be endured with patience. Ridicule is, however, not the teſt of truth, tho' when joined to ſatyr, it ſeldom fails of ſeducing the many who would rather laugh than think. Should this prove the caſe in the preſent inſtance, let the author be allowed to ſkreen himſelf from cenſure, by ſaying he writes not to the many, but the few ; to thoſe alone who can examine the parts with a view to the *whole*, and who ſcorn to deſpiſe even the moſt deformed, or the moſt minute work of an all-wiſe CREATOR.

G E N E R A.

I.   T O R T O I S E.
II.   F R O G.
III.   L I Z A R D.
IV.   S E R P E N T.

Pl I.

CORIACEOUS TORTOISE.

C L A S S III.

R E P T I L E S.

Body covered either with a fhell or ftrong hide, I. TORTOISE. divided by futures; four fin-like feet; a fhort tail.

Teftudo coriacea five Mercurii. *Rondel.* 450? *Gefner pifc.* 946 ? Teftudo coriacea ? Teftudo pedibus pinniformibus mu-	ticis, tefta coriacea, cauda angulis feptem exaratis. *Lin. fyft.* 350. Turtle. *Borlafe Cornwall,* 285. *Plate* 27.

I. CORIACEOUS.

THIS fpecies is common to the *Mediterranean*, and to our fouthern feas, and is not, as far as we know, difcovered in any other.

Two were taken on the coaft of *Cornwall* in the mackrel nets, of a vaft fize, a little after *Midfummer* 1756; the largeft weighed eight hundred pounds, the leffer near feven hundred. A third, of equal weight with the firft, was caught on the

B 4   coaft

-15-

coaſt of *Dorſetſhire*, and depoſited in the *Leverian Muſeum*.

The length of the body is four feet ten inches; of the head nine inches and a half; of the neck three; or of the whole five feet twelve. The upper jaw bifurcated at the end: the extremity of the lower ſharp, claſping into the fork of the upper. The noſtrils ſmall and round.

The breadth of the body in the largeſt part is three feet. The length of the fore fins two feet ſeven: of the hind thirteen inches and a half: are ſmooth, grow pointed to the extremity, and are deſtitute of toes. Theſe fins are ſtuffed: perhaps the bones might have been taken out; for in the figure given by *Rondeletius*, which agrees in all other reſpects with this ſpecies, there is appearance of toes, and even nails.

The body is covered with a ſtrong hide, exactly reſembling black leather, deſtitute of ſcales, but marked with the appearance. The back is divided into five longitudinal flutings or grooves, with as many ſharp but ſmooth riſings.

This ſpecies is ſaid to be extremely fat: but the fleſh coarſe and bad *, according to the report made by writers who had opportunity of taſting them in the *Mediterranean* ſea. I am informed that the *Carthuſians* will eat no other than this ſpecies.

---

* *Rondeletius. Boſſuet.*

Body

Body naked.
Four legs, the feet divided into toes.
No tail.

Βατραχ☉. *Arift. Hift. an.* *Lib.* IV. c. 9.

La Grenoille. *Belon poiffons,* 48.

Rana fluviorum. *Rondel.* 217,

Rana aquatica innoxia. *Gefner quad. ovip.* 46. *Aquatil* 805.

Rana aquatica. *Raii fyn. quad.* 447.

Waffer Frofche. *Meyer an.* I. Tab. 52.

Rana temporaria. R. dorfo planiufculo fubangulato. *Lin. fyft.* 357.

Groda, Fro, Klaffa. *Faun. Suec.* No. 102.

Rana. *Gronov. Zooph.* No. 62.

2. COMMON.

SO common and well-known an animal requires no defcription; but fome of its properties are fo fingular, that we cannot pafs them unnoticed.

Its fpring or power of taking large leaps is remarkably great, and it is the beft fwimmer of all four-footed animals. Nature hath finely adapted its parts for thofe ends, the fore members of the body being very lightly made, the hind legs and thighs very long, and furnifhed with very ftrong mufcles.

While in a tadpole ftate, it is entirely a water animal; the work of generation is performed in that element, as may be feen in every pond during

fpring;

ſpring; when the female remains oppreſſed by the male for a number of days.

GENERA-
TION.

The work of propagation is extremely ſingular, it being certain that the frog has not a *penis intrans*; there appears a ſtrong analogy in this caſe between a certain claſs of the vegetable kingdom and thoſe animals; for it is well known, that when the female frog depoſits its ſpawn, the male inſtantaneouſly impregnates it with what we may call a *farina fœcundans,* in the ſame manner as the male *Palm* tree conveys fructification to the flowers of the female, which would otherwiſe be barren *.

As ſoon as the frogs are releaſed from their tadpole ſtate, they immediately take to land; and if the weather has been hot, and there fall any refreſhing ſhowers, you may ſee the ground for a conſiderable ſpace perfectly blackened by myriads of theſe animalcules, ſeeking for ſome ſecure lurking places. Some philoſophers † not giving themſelves time to examine into this phænomenon, imagined them to have been generated in the clouds, and ſhowered on the earth; but had they, like our *Derham* ‡, but traced them to the next pool, they would have found a better ſolution of the difficulty.

As frogs adhere cloſely to the backs of their own ſpecies, ſo we know they will do the ſame by fiſh: *Walton* § mentions a ſtrange ſtory of their deſtroy-

* *Shaw's Travels,* 224. *Haſſelquiſt Trav. Engl. Ed.* 416.
† *Rondeletius,* 216. *Wormii Muſ.* 327.
‡ *Ray's Wiſdom Creat.* 316.   § *Complete Angler,* 161.

ing

ing pike; but that they will injure, if not entirely kill carp, is a fact indisputable, from the following relation: a very few years ago, on fishing a pond belonging to Mr. *Pit*, of *Encomb*, *Dorsetshire*, great numbers of the carp were found each with a frog mounted on it, the hind legs clinging to the back, the fore legs fixed in the corner of each eye of the fish, which were thin and greatly wasted, teized by carrying so disagreeable a load. These frogs we imagine to have been males disappointed of a mate.

The croaking of frogs is well known, and from that in fenny countries they are distinguished by ludicrous titles, thus they are stiled *Dutch Nightingales* and *Boston Waites*; even the *Stygian* frogs have not escaped notice, for *Aristophanes* hath gone farther, and formed a chorus of them.

Βρεκεκεξ, κοάξ, κοάξ,
Βρεκεκεξ, κοαξ, κοαξ,
Λιμναῖα κρηνῶν τεκνα *.

Brekekex, coax, coax,
Brekekex, coax, coax,
The offspring of the pools and fountains.

Yet there is a time of year when they become mute, neither croaking nor opening their mouths for a whole month: this happens in the hot season, and that is in many places known

<small>PERIODICAL SILENCE.</small>

---

* *Comedy of the Frogs.*

to

to the country people by the name of the *Paddock Moon*. I am informed that for that period, their mouths are fo clofed, that no force (without killing the animal) will be capable of opening them.

Morton * endeavours to find a reafon for their filence, but tho' his facts are true, he is unfortunate in his philofophy. Frogs are certainly endued (as he well obferved) with a power of living a good while under water without refpiration, which is owing to their lungs being compofed of a feries of bladders: but he miftakes the nature of air, when he affirms that they receive a quantity of cool air, and dare not open their mouths for a month, from a dread of admitting a warmer into their lungs. It is hardly neceffary to fay, that in whatever ftate the air was received, it would affimilate itfelf to the external atmofphere in a fhort time. We muft leave the fact to be accounted for by farther experiments. But from what we do know, we may partly vindicate *Theophraftus*, and other antients, about the filence of the frogs at *Seriphus*. That philofopher affirms it, but afcribes it to the coldnefs of the waters in that ifland: Now when Monfieur *Tournefort* was there, the waters were lukewarm, and the frogs had recovered their voices †. Is it not probable that *Theophraftus* might be at *Seriphus* at that feafon when the frogs were mute, and having never obferved it elfewhere,

* *Hift. Northampt.* 441.
† *Tournefort's voy.* I. 142.

might

might conclude their filence to be general as to the time, but particular as to the place. *Ælian**, who quotes *Theophraftus* for the laft paffage, afcribes the fame filence to the frogs of the lake *Pierus* in *Theffaly*, and about *Cyrene* in *Africa*: but he is fo uncertain a writer, that we cannot affirm whether the fpecies of the *African* frogs is the fame with ours.

Thefe, as well as other reptiles, feed but a fmall fpace of the year. The food of this genus is flies, infects, and fnails. Toads are faid to feed alfo on bees, and to do great injury to thofe ufeful infects.

FOOD.

During winter frogs and toads remain in a torpid ftate : the laft of which will dig into the earth, and cover themfelves with almoft the fame agility as the mole.

Rana gibbofa. *Gefner pifc.* 809.
Rana efculenta. *Lin. fyft.* 357. *Faun. Suec.* No. 279.

R. corpore angulato, dorfo tranfverfe gibbo, abdomine marginato. *Ibid.*

3. EDIBLE.

THIS differs from the former in having a high protuberance in the middle of the back, forming a very fharp angle. Its colors are alfo more vivid, and its marks more diftinct; the ground

* *Ælian, Lib.* III. *ch.* 35, 37.

COLOR

color being a pale or yellowiſh green, marked with rows of black ſpots from the head to the rump.

This and, we think, the former, are eaten. We have ſeen in the markets at *Paris* whole hampers full, which the venders were preparing for the table, by ſkinning and cutting off the foreparts, the loins and legs only being kept. Our ſtrong diſlike to theſe reptils, prevented a cloſe examination into the ſpecies.

4. Toad.

Φϱυνῶ. *Ariſt. Hiſt. an. lib.* ix. c. I. 40.
Bufo. *Virg. Georg.* I. 184.
Rubeta. *Plin. lib.* VIII. c. 31.
Rubeta ſc. Phrynum. *Geſner piſc.* 807. *Rondel,* 222.
Bufo five Rubeta. *Raii ſyn. quad.* 252.

Bufo rubetarum. *Klein quad.* 122.
Rana Bufo. R. corpore ventricoſo, verrucoſo lurido fuſcoque. *Lin. ſyſt.* 354.
Padda, Taſſa. *Faun. ſuec.* No. 275.
Gronov. Zooph. No. 64.

THE moſt deformed and hideous of all animals; the body broad, the back flat, and covered with a pimply duſky hide; the belly large, ſwagging, and ſwelling out; the legs ſhort, and its pace labored and crawling: its retreat gloomy and filthy: in ſhort, its general appearance is ſuch as to ſtrike one with diſguſt and horror; yet we have been told by thoſe who have reſolution to view it with attention, that its eyes are fine: to this it

feems

seems that *Shakespear* alludes, when he makes his *Juliet* remark,

> Some say the lark and loathed toad change eyes.

As if they would have been better bestowed on so charming a songster than on this raucous reptile.

But the hideous appearance of the toad is such as to make this one advantageous feature overlooked, and to have rendered it in all ages an object of horor, and the origin of most tremendous inventions. *Ælian* * makes its venom so potent, that *Basilisk*-like, it conveyed death by its very look and breath; but *Juvenal* is content with making the *Roman* ladies, who were weary of their husbands, form a potion from its intrails †, in order to get rid of the good man.

> Occurrit Matrona potens, quæ molle Calenum
> Porrectura viro miscet sitiente rubetam.   *Sat.* I. 68.

> To quench the husband's parching thirst, is brought
> By the great Dame, a most deceitful draught;
> In rich *Calenian* wine she does infuse,
> (To ease his pains) the toad's envenom'd juice.

This opinion begat others of a more dreadful nature; for in after-times superstition gave it preternatural powers, and made it a principal ingredient in the incantations of nocturnal hags:

* *Hist. an. lib.* ix. *c.* 11.
† *Sat.* vi. 658. Vide *Ælian Hist. an. lib.* xvii. *c.* 12. and 15.

Toad.

Toad that under the cold ſtone,
Days and nights has, thirty-one,
Swelter'd venom ſleeping got,
Boil thou, *firſt* i'th' charmed pot.

We know by the poet that this charm was intended for a deſign of the firſt conſideration, that of raiſing the dead from their repoſe, and bringing before the eyes of *Macbeth* a hateful ſecond-ſight of the proſperity of *Banquo*'s line.

This ſhews the mighty powers attributed to this animal by the dealers in the magic art; but the powers our poet indues it with, are far ſuperior to thoſe that *Geſner* aſcribes to it: *Shakeſpear*'s witches uſed it to diſturb the dead; *Geſner*'s, only to ſtill the living, *Ut vim coeundi ni fallor, in viris tollerent* *.

TOAD-
STONE.

We may add here another ſuperſtition in reſpect to this animal: it was believed by ſome old writers to have a ſtone in its head, fraught with great virtues medical and magical: it was diſtinguiſhed by the name of the Reptile, and called the *Toad-Stone, Bufonites, Crapaudine, Krottenſtein* †; but all its fancied powers vaniſhed on the diſcovery of its being nothing but the foſſil tooth of the *ſea-wolf*, or of ſome flat-toothed fiſh, not unfrequent in our iſland, as well as ſeveral other countries; but we may well excuſe this tale, ſince *Shakeſpear* has extracted from it a ſimile of uncommon beauty:

* *Hiſt. quad. ovip.* 72.
† *Boet. de Boot. de Lap. et Gem.* 301. 303.

Sweet

Sweet are the ufes of adverfity,
Which, like the toad, ugly and venomous,
Wears yet a precious jewel in his head.

But thefe fables have been long exploded : we fhall now return to the notion of its being a poifonous animal, and deliver, as our opinion, that its exceffive deformity, joined to the faculty it has of emitting a juice from its pimples, and a dufky liquid from its hind parts, is the foundation of the report.

That it has any noxious qualities we have been unable to bring proofs in the fmalleft degree fatisfactory, though we have heard many ftrange relations on that point.

On the contrary, we know feveral of our friends who have taken them in their naked hands, and held them long without receiving the left injury : It is alfo well known that quacks have eaten them, and have befides fqueezed their juices into a glafs, and drank them with impunity. NOT POISONOUS.

We may fay alfo, that thefe reptiles are a common food to many animals ; to *buzzards, owls, Norfolk plovers, ducks,* and *fnakes,* who would not touch them were they in any degree noxious.

So far from having venomous qualities, they have of late been confidered as if they had beneficent ones. We wifh, for the benefit of mankind, that we could make a favorable report of the many attempts of late to cure the moft terrible of difeafes

VOL. III.   C   the

the *cancer*, by the application of live toads; but, alas, they feem only to have rendered a horrible complaint more loathfome. My enquiries on this fubject, and fome further particulars relating to the hiftory of this animal, may be found in the Appendix.

In a word, we may confider the toad as an animal that has neither good nor harm in it; that being a defencelefs creature, nature had furnifhed it, inftead of arms, with a moft difgufting deformity, that ftrikes into almoft every being capable of annoying it, a ftrong repugnancy to meddle with fo hideous and threatening an appearance.

GENERA-
TION.

The time of their propagation is very early in the fpring: at that feafon the females are feen crawling about oppreffed by the males, who continue on them for fome hours, and adhere fo faft as to tear the very fkin from the parts they ftick to. They fpawn like frogs; but what is fingular, the male affords the female obftetrical aid, in a manner that will be defcribed in the Appendix.

To conclude this account with the marvellous, this animal is faid to have often been found in the midft of folid rocks, and even in the centre of growing trees, imprifoned in a fmall hollow, to which there was not the leaft adit or entrance*: how the animal breathed, or how it fubfifted (fuppofing the poffibility of its confinement) is paft

* *Plot's Hift. Staff.* 247.

our

our comprehenſion. *Plot*'s* ſolution of this phænomenon is far from ſatisfactory; yet as we have
the great *Bacon*'s † authority for the fact, we do
not entirely deny our aſſent to it.

Rana Rubeta? *Lin. ſyſt.* 355.   obtuſo   ſubtus   punctato. 5. NATTER
  Faun. Suec. No. 101.   *Ibid.*   JACK.
R. corpore verrucoſo, ano

THIS ſpecies frequents dry and ſandy places:
    it is found on *Putney Common*, and alſo
near *Reveſby Abby, Lincolnſhire*, where it is called
the *Natter Jack*. It never leaps, neither does it
crawl with the ſlow pace of a toad, but its motion
is liker to running. Several are found commonly
together, and, like others of the genus, they appear in the evenings.

The upper part of the body is of a dirty yellow, clouded with brown, and covered with po-
rous pimples, of unequal ſizes: on the back is a
yellow line.

The upper ſide of the body is of a paler hue,
marked with black ſpots, which are rather rough.

On the fore feet are four divided toes; on the
hind five, a little webbed.

The length of the body is two inches and a quar-

* P. 249.
† *Nat. Hiſt. Cent.* vi. *Exp.* 570.
        C 2   ter;

ter; the breadth, one and a quarter : the length of the fore legs one inch one-fixth; of the hind legs, two inches.

We are indebted to *Jofeph Banks*, Efq; for this account.

6. GREAT.   INHABITS the woods near *Loch Ranfa*, in the *Ifle of Arran*.

Is double the fize of the common frog : the body fquare : belly great : legs fhort : four toes on the fore-feet, four and a thumb to the hind ; the fecond outmoft toe the longeft. The color above, is a dirty olive, marked with great warty fpots; the head alone plain. The color beneath whitifh.

It leaped flowly.

Slender

Pl. II.

BROWN LIZARD.

SCALY LIZARD.

P. Mazell Del. & Sculp.

Nº 7.

Nº 9.

Slender naked body : four legs :
Divided toes on each :
Very long tail.

Lacertus terreftris lutea fqua-
　mofa anglica.　*Raii fyn.*
　quad. 264.
Plott's Hiſt. Staff. 252. *tab.*
　22.
Lacerta agilis ? L. cauda ver-
　ticillata longiufcula fquamis
　acutis, collari fubtus fquamis
　conftructo. *Lin. fyft.* 363.

Odla, Fyrfot.　*Faun. Suec.*
　No. 284.
Lacerta, *Gronov. Zooph.* No.
　60.
Little Brown Lizard. *Edw.*
　225.
Padzher pou.　*Borlafe Corn-*
　wall, 284. *tab.* 28.

7. SCALY.

THOSE we have feen differ in color, but
agree in all other refpects with the fpecies
defcribed by Doctor *Plot.*　Their length from the
nofe to the hind-legs was three inches; from
thence to the end of the tail three and three quarters.

Along the back was a black lift; each fide of
that a brown one : then fucceeded a narrow ftripe,
fpotted alternately yellow and brown ; beneath that
a broad black one; thofe ended a little beyond
the hind-legs.　The belly was yellow, and the
fcales large but even. The fcales on the back fmall ;
on the tail the ends projected : thofe on the latter
were varied with black and brown.

C 3　　　　　　　　The

The legs and feet were dufky; on each foot were five toes, furnifhed with claws.

This fpecies is extremely nimble: in hot weather it bafks on the fides of dry banks, or of old trees; but on being obferved immediately retreats to its hole.

The food of this fpecies, as of all the other *Englifh* lizards, is infects: they themfelves of birds of prey. Each of our lizards are perfectly harmlefs; yet their form is what ftrikes one with difguft, and has occafioned great obfcurity in their hiftory.

OTHER SPECIES.

Related to this fpecies is the *Guernfey* lizard, which we are informed has been propagated in *England* from fome originally brought from that ifland. We have alfo heard of a green lizard frequent near *Farnham*, which probably may be of that kind: but the moft uncommon fpecies we ever met with any account of, is that which was killed near *Wofcot*, in the parifh of *Swinford*, *Worcefterfhire*, in 1741, which was two feet fix inches long, and four inches in girth. The fore-legs were placed eight inches from the head; the hind legs five inches beyond thofe: the legs two inches long: the feet divided into four toes, each furnifhed with a fharp claw. Another was killed at *Penbury*, in the fame county. Whether thefe are not of exotic defcent, and whether the breed continues, is what we are at prefent uninformed of.

Lacertus

Pl. III.

II.

I WARTY LIZARD.

II.

Lacertus aquaticus. *Gefner*   *Lin. fyft.* 370.   8. WARTY.
    quad. ovip. 31.   Skrot-abborre, Gruffgrabbe.
Salamandra aquatica. *Raii*   *Faun. Suec.* No. 281.
    fyn. quad. 273.   Lacerta Americana. *Seb. Muf.*
Lacerta paluftris. L. cauda   I. *tab.* 89. *fig.* 4, 5
    lanceolata mediocri, pedibus Salamandra alepidota verruco-
    muticis palmis tetradactylis.   fa. *Gronov. Zooph.* No. 47.

THE length of this fpecies was fix inches and
    an half, of which the tail was three and a
quarter.

The irides yellow : the head and beginning of
the back flat, and covered with fmall pimples or
warts, of a dark dufky color ; the fides with white
ones : the belly, and the fide of the tail, was of
a bright yellow ; the firft fpotted with black.

The tail was compreffed fideways, and very thin
towards the upper edge, and flender towards the
end.

The fore-feet divided into four toes ; the hind
into five ; all without nails, dufky fpotted with
yellow.

Its pace is flow and crawling.

This fpecies we have frequently feen in the
ftate we defcribe, but are uncertain whether we
ever met with it under the form of a *larve*. We
have more than once found under ftones and old
logs, fome very minute young lizards that had
much the appearance of this kind : they were

C 4   perfectly

perfectly formed, and had not the least vestiges of fins; so that circumstance, joined to their being found in a dry place, remote from water, makes us imagine them to have never been inhabitants of that element, as it is certain many of our lizards are in their first state.

At that period they have a fin above and below their tail; that on the upper part extends along the back as far as the head, but both drop off as soon as the animal takes to the land, being then no longer of any use.

Besides these circumstances that attend them in form of a *larve*, Mr. *Ellis* * has remarked certain pennated fins at the gills of one very common in most of our stagnating waters, and which is frequently observed to take a bait like a fish.

9. Brown.   Lacertus vulg. terrestris ventre nigro maculato. *Raii syn. quad.* 264.
L. vulgaris. L. cauda tereti mediocri, pedibus unguiculatis, palmis tetradactylis, dorso linea duplici fusca. *Lin. syst.* 370. *Faun. Suec.* No. 283.

THIS is three inches long: the body slender; the tail long, slightly compressed, small and taper; that and the upper part of the body of a pale brown, marked on each side the back with a

* *Phil. Tran.* Vol. LVI. P. 191.

narrow

narrow black line reaching to the end of the tail:
the belly of a pale yellow, marked with small dusky spots; the toes formed like those of the preceding.

Lacertus parvus terrestris fuscus oppido rarus. *Raii syn.*   10. LITTLE.
quad. 264.

THIS species is mentioned by Mr. *Ray* in his
list of the *English* lizards, without any other
description than is comprehended in the *synonym*.

Lacertus terrestris anguiformis in ericetis. *Raii syn. quad.* 264.   11. AN-
GUINE.

WE remain also in the same obscurity in respect
to this species.   It seems to be of that kind
which connects the serpent and lizard genus, having a long and very slender body, and very small
legs.   Such are the *Seps,* or *Lacerta Chalcidica* of
Raii syn. quad. 272, the *Lacerta anguina* of *Linnæus,*
371, or that figured by *Seba, tom.* ii. *tab.* 68. under the name of *Vermis serpentiformis.*

Long

IV.
SERPENT.   Long and flender bodies, covered with fcaly plates:
No feet.

12. VIPER.   Ἔχις. *Arift. Hiſt. an. lib.* iii.   *tab.* 28.
c. 1.   Coluber Berus. *Lin ſyſt.* 377.
Vipera. *Virg. Georg.* iii. 417.   Huggorm *Faun. Suec.* No.
Plinii, lib. x. *c.* 42.   286. C. Berus fcutis ab-
Vipera. *Geſner Serp.* 71.   dom. 146. fquamis caudæ.
Viper, or Adder. *Raii ſyn.*   39. *Ibid.*
quad. 285. *Borl. Corn.* 282.   *Amœn. Acad.* I. 527.

V IPERS are found in many parts of this ifland,
but the dry, ftony, and, in particular, the
chalky countries abound with them. They fwarm
in many of the *Hebrides.*

They are viviparous, not but that they are hatched from an internal egg; being of that clafs of
animals, of whofe generation *Ariſtotle* * fays,
Εν αυτοις μέν ωοτοκει τό τέλειον ωόν, ἔξω δε ζωοτοκει, *i. e.* They
conceive a perfect egg within, but bring forth their
young alive.

Providence is extremely kind in making this fpecies far from being prolific, we having never heard
of more than eleven eggs being found in one viper,
and thofe are as if chained together, and each about
the fize of a blackbird's egg.

* *De Gen. an. Lib.* III. *c.* 2.

two

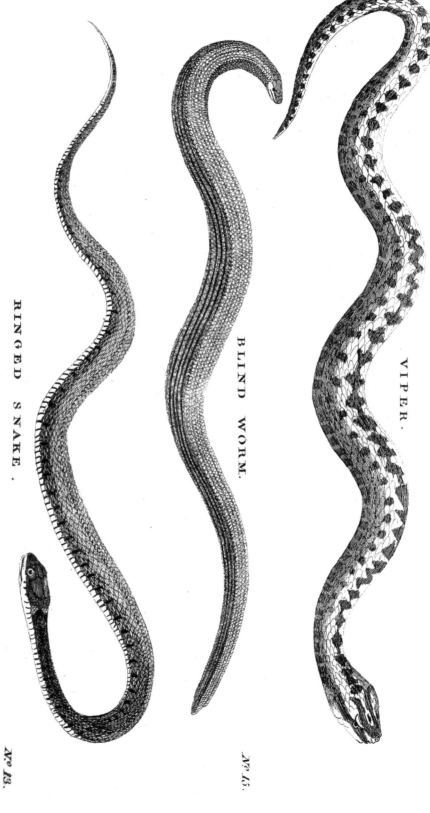

PL. IV.

VIPER.

N° 12.

BLIND WORM.

N° 15.

RINGED SNAKE.

N° 13.

The viper grows feldom to a greater length than two feet; though once we faw a female (which is nearly a third larger than the male) which was almoft three feet long.

The ground-color of this ferpent is a dirty yellow; that of the female deeper. Its back is marked the whole length with a feries of rhomboid black fpots, touching each other at the points; the fides with triangular ones; the belly entirely black.

Descrip.

There is a variety wholly black; but the rhomboid marks are very confpicuous even in this, being of a deeper and more gloffy hue than the reft. *Petiver* calls it the *Vipera Anglica Nigricans. Pet. Muf.* No. 204 *.

The head of the viper is inflated, which diftinguifhes it from the common fnake. The tongue forked; the teeth fmall; the four canine teeth are placed two on each fide the upper jaw: thefe inftruments of poifon are long, crooked, and moveable, and can be raifed and depreffed at pleafure; they are hollow from near the point to their bafe, near which is a gland that fecretes, prepares, and lodges the poifon; and the fame action that gives the wounds, forces from this gland, through the tooth, the fatal juice into it.

Teeth

Thefe iflands may be particularly thankful for the bleffing they enjoy, in being poffeffed of only

* Coluber Prefter. *Lin. fyft.* 377. Bofe. *Faun. Suec.* No. 287.

one

one venomous animal, and that of a kind which encreafes fo little.

They copulate in *May*, and are fuppofed to be about three months before they bring forth.

They are faid not to arrive at their full growth in lefs than fix or feven years; but that they are capable of engendering at two or three.

We have been often affured by intelligent people of the truth of a fact mentioned by Sir *Thomas Brown**, who was far from a credulous writer, that the young of the viper, when terrified, will run down the throat of the parent, and feek fhelter in its belly in the fame manner as the young of the *oppoffum* retire into the ventral pouch of the old one.

Food.

From this fome have imagined that the viper is fo unnatural as to devour its own young; we difbelieve the fact, it being well known that the food of thefe ferpents is frogs, toads, lizards, mice, and, according to Doctor *Mead*, even an animal fo large as a mole. Thefe they fwallow entire; which, if we confider the narrownefs of their neck, fhews it is capable of a diftenfion hardly credible, had we not ocular proofs of the fact.

It is alfo faid, from good authority, that they will prey on young birds; whether on fuch as neftle on the ground, or whether they climb up trees for them as the *Indian* ferpents do, we are quite un-

* *Vulgar errors*, 114.

certain;

certain; but we are well affured that this difcovery is far from a recent one:

Ut affidens implumibus pullis avis
*Serpentium allapfus timet *.*

Thus, for its young the anxious bird
The gliding ferpent fears.

The viper is capable of fupporting very long
abftinence, it being known that fome have been
kept in a box fix months without food, yet did
not abate of their vivacity. They feed only a
fmall part of the year, but never during their confinement; for if mice, their favorite diet, fhould
at that time be thrown into their box, tho' they
will kill, yet they never will eat them.

The poifon decreafes in violence in proportion
to the length of their confinement: it muft be alfo
added, the virtues of its flefh (whatfoever they be)
are at the fame time confiderably leffened.

Thefe animals, when at liberty, remain torpid
throughout the winter; yet when confined have
never been obferved to take their annual repofe.

The method of catching them is by putting a
cleft ftick on or near their head; after which they
are feized by the tail, and put inftantly into a bag.

The viper-catchers are very frequently bit by
them in the purfuit of their bufinefs, yet we very

* *Hor. Epod.* I.

rarely

rarely hear of the bite being fatal. The remedy, if applied in time, is very certain, and is nothing elfe but fallad oil, which the viper-catchers feldom go without. The *axungia viperina,* or the fat of vipers, is alfo another. Doctor *Mead* fufpects the efficacy of this laft, and fubftitutes one of his own in its place *; but we had rather truft to vulgar receipts which perpetual trials have fhewn to be infallible.

The fymptoms of the venom, if the wound is neglected, are very terrible: it firft caufes an acute pain in the place affected, attended with a fwelling, firft red, afterwards livid, which by degrees fpreads to the neighboring parts; great faintnefs, and a quick tho' low and interrupted pulfe enfue ; great ficknefs at the ftomach, bilious convulfive vomitings, cold fweats, and fometimes pains about the navel; and in confequence of thefe, death itfelf. But the violence of the fymptoms depends much on the feafon of the year, the difference of the climate, the fize or rage of the animal, or the depth or fituation of the wound.

Dreadful as the effects of its bite may be, yet its flefh has been long celebrated as a noble medicine. Doctor *Mead* cites from *Pliny, Galen,* and other antients, feveral proofs of its efficacy in the cure of *ulcers,* the *elephantiafis,* and other bad complaints. He even fays he has feen good effects

Uses.

* *Effay on Poifons,* 47.

from

from it in an obſtinate *lepra:* it is at preſent uſed, as a reſtorative, tho' we think the modern phyſicians have no great dependence on its virtues. The antients preſcribed it boiled, and to be eaten as fiſh; for when freſh, the medicine was much more likely to take effect than when dried, and given in form of a powder or troche. Mr. *Keyſler* relates that Sir *Kenelm Digby* uſed to feed his wife, who was a moſt beautiful woman, with capons fattened with the fleſh of vipers.

The antient *Britons* had a ſtrange ſuperſtition in reſpect to theſe animals, and of which there ſtill remains in *Wales* a ſtrong tradition. The account *Pliny* gives of it is as follows: we ſhall not attempt a tranſlation, it being already done to our hands in a ſpirited manner by the ingenious Mr. *Maſon,* which we ſhall take the liberty of borrowing.

Præterea eſt ovorum genus in magna Galliarum *fama, omiſſum* Græcis. *Angues innumeri æſtate convoluti, ſalivis faucium corporumque ſpumis artifici complexu glomerantur;* anguinum *appellatur.* Druidæ *ſibilis id dicunt in ſublime jactari, ſagoque oportere intercipi, ne tellurem attingat: profugere raptorem equo: ſerpentes enim inſequi, donec arceantur amnis alicujus interventu* *.

* *Lib.* XXIX. *c.* 3.

But

> But tell me yet
> From the grot of charms and fpells,
> Where our matron fifter dwells,
> *Brennus,* has thy holy hand
> Safely brought the Druid wand,
> And the potent *Adder-ftone,*
> Gender'd 'fore the autumnal moon ?
> When in undulating twine,
> The foaming fnakes prolific join ;
> When they hifs, and when they bear
> Their wond'rous egg aloof in air ;
> Thence before to earth it fall,
> The *Druid* in his hallow'd pall,
> Receives the prize,
> And inftant flies,
> Follow'd by the envenom'd brood,
> 'Till he crofs the cryftal flood *.

This wondrous egg feems to be nothing more than a bead of glafs, ufed by the *Druids* as a charm to impofe on the vulgar, whom they taught to believe, that the poffeffor would be fortunate in all his attempts, and that it would gain him the favor of the great.

Our modern *Druideffes* give much the fame account of the *ovum anguinum, Glain Neidr,* as the *Welch* call it, or the *Adder-Gem,* as the *Roman* philofopher does, but feem not to have fo exalted an opinion of its powers, ufing it only to affift children in cutting their teeth, or to cure the chincough, or to drive away an ague.

* *Mafon's Caractacus.* The perfon fpeaking is a Druid.

We

We have some of these beads in our cabinet:
they are made of glass, and of a very rich blue color; some are plain, others streaked: we say nothing of the figure, as the annexed plate will convey a stronger idea of it than words.

Ευυδρις. *Arist. Hist. an.* I.
c. 1.
Natrix torquata. *Gesner Ser-*
pent. 63.
Natrix torquata. *Raii syn.*
quad. 334.
Anguis vulgaris fuscus collo
flavescente, ventre albis ma-
culis distinctus. *Pet. Mus.*
XVII. No. 101.
Coluber natrix. *Lin. syst.* 380.
Tomt-Orm, Snok, Ring-Orm.
Faun. Suec. No. 288.
C. natrix scutis abdom. 170.
Squamis caudæ, 60. *Ibid.*

13. RINGED.

THE snake is the largest of the *English* serpents, sometimes exceeding four feet in
length: the neck is slender; the middle of the

body thickeſt; the back and ſides covered with ſmall ſcales, the belly with oblong, narrow, tranſverſe plates. The firſt *Linnæus* diſtinguiſhes by the name of *ſquamæ*, the laſt he calls *ſcuta*, and from them forms his genera of ſerpents.

Thoſe that have both *ſquamæ* and *ſcuta* he calls *Colubri*; thoſe that have only *ſquamæ*, *Angues*. The viper and ſnake are comprehended in the firſt genus, the blind-worm under the ſecond; but we chuſe (to avoid multiplying our genera) to take in the few ſerpents we have by a ſingle genus, their marks being too evident to be confounded.

Descrip.   The color of the back and ſides of the ſnake are duſky or brown; the middle of the back marked with two rows of ſmall black ſpots running from head to tail; and from them are multitudes of lines of ſpots croſſing the ſides; the plates on the belly are duſky, the ſcales on the ſides of a bluiſh white.

On each ſide the neck is a ſpot of pale yellow, and the baſe of each is a triangular black ſpot, one angle of which points towards the tail.

The teeth are ſmall and ſerrated, lying each ſide the jaw in two rows.

This ſpecies is perfectly inoffenſive; it frequents and lodges itſelf among buſhes in moiſt places, and will readily take the water, ſwimming very well.

It preys on frogs, inſects, worms, and mice, and, conſidering the ſmallneſs of the neck, it is amazing how large an animal it will ſwallow.

                        The

The fnake is oviparous: it lays its eggs in
dung hills, and in hot-beds, whofe heat, aided by
that of the fun, promotes the exclufion of the
young.

During winter it lies torpid in banks of hedges,
and under old trees.

EGGS.

Anguis Eryx. *Lin. fyft.* 392.
A new Snake. *Tour in Scotl.* 1769. App.

14. ABER-
DEEN.

LENGTH fifteen inches. Tongue broad and
forked. Noftrils fmall, round, and placed
near the tip of the nofe. Eyes lodged in oblong
fiffures above the angle of the mouth. Belly of
a bluifh lead color, marked with fmall white fpots
irregularly difpofed. The reft of the body of a
greyifh brown, with three longitudinal dufky lines,
one extending from the head along the back to
the point of the tail; the others broader, and
extending the whole length of the fides. It had
no *fcuta*; but was entirely covered with fmall
fcales; largeft on the upper part of the head.

Inhabits *Aberdeenfhire.* Communicated to me by
the late Doctor *David Skene.*

D 2   The

15. Blind-
worm.

The Blind-worm, or flow-
worm, *Cæcilia Typhline*
Græcis. *Raii syn. quad.*
289. *Grew's Muf.* 48.
Cæcilia anglica cinerea fquamis parvis mollibus, com-
pactis. *Pet. Muf.* xvii. No.
102.

Long Cripple. *Borlafe Cornw.*
284. *tab.* 28.
Anguis fragilis. *Lin. syft.*
392.
Ormfla, Koppar-Orm. *Faun.*
Suec. 289.
A. fragilis fquamis abdomi-
nis caudæque, 135. *Ibid.*

Descrip.

THE ufual length of this fpecies is eleven
inches: the irides are red; the head fmall;
the neck ftill more flender; from that part the
body grows fuddenly, and continues of an equal
bulk to the tail, which ends quite blunt.

The color of the back is cinereous, marked with
very fmall lines compofed of minute black fpecks;
the fides are of a reddifh caft; the belly dufky,
both marked like the back.

The tongue is broad and forky; the teeth minute, but numerous; the fcales fmall.

The motion of this ferpent is flow, from which,
and from the fmallnefs of the eyes, are derived its
names. Like others of the genus, they lie torpid
during winter, and are fometimes found in vaft
numbers twifted together.

Like the former it is quite innocent. Doctor
'*Borlafe* mentions a variety of this ferpent with a
pointed tail; and adds, that he was informed that
a man loft his life by the bite of one in *Oxfordfhire.*
We

We are inclined to think that his informant mif-
took the black or dufky viper for this kind; for,
excepting the viper, we never could learn that
there was any fort of poifonous ferpent in thefe
kingdoms.

In *Sweden* is a fmall reddifh ferpent, called there
Afping, the *Coluber Cherfea* of *Linnæus*, whofe bite
is faid to be mortal. Is it poffible that this could
be the fpecies which has hitherto efcaped the notice of our naturalifts? I the rather fufpect it, as
I have been informed, that there is a fmall fnake
that lurks in the low grounds of *Galloway*, which
bites and often proves fatal to the inhabitants.

D 3   CLASS

C L A S S IV.

F I S H.

Oh Deus! ampla tuæ, quam ſunt miracula dextræ!
  O quam ſolerti ſingula mente regis!
Divite tu gazâ terras, et meſſibus imples;
  Nec minus eſt vaſti fertilis unda maris:
Squammiger hunc peragrat populus, proleſque parentum
  Stipat, et ingentes turba minuta duces.

JONSTON. PSALMUS CIV.

F I S H.

Div. I. CETACEOUS FISH.

NO gills; an orifice on the top of the head,
thro' which they breathe, and eject water; a
flat or horizontal tail; exemplified in the explanatory plate, *fig.* 1. by the BEAKED WHALE, borrowed from *Dale's Hift. Harw.* 411. *Tab.* 14.

G E N E R A.

I.   W H A L E.
II.   C A C H A L O T.
III.   D O L P H I N.

Div. II. CARTILAGINOUS FISH.

BREATHING thro' certain apertures, generally placed on each fide the neck, but in
fome inftances beneath, in fome above, and from
one to feven in number on each part, except in
the PIPE FISH, which has only one.

The mufcles fupported by cartilages, inftead of
bones.

Explan. Pl. *fig.* 2. the PICKED DOG FISH.
     a. The lateral apertures.

IV.   L A M-

F I S H.

D ɪ v. III. B O N Y F I S H.

THIS divifion includes thofe whofe mufcles are fupported by bones or fpines, which breathe thro' gills covered or guarded by thin bony plates, open on the fide, and dilatable by means of a certain row of bones on their lower part each feparated by a thin web, which bones are called the *Radii Branchioftegi*, or the *Gillcovering Rays*.

The tails of all the fifh that form this divifion, are placed in a fituation perpendicular to the body, and this is an invariable character.

The later Icthyologifts have attempted to make the number of the branchioftegous rays a character of the *genera*; but I found (yet too late in fome inftances, where I yielded an implicit faith) that their rule was very fallible, and had induced me into error; but as I borrowed other definitions, it is to be hoped the explanation of the *genera* will be intelligible. I fhould be very difingenu-ous,

ous, if I did not own my obligations in this refpect to the works of ARTEDI, Dr. GRONOVIUS,
and LINNÆUS.

It is from the laft I have copied the great fections
of the BONY FISH into

APODAL,	JUGULAR,
THORACIC,	ABDOMINAL*.

He founds this fyftem on a comparifon of the
ventral fins to the feet of land animals or reptiles;
and either from the want of them, or their particular fituation in refpect to the other fins, eftablifhes
his fections.

In order to render them perfectly intelligible,
it is neceffary to refer to thofe feveral organs of
movement, and fome other parts, in a perfect fifh,
or one taken out of the three laft fections.

The HADOCK.   Expl. Pl. *fig.* 4.

a.	The pectoral fins.
b.	ventral fins.
c.	anal fins.
d.	caudal fin, or the tail.
e. e. e.	dorfal fins.
f.	bony plates that cover the gills.
g.	branchioftegous rays, and their mem-branes.
h.	lateral, or fide line.

* *Vide Syft. Nat.* 422.

S E C T,

S ᴇᴄᴛ. I.   A P O D A L.

THE moſt imperfect, wanting the ventral fins; illuſtrated by the Conger, *fig.* 3. This alſo expreſſes the union of the dorſal and anal fins with the tail, as is found in ſome few fiſh.

    XII.   E E L.
    XIII.   W O L F   F I S H.
    XIV.   L A U N C E.
    XV.   M O R R I S.
    XVI.   S W O R D   F I S H.

S ᴇᴄᴛ. II.   J U G U L A R.

THE ventral fins *b*, placed before the pectoral fins *a*, as in the Hadock, *fig.* 4.

    XVII.   D R A G O N E T.
    XVIII.   W E E V E R.
    XIX.   C O D F I S H.
    XX.   B L E N N Y.

Sᴇᴄᴛ.

SECT. III. THORACIC.

THE ventral fins *a*, placed beneath the pectoral fins *b*, as in the FATHER LASHER,
fig. 5.

XXI.	GOBY.
XXII.	BULL-HEAD.
XXIII.	DOREE.
XXIV.	FLOUNDER.
XXV.	GILT-HEAD.
XXVI.	WRASSE.
XXVII.	PERCH.
XXVIII.	STICKLEBACK.
XXIX.	MACKREL.
XXX.	SURMULLET.
XXXI.	GURNARD.

SECT.

Sect. IV. ABDOMINAL.

THE ventral fins placed behind the pectoral fins, as in the Minow, *fig.* 6.

XXXII.	L O C H E.
XXXIII.	S A L M O N.
XXXIV.	P I K E.
XXXV.	A R G E N T I N E.
XXXVI.	A T H E R I N E.
XXXVII.	M U L L E T.
XXXVIII.	F L Y I N G F I S H.
XXXIX.	H E R R I N G.
XL.	C A R P.

D I v.

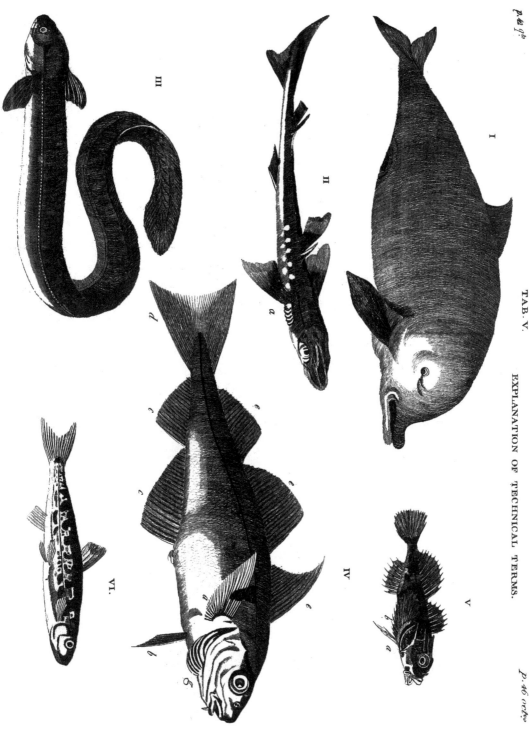

TAB. V. EXPLANATION OF TECHNICAL TERMS.

Div. I. CETACEOUS FISH.

NATURE on this tribe hath bestowed an internal structure in all respects agreeing with that of quadrupeds; and in a few other the external parts in both are similar.

Cetaceous Fish, like land animals, breathe by means of lungs, being destitute of gills. This obliges them to rise frequently to the surface of the water to respire, to sleep on the surface, as well as to perform several other functions.

They have the power of uttering sounds, such as bellowing and making other noises, a faculty denied to genuine fish *.

Like land animals they have warm blood, are furnished with organs of generation, copulate, bring forth, and suckle their young, shewing a strong attachment to them.

Their bodies beneath the skin are entirely surrounded with a thick layer of fat (blubber) analogous to the lard on hogs.

The number of their fins never exceeds three,

* *Pontop. Hift. Norw.* II. 123. *Blafius Anat. Animal,* 288.

viz.

viz. two pectoral fins, and one back fin; but in some species the last is wanting.

Their tails are placed horizontally or flat in respect to their bodies; contrary to the direction of those of all other fish, which have them in a perpendicular site. This situation of the tail enables them to force themselves suddenly to the surface of the water to breathe, which they are so frequently constrained to do.

Many of these circumstances induced *Linnæus* to place this tribe among his *Mammalia*, or what other writers style quadrupeds.

To have preserved the chain of beings entire, he should in this case have made the genus of *Phocæ*, or *Seals*, and that of the *Trichecus* or *Manati*, immediately precede the whale, those being the links that connect the *Mammalia* or quadrupeds with the fish; for the *Seal* is, in respect to its legs, the most imperfect of the former class; and in the *Manati* the hind feet coalesce, assuming the form of a broad horizontal tail.

Notwithstanding the many parts and properties which cetaceous fish have in common with land animals, yet there still remain others, that in a natural arrangement of the animal kingdom, must determine us after the example of the illustrious *Ray* *, to place them in the rank of fish; and for

---

* Who makes two divisions of fish.

     1. *Pulmone respirantes.*

     2. *Branchiis respirantes.*

the

the fame reafons, that firft of fyftematic writers affigns,

That the form of their bodies agrees with that of fifh.

They are entirely naked, or covered only with a fmooth fkin.

They live entirely in the water, and have all the actions of fifh.

I.
WHALE.

Cetaceous Fiſh without teeth, with horny laminæ
in their mouths.

16. COMMON. Μυσικητος. *Ariſt. hiſt. an.*
Lib. III. *c.* 12.
Muſculus *Plinii, Lib.* XI. *c.*
37.
Balæna. *Rondel.* 475. *Geſner*
Piſc. 114.
Balæna major, laminas cor-
neas in ſuperiore maxillas
habens, fiſtula donata, bi-
pinnis. *Sib. Phalæn.* 28.
Balæna vulgaris edentula, dor-
ſo non pinnato. *Raii ſyn.*
piſc. 6.
Balæna. *Rondel. Wil. Iĉth.* 35.

The Whale. *Marten's Spitz-*
berg. 130. *Crantz's Greenl.*
I. 107.
La Baleine ordinaire. *Briſſon*
Cet. 218.
Balæna fiſtula in medio capite,
dorſo caudum verſus, acu-
minato. *Arted. ſyn.* 106.
Sp. 106.
Balæna myſticetus. *Lin. ſyſt.*
105. Gronlands Walfiſk.
Faun. Suec. No. 49.
Balæna. *Gronov. Zooph.* 29.

SIZE.

THIS ſpecies is the largeſt of all animals: it
is even at preſent ſometimes found in the
northern ſeas ninety feet in length; but formerly
they were taken of a much greater ſize, when the
captures were leſs frequent, and the fiſh had time
to grow. Such is their bulk within the *arĉtic* circle, but in thoſe of the *torrid* zone, where they are
unmoleſted, whales are ſtill ſeen one hundred and
ſixty feet long*.

The

* *Adanſon's voy.* 174. From this account we find no reaſon to diſbelieve the vaſt ſize of the *Indian* whales, of whoſe
bones

The head is very much difproportioned to the fize of the body, being one-third the fize of the fifh : the under lip is much broader than the upper. The tongue is compofed of a foft fpongy fat, capable of yielding five or fix barrels of oil. The gullet is very fmall for fo vaft a fifh, not exceeding four inches in width. In the middle of the head are two orifices, thro' which it fpouts water to a vaft height, and with a great noife, efpecially when difturbed or wounded.

The eyes are no larger than thofe of an ox.

On the back there is no fin, but on the fides, beneath each eye, are two large ones.

The penis is eight feet in length, inclofed in a ftrong fheath. The teats in the female are placed in the lower part of the belly.

The tail is broad and femilunar.

This whale varies in color : the back of fome being red, the belly generally white. Others are black, fome mottled, others quite white, according to the obfervation of *Marten*, who fays, that their colors in the water are extremely beautiful, and that their fkin is very fmooth and flippery.

What is called *whalebone* adheres to the upper jaw, and is formed of thin parallel laminæ, fome

WHALE-BONE.

bones and jaws, both *Strabo*, *Lib.* XV. and *Pliny*, *Lib.* IX. *c.* 3. relate, that the natives made their houfes, ufing the jaws for door-cafes. This method of building was formerly practifed by the inhabitants of *Greenland*, as we find from *Frobifher*, in his fecond voyage, p. 18, publifhed in 1587.

E 2

of

of the longeft four yards in length; of thefe there are commonly 350 on each fide, but in very old fifh more; of thefe about 500 are of a length fit for ufe, the others being too fhort. They are furrounded with long ftrong hair, not only that they may not hurt the tongue, but as ftrainers to prevent the return of their food when they difcharge the water out of their mouths.

It is from thefe hairs that *Ariftotle* gave the name of Μυστικητος, or the *bearded whale*, to this fpecies, which he tells us had in its mouth hairs inftead of teeth*; and *Pliny* defcribes the fame under the name of *Mufculus* †. Though the antients were acquainted with this animal, yet as far as we recollect, they were ignorant of their ufes as well as capture.

Aldrovand ‡ indeed defcribes from *Oppian*, what he miftakes for whale fifhing: he was deceived by the word κητος, which is ufed not only to exprefs whale in general, but any great fifh. The *poet* here meant the *fhark*, and fhews the way of taking it in the very manner practifed at prefent, by a ftrong hook baited with flefh. He defcribes too its three-fold row of teeth, a circumftance that at once difproves its being a whale:

---

* ετι δε και ὁ μυστικητος οδοντας μεν εν τω στοματι εκ εχει, τριχας δε ὁμοιας ὑειας. *Hift. an. Lib.* III. *c.* 12.

† *Lib.* XI. *c.* 37.

‡ *De Cetis.* 261.

Δεινυς

Δεινὸς χαυλιοδοντας ἀναιδέας ἠΰτ᾽ ἀκοντας,
Τριστοιχεὶ πεφυῶτας ἐπασσυτέρῃσιν ἀκωκαῖς.

Halieut. V. *lin.* 526.

Whofe dreadful teeth in triple order ftand,
Like fpears out of his mouth.

The whale, though fo bulky an animal, fwims
with vaft fwiftnefs, and generally againft the wind.

It brings only two young at a time, as we believe is the cafe with all other whales.

Its food is a certain fort of fmall fnail, and as
Linnæus fays, the *medufa,* or fea blubber.

Food.

The great refort of this fpecies is within the
arctic circle, but they fometimes vifit our coafts.
Whether this was the *Britifh* whale of the antients
we cannot pretend to fay, only we find, from a line
in *Juvenal,* that it was of a very large fize;

Quanto Delphinis Balæna Britannica *major.*

Sat. X.

As much as *Britifh* whales in fize furpafs
The dolphin race.

To view thefe animals in a commercial light, we
muft add, that the *Englifh* were late before they
engaged in the whale-fifhery: it appears by a fet of
queries, propofed by an honeft merchant in the
year 1575, in order to get information in the
bufinefs, that we were at that time totally igno-

E 3
rant

rant of it, being obliged to fend to *Bifkaie for men fkilful in the catching of the whale, and ordering of the oil, and one cooper fkilful to fet up the ftaved cafk* *. This feems very ftrange; for by the account O*Æther* gave of his travels to King *Alfred*, near 700 years † before that period, it is evident that he made that monarch acquainted with the *Nor-wegians* practifing the whale-fifhery; but it feems all memory of that gainful employ, as well as of that able voyager O*Æther*, and all his important difcoveries in the North were loft for near feven centuries.

It was carried on by the *Bifcayeners* long before we attempted the trade, and that for the fake not only of the oil, but alfo of the whalebone, which they feem to have long trafficked in. The earlieft notice we find of that article in our trade is by *Hackluyt* ‡, who fays it was brought from the *Bay of St. Laurence* by an *Englifh* fhip that went there for the *barbes* and *fynnes* of whales and train oil, A. D. 1594, and who found there feven or eight hundred *whale fynnes*, part of the cargo of two great *Bifkaine* fhips, that had been wrecked there three years before. Previous to that, the ladies ftays muft have been made of fplit cane, or fome tough wood, as Mr. *Anderfon* obferves in his

---

* *Hackluyt's Col. voy.* I. 414.
† *Idem*, I. 4.
‡ *Idem*, III. 194.

Dictionary

Dictionary of Commerce *, it being certain that the whale fishery was carried on, for the fake of the oil, long before the difcovery of the ufe of whale bone.

The great refort of thefe animals was found to be on the inhofpitable fhores of *Spitzbergen*, and the *European* fhips made that place their principal fifhery, and for numbers of years were very fuccefsful: the *Englifh* commenced that bufinefs about the year 1598, and the town of *Hull* had the honor of firft attempting that profitable branch of trade. At prefent it feems to be on the decline, the quantity of fifh being greatly reduced by the conftant capture for fuch a vaft length of time: fome recent accounts inform us, that the fifhers, from a defect of whales, apply themfelves to feal fifhery, from which animals they extract an oil. This we fear will not be of any long continuance; for thefe fhy and timid creatures will foon be induced to quit thofe fhores by being perpetually harraffed, as the *morfe* or *walrus* has already in a great meafure done. We are alfo told, that the poor natives of *Greenland* begin even now to fuffer from the decreafe of the feal in their feas, it being their principal fubfiftence; fo that fhould it totally defert the coaft, the whole nation would be in danger of perifhing through want.

In old times the whale feems never to have been

ROYAL FISH.

* *Vol.* I. 442.

E 4   taken

taken on our coafts, but when it was accidentally flung afhore: it was then deemed a royal fifh *, and the king and queen divided the fpoil; the king afferting his right to the head, her majefty to the tail †.

17. PIKE-
HEADED.

Balæna tripinnis nares habens cum roftro acuto, et plicis in ventre. *Sib. Phalain* 29. *tab.* 1.
Idem. *Raii fyn. pifc.* 16.
Pike-headed Whale. *Dale Harwich*, 410. No. 3.

La Baleine a mufeau pointu. *Briffon Cet.* 224.
Balæna fiftula duplici in roftro, dorfo extremo protuberantia cornuiformi. *Arted. fyn.* 107.
Balæna Boops. *Lin. fyft.* 106.

SIZE.

THE length of that taken on the coaft of *Scotland*, as remarked by Sir *Robert Sibbald*, was forty-fix feet, and its greateft circumference twenty.

DESCRIP.

The head of an oblong form, floping down, and growing narrower to the nofe; fix feet eight inches from the end of which were two fpout-holes, feparated by a thin divifion: the eyes fmall.

The pectoral fins five feet long, and one and a half broad: on the back, about eight feet and an half from the tail, in lieu of a back fin, was a hard horny protuberance: the tail was nine feet and a half broad.

* Item habet warectum maris per totum regnum *Ballenas* et *Sturgiones* captos, &c. *Edwardi* II. anno 17mo.
† *Blackftone's Com.* I. c. 4.

The

The belly was uneven, and formed into folds running length-ways.

The ſkin extremely ſmooth and bright; that on the back black; that on the belly white.

This ſpecies takes its name from the ſhape of its noſe, which is narrower and ſharper pointed than that of other whales.

Balæna edentula corpore ſtrictiore, dorſo pinnato. *Raii ſyn. piſc.* 9. *Dale Harwich,* 410. No. 2.
Fin Fiſh. *Marten's Spitzberg.* 165.
Egede Greenl. 65. *Crantz Greenl.* I. 110.

Le Gibbar. *Briſſon Cet.* 222. 18. Fin Fiſh, Balæna fiſtula in medio capite tubero penniformi in extremo dorſo. *Arted. ſyn.* 107.
Balæna Phyſalus. *Lin. ſyſt.* 106.

THIS ſpecies is diſtinguiſhed from the common whale by a fin on the back, placed very low and near the tail.

The length is equal to that of the common DESCRIP. kind, but much more ſlender. It is furniſhed with whale-bone in the upper jaw, mixed with hairs, but ſhort and knotty, and of little value. The blubber alſo on the body of this kind is very inconſiderable; theſe circumſtances, added to its extreme fierceneſs and agility, which renders the capture very dangerous, cauſe the fiſhers to neglect it. The natives of *Greenland* though hold it in great eſteem, as it affords a quantity of fleſh, which to their palate is very agreeable.

The

The lips are brown, and like a twisted rope: the spout hole is as it were split in the top of its head, through which it blows water with much more violence, and to a greater height, than the common whale. The fishers are not very fond of seeing it, for on its appearance the others retire out of those seas.

Some writers conjecture this species to have been the Φυσαλος, and *Physeter*, or blowing whale of *Oppian*, *Ælian*, and *Pliny* *; but since those writers have not left the left description of it, it is impossible to judge which kind they meant; for in respect to the faculty of spouting out water, or blowing, it is not peculiar to any one species, but common to all the whale kind.

19. Round-
lipped.

Balæna tripinnis maxillam inferiorem rotundam et superiore multo latiorem habens. *Sib. Phalain.* 33. *tab.* T. 3. Idem. *Raii syn. pisc.* 16. La Baleine a museau rond.

Brisson Cet. 222. B. fistula duplici in fronte maxilla inferiore multo latiore. *Arted. syn.* 107. Balæna musculus. *Lin. syst.* 106.

THE character of this species is to have the lower lip broader than the upper, and of a semicircular form.

That taken in 1692 near *Abercorn-Castle*, was seventy-eight feet long, the circumference thirty-

* *Oppian, Halieut,* I. *Lin.* 368. *Ælian Hist. an.* ix. *c.* 49. *Plin. lib.* ix. *c.* 5.

five;

five; the *rictus* or gape very wide; the tongue fifteen feet and a half long; the mouth furnished with short whale-bone, about three feet in length. On the forehead were two spout holes of a pyramidal form.

The eyes were placed thirteen feet from the end of the nose: the pectoral fins ten feet long: the back fin about three feet high, placed near the tail, which was eighteen feet broad: the belly was full of folds.

This species is said to feed on herrings.

Butskopf. *Marten's Spitzberg.* 124.
Bottle-head, or Flounders-head. *Dale Harwich,* 411.

tab. 14.
Nebbe-haul, or beaked Whale. *Pontop. Norway,* I. 123.

20. BEAKED.

THIS species was taken near *Maldon,* 1717, and thus described by *Dale* and *Marten.*

The length was fourteen feet, the circumference seven and an half; the body very thick, the forehead high, the nose depressed, and of the same thickness its whole length, not unlike the beak of a bird: in the mouth were no teeth.

The eyes large, the eyelids small, and placed a little above the line of the mouth. The spout hole was on the top of the head semicircular, with the corners pointed towards the tail.

The pectoral fins were seventeen inches long.
The

The back fin was placed rather nearer the tail than the head, and was a foot long: the breadth of the tail was three feet two inches.

These fish sometimes grow to the length of twenty feet; they make but little noise in blowing, are very tame, come very near the ships, and will accompany them for a great way.

Belon describes and figures a fish very much resembling, if not the same with this: he says it furnished whale-bone, *Dont les Dames font aujourdhuy leurs buftes et arrondiffent leurs verdugades**, by which it appears, that the commodity was but newly known at that time in *France*. He adds, that the tongue was very good eating, and both that and the flesh used to be salted for provision.

---

* *Belon de la nat. &c. des Poiffons*, 1555, p. 6, by which it appears that the *French* were acquainted with that article at left forty years before we were.

Cetaceous

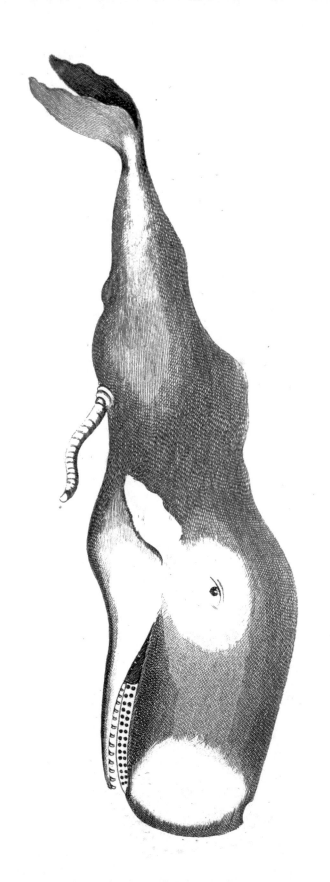

Cetaceous Fiſh, with teeth in the lower jaws only. II. CACHA-LOT.

Trumpa. *Purchaſs's Pilgrimes* III. 471.

Balæna major in inferiore tantum maxilla dentata dentibus arcuatis falciformibus, pinnam ſive ſpinam in dorſo habens. *Sib. Phalian.* 13. *tab.* A. 1. *Raii ſyn. piſc.* 15.

Le cachalot a dents en faucil-les. *Briſſon Cet.* 229.

The Parmacitty Whale, or Pot Wal Fiſh. *Dale Harwich,* 413.

Phyſeter microps. *Lin. ſyſt.* 107. *Arted. ſyn.* 104.

Caſhalot, Catodon, or Pot Fiſh. *Crantz Greenl.* I. 112.

21. BLUNT-HEADED.

A FISH of this kind was caſt aſhore on *Cramond Iſle*, near *Edinburgh, December* 22d, 1769; its length was fifty-four feet, the greateſt circumference, which was juſt beyond the eyes, thirty: the upper jaw was five feet longer than the lower, whoſe length was ten feet.

SIZE.

The head was of a moſt enormous ſize, very thick, and above one-third the ſize of the fiſh: the end of the upper jaw was quite blunt, and near nine feet high: the ſpout hole was placed near the end of it.

DESCRIP.

The teeth were placed in the lower jaw, twenty-three on each ſide, all pointing outwards; in the upper jaw, oppoſite to them, were an equal number of cavities, in which the ends of the teeth lodged when the mouth was cloſed. The tooth,

TEETH.

figured

figured in *plate* iii. *No.* 2. was eight inches long, the greateft circumference the fame. It is hollow within fide for the depth of three inches, and the mouth of the cavity very wide : it is thickeft at the bottom, and grows very fmall at the point, bending very much; but in fome the flexure is more than in others. Thefe, as well as the teeth of all other whales we have obferved, are very hard, and cut like ivory.

The eyes very fmall, and remote from the nofe.

The pectoral fins placed near the corners of the mouth, and were only three feet long: it had no other fin, only a large protuberance on the middle of the back.

The tail a little forked, and fourteen feet from tip to tip.

The penis feven feet and a half long.

The figure, *plate* ii. we borrowed from a print in the LX. vol. of the *Ph. Tr. p.* where there is a very good account of this fpecies by Mr. *James Robertfon*, furgeon.

This is one of the fpecies which yield what is SPERMACETI improperly called *fperma ceti*; that fubftance being found lodged in the head of the fifh that form this genus, which the *French* call *Cachalot*, a name we have adopted, having no general term for it in our tongue.

Linnæus informs us, that this fpecies purfues and terrifies the Porpeffes to fuch a degree as often to drive them on fhore.

Belæna

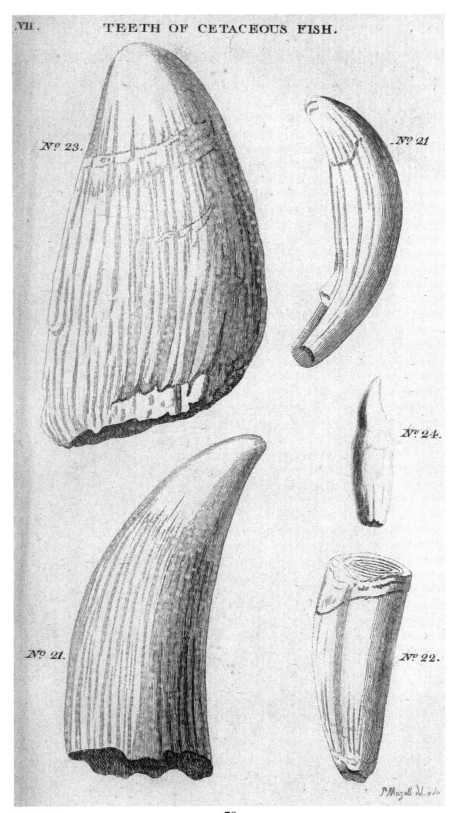

N.º 23.

N.º 21

N.º 24.

N.º 21.

N.º 22.

P. Mazell del. & Sc.

Balæna minor in inferiore maxilla tantum dentata fine fpina aut pinna in dorfo. *Sib. Phalain.* 9. *Raii fyn. pifc.* 15.
Le petit Cachalot. *Briffon*

Cet. 228.
Phyfeter Catodon. *Lin. fyft.* 107.
Catodon fiftula in roftro. *Arted. fynon.* 108.

THIS fpecies was taken on one of the *Orkney* Ifles, a hundred and two of different fizes being caft afhore at one time, the largeft twenty-four feet in length.

The head was round, the opening of the mouth fmall: *Sibbald* fays it had no fpout hole, but only noftrils. We rather think, that the former being placed at the extremity of the nofe was miftaken by him for the latter.

The teeth we have in our cabinet of this fpecies (*plate* iii. *No.* 4.) are an inch and three-quarters long, and in the largeft part, of the thicknefs of one's thumb. The top is quite flat, and marked with concentric lines; the bottom is more flender than the top, and pierced with a fmall orifice.

The back fin was wanting; inftead was a rough fpace.

Belæna

23. High-
FINNED.
Balæna macrocephala tripin-   nentes. *Sib. Phalain.* 18.
nis, quæ in mandibula in-   *Raii fyn. pifc.* 16.
feriore dentes habet minus   Le Cachalot a dents plattes.
inflexos et in planum defi-   *Briffon Cet.* 230.

ONE of this fpecies was caft on the *Orkney* Ifles in 1687. The fpout hole was placed in front, and on the middle of the back was a high fin, which *Sibbald* compares to the mizen maft of a fhip. The head abounded with *fperma ceti* of the beft fort.

TEETH.   The teeth of this kind are very flightly bent; that which we have figured, *plate* iii. *No.* 1. is feven inches three-quarters in length; the greateft circumference nine: it is much compreffed on the fides; the point rather blunt than flat; the bottom thin, having a very narrow but long orifice, or flit, hollowed to the depth of five inches and a quarter, and the tooth was immerfed in the jaw as far as that hollow.

Cetaceous

Cetaceous Fish, with teeth in both jaws.

Δελφίς. *Arist. Hist. an. lib.* vi. *c.* 12. Δελφίν. *Ælian lib.* I. *c.* 18.
Delphinus *Plinii, lib.* ix. *c.* 8. Le Daulphin, ou oye de mer *Belon Piscis* 7.
Delphinus. *Rondel.* 459. *Gefner pisc.* 319. *Caii opusc.* 113.
Delphinus Antiquorum. *Wil.*

Icth. 28. *Raii syn pisc.* 12
Delphinus corpore longo subtereti, rostro longo acuto. *Arted. syn.* 105.
Le Dauphin. *Brisson Cet.* 233.
Delphinus Delphis. *Lin. syst.* 108.
Dolphin. *Borlase Cornwall,* 264. *tab.* 27. *Crantz Greenl.* I. 115.

24. DOL-PHIN.

HISTORIANS and philosophers seem to have contended who should invent most fables concerning this fish. It was consecrated to the Gods, was celebrated in the earliest time for its fondness of the human race, was honored with the title of the *Sacred Fish**, and distinguished by those of *Boy-loving*, and *Philanthropist*. It gave rise to a long train of inventions, proofs of the credulity and ignorance of the times.

Aristotle steers the clearest of all the antients from these fables, and gives in general so faithful a natural history of this animal, as evinces the superior judgment of that great philosopher, in comparison to those who succeeded him. But the elder

---

* *Athenæus,* 281.

Pliny, *Ælian*, and others, feem to preferve no bounds in their belief of the tales related of this fifh's attachment to mankind.

Pliny * the younger, (apologizing for what he is going to fay) tells the ftory of the enamoured dolphin of *Hippo* in a moft beautiful manner. It is too long to be tranfcribed, and would be injured by an abridgement; therefore we refer the reader to the original, or to Mr. *Melmoth*'s elegant tranflation.

Scarce an accident could happen at fea but the dolphin offered himfelf to convey to fhore the unfortunate. *Arion*, the mufician, when flung into the ocean by the pyrates, is received and faved by this benevolent fifh.

Inde (fide majus) tergo Delphina recurvo,
  Se memorant oneri fuppofuiffe novo.
Ille fedens citharamque tenens, pretiumque vehendi
  Cantat, et æquoreas carmine mulcet aquas.

Ovid. Fafti, lib. ii. 113.

But (paft belief) a Dolphin's arched back,
Preferved *Arion* from his deftined wrack;
Secure he fits, and with harmonious ftrains,
Requites his bearer for his friendly pains.

We are at a lofs to account for the origin of thofe fables, fince it does not appear that the dolphin fhews a greater attachment to mankind than

* *Epift. lib.* ix. *ep.* 33.

the

the reft of the cetaceous tribe. We know that at prefent the appearance of this fifh, and the porpeffe, are far from being efteemed favorable omens by the feamen; for their boundings, fprings and frolics in the water, are held to be fure figns of an approaching gale.

It is from their leaps out of that element that they affume a temporary form that is not natural to them, but which the old painters and fculptors have almoft always given them. A dolphin is fcarce ever exhibited by the antients in a ftrait fhape, but almoft always incurvated : fuch are thofe on the coin of *Alexander the Great*, which is preferved by *Belon*, as well as on feveral other pieces of antiquity. The poets defcribe them much in the fame manner, and it is not improbable but that the one had borrowed from the other :

Tumidumque pando tranfilit dorfo mare
Tyrrhenus omni pifcis exfultat freto,
Agitatque gyros.

Senec. Trag. Agam. 450.

Upon the fwelling waves the dolphins fhew
Their bending backs, then fwiftly darting go,
And in a thoufand wreaths their bodies throw.

The natural fhape of the dolphin is almoft ftrait, DESCRIP. the back being very flightly incurvated, and the body flender : the nofe is long, narrow, and pointed

F 2

ed

ed, not much unlike the beak of some birds, for which reason the *French* call it *L' oye de mer.*

It has in all forty-two teeth, twenty-one in the upper jaw, and nineteen in the lower, a little a-bove an inch long, conic at their upper end, sharp pointed*, bending a little in. They are placed at small distances from each other, so that when the mouth is shut, the teeth of both jaws lock into one another: a single one is figured *plate* iii. *No. 5.*

The spout hole is placed in the middle of the head.

The back fin is high, triangular, and placed rather nearer to the tail than to the head; the pecto-ral fins situated low.

The tail is semilunar.

The skin is smooth, the color of the back and sides dusky; the belly whitish.

It swims with great swiftness: its prey is fish.

It was formerly reckoned a great delicacy: Doctor *Caius* says, that one which was taken in his time, was thought a present worthy the Duke of *Norfolk,* who distributed part of it among his friends. It was roasted and dressed with porpesse sauce, made of crumbs of fine white bread, mixed with vinegar and sugar.

This species of dolphin must not be confound-ed with that to which seamen give the name, the

* *Plate* iii. *fig.* 5.

latter

latter being quite another kind of fish, the *Coryphæna, Hippuris* of *Linnæus, p.* 446. and the *Dorado* of the *Portuguese,* described by *Willughby, p.* 213.

Φώαινα. *Arist. hist. an. Lib.* VI. *c.* 12. Tursio *Plinii, Lib.* IX. *c.* 9.
Le Marsouin. *Belon.*
Tursio. *Rondel.* 474. *Gesner pisc.* 711.
Porpesse. *Wil. Icth.* 31. *Raii syn. pisc.* 13. *Crantz's Greenl.* I. 114. *Kolben's Hist. Cape,* II. 200.

Le Marsouin.   *Brisson Cet.* 234.
Delphinus corpore fere coniformi, dorso lato, rostro subacuto. *Arted. synon.* 104.
Delphinus Phocæna. *Lin. syst.* 108.
Marswin, Tumbiare. *Faun. Succ.* No. 51.

25. POR-PESSE.

THESE fish are found in vast multitudes in all parts of the sea that wash these islands, but in greatest numbers at the time when fish of passage appear, such as mackrel, herrings, and salmon, which they pursue up the bays with the same eagerness as a pack of dogs does a hare. In some places they almost darken the sea as they rise above water to take breath: but porpesses not only seek for prey near the surface, but often descend to the bottom in search of sand eels, and sea worms, which they root out of the sand with their noses in the same manner as hogs do in the fields for their food.

Their bodies are very thick towards the head, DESCRIP.

F 3   but

but grows flender towards the tail, forming the figure of a cone.

The nofe projects a little, is much fhorter than that of the dolphin, and is furnifhed with very ftrong mufcles, which enables it the readier to turn up the fand.

TEETH.

In each jaw are forty-eight teeth, fmall, fharp pointed, and a little moveable: like thofe of the dolphin, they are fo placed as that the teeth of one jaw locks into thofe of the other when clofed.

The tongue is flat, pectinated at the edges, and faftened down to the bottom of the mouth.

The eyes fmall; the fpout hole on the top of the head.

On the back is one fin placed rather below the middle; on the breaft are two fins. The tail femilunar.

The color of the porpeffe is generally black, and the belly whitifh, not but they fometimes vary; for in the river *St. Laurence* there is a white kind; and Doctor *Borlofe*, in his voyage to the *Scilly* ifles, obferved a fmall fpecies of cetaceous fifh, which he calls *thornbacks*, from their broad and fharp fin on the back, fome of thefe were brown, fome quite white, others fpotted: but whether they were only a variety of this fifh, or whether they were fmall *grampufes*, which are alfo fpotted, we cannot determine.

FAT.

The porpeffe is remarkable for the vaft quantity of the fat or lard that furrounds the body, which

which yields a great quantity of excellent oil : from this lard, or from their rooting like fwine, they are called in many places *fea hogs*; the *Germans* call them *meerfchwein*; the *Swedes*, *marfuin*; and the *Englifh*, *porpeffe*, from the *Italian*, *porco pefce*.

It would be curious to trace the revolutions of fafhion in the article of eatables; what epicure firft rejected the Sea-Gull and Heron; and what delicate ftomach firft naufeated the greafy flefh of the Porpeffe. This latter was once a royal difh, even fo late as the reign of *Henry* VIII. and from its magnitude muft have held a very refpectable ftation at the table; for in a houfhold book of that prince, extracts of which are publifhed in the third volume of the *Archaelogia*, it is ordered that if a Porpeffe fhould be too big for a horfe-load, allowance fhould be made to the purveyor. I find that this fifh continued in vogue even in the reign of *Elizabeth*, for Doctor *Caius* * on mentioning a Dolphin (that was taken at *Shoreham*, and brought to *Thomas* Duke of *Norfolk*, who divided, and fent it as a prefent to his friends) fays, that it eat beft with *Porpeffe fauce*, which was made of vinegar, crums of fine bread, and fugar.

* *Opufcula*, 116.

F 4 Orca

26. GRAM-
   PUS.

Orca *Plinii, Lib.* IX. *c.* 6.
L'oudre ou grand marfouin.
   Belon, 13.
Orca. *Rondel.* 483. *Gefner*
   pifc. 6;5. Leper, Springer.
   Schonevelde, 53.
Butſkopf. *Marten's Spitzberg.*
   124.
Balæna minor utraque maxilla

dentata. *Sib. Phalæn.* 7, 8.
Wil. Icth. 40. *Raii fyn.*
   pifc. 15.
L'Epaulard. *Briſſon Cet.* 236.
Delphinus orca. *Lin. fyſt.* 108.
Lopare, Delphinus roſtro fur-
   ſum repando, dentibus la-
   tis ferratis. *Arted. fyn.* 106.

THIS ſpecies is found from the length of fifteen feet to that of twenty-five. It is remarkably thick in proportion to its length, one of
eighteen feet being in the thickeſt place ten feet
diameter. With reaſon then did *Pliny* call this
an immenſe heap of fleſh, armed with dreadful
teeth *.

It is extremely voracious, and will not even
ſpare the porpeſſe, a congenerous fiſh. It is ſaid
to be a great enemy to the whale, and that it will
faſten on it like a dog on a bull, till the animal
roars with pain.

TEETH.

The noſe is flat, and turns up at the end.
There are thirty teeth in each jaw; thoſe before
are blunt, round, and ſlender; the fartheſt ſharp
and thick: between each is a ſpace adapted to receive the teeth of the oppoſite jaw when the mouth
is cloſed.

* Cujus imago nulla repreſentatione exprimi poſſit alia,
quam carnis immenſæ dentibus truculentis. *Lib.* IX. *c.* 6.

The

The spout hole is in the top of the neck. In respect to the number and site of the fins, it agrees with the dolphin.

The color of the back is black, but on each shoulder is a large white spot, the sides marbled with black and white, the belly of a snowy whiteness.

These sometimes appear on our coasts, but are found in much greater numbers off the *North Cape* in *Norway*, whence they are called the *North Capers*. These and all other whales are observed to swim against the wind, and to be much disturbed, and tumble about with unusual violence at the approach of a storm.

Linnæus and *Artedi* say, that this species is furnished with broad serrated teeth, which as far as we have observed, is peculiar to the *shark* tribe. We therefore suspect that those naturalists have had recourse to *Rondeletius*, and copied his erroneous account of the teeth: Sir *Robert Sibbald*, who had opportunity of examining and figuring the teeth of this fish, and from whom we take that part of our description, giving a very different account of them.

It will be but justice to say, that no one of our countrymen ever did so much towards forming a general natural history of this kingdom as Sir *Robert Sibbald*: he sketched out a fine outline of the Zoology of *Scotland*, which comprehends the greatest part of the *English* animals, and, we are told,

had

had actually filled up a confiderable part of it : he publifhed a particular hiftory of the county of *Fife*, and has left us a moft excellent hiftory of the whales which frequent the coaft of *Scotland.* We acknowledge ourfelves much indebted to him for information in refpect to many of thofe fifh, few of which frequent the fouthern feas of thefe kingdoms, and thofe that are accidentally caft afhore on our coafts, are generally cut up by the country people, before an opportunity can be had of examining them.

Div.

Div. II. CARTILAGINOUS FISH.

THIS title is given to all fish whose muscles are supported by cartilages instead of bones, and comprehends the same genera of fish to which *Linnæus* has given the name of *amphibia nantes:* but the word *amphibia,* ought properly to be confined to such animals who inhabit both elements, and can live without any inconvenience for a considerable space, either in land or under water; such as *tortoises, frogs,* and several species of *lizards*; and among the quadrupeds, hippopotami, seals, &c. &c. This definition therefore excludes all that form this division.

Many of the cartilaginous fish are viviparous, being excluded from an egg, which is hatched within them. The egg consists of a white and a yolk, and is lodged in a case, formed of a thick tough substance, not unlike softened horn: such are the eggs of the *Ray* and *Shark* kinds.

Some again differ in this respect, and are oviparous; such is the *Sturgeon,* and others.

They breathe either through certain apertures beneath, as in the *Rays*; on their sides as in the *Sharks,* &c. or on the top of the head, as in the *Pipe-fish*; for they have not covers to their gills like the bony fish.

Slender

IV.
LAMPREY.

Slender Eel-fhaped body.
Seven apertures on each fide;
One on the top of the head.
No pectoral or ventral fins.

27. SEA.

La Lamproye de mer. *Belon*, 66.
Lampetra. *Rondel*. 398.
Lampreda. *Gefner*. *Paralip*, 22. *Pifc.* 590.
Lamprey, or Lamprey Eel. *Wil. Ich.* 105.
Lampetra. *Raii fyn. pifc.* 35.

Petromyzon maculofus ordinibus dentium circiter viginti. *Arted. fynon.* 90.
Petromyzon marinus. P. ore intus papillofo, pinna dorfali pofteriori a cauda diftincta. *Lin. fyft.* 394. *Faun. Suec.* No. 292.

PLACE.

LAMPREYS are found at certain feafons of the year in feveral of our rivers, but the *Severn* is the moft noted for them*. They are fea fifh, but like falmon, quit the falt waters, and afcend the latter end of the winter, or beginning of fpring, and after a ftay of a few months return again to the ocean, a very few excepted. The beft feafon for them is in the months of *March*, *April*, and *May*; for they are more firm when juft arrived out of the falt water than they are afterwards, being obferved to be much wafted, and very flabby at the approach of hot weather.

* They are alfo found in the moft confiderable of the *Scotch* and *Irifh* rivers.

They

Pl. VIII.

LAMPRIE S.

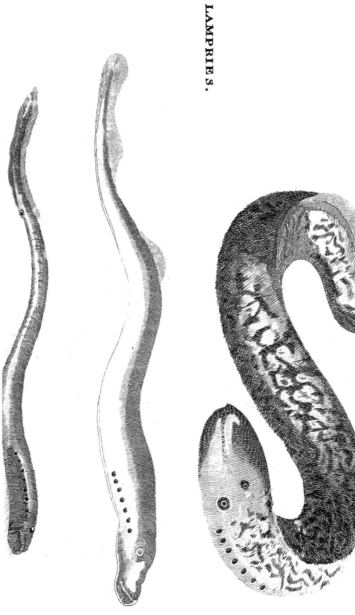

N.º 27.

N.º 28.

N.º 29.

They are taken in nets along with falmon and fhad, and fometimes in weels laid in the bottom of the river.

It has been an old cuftom for the city of *Gloucefter*, annually, to prefent his majefty with a *lamprey* pye, covered with a large raifed cruft. As the gift is made at *Chriftmas*, it is with great difficulty the corporation can procure any frefh lampreys at that time, though they give a guinea a-piece for them, fo early in the feafon.

They are reckoned a great delicacy, either when potted or ftewed, but are a furfeiting food, as one of our monarchs fatally experienced, *Henry* the Firft's death being occafioned by a too plentiful meal of thefe fifh.   It appears that notwithftanding this accident, they continued in high efteem; for *Henry* the Fourth granted protections to fuch fhips as brought over lampreys for the table of his royal confort*.   His fucceffor iffues out a warrant to *William* of *Nantes*, for fupplying him and his army with lampreys, wherefoever they happen to march †. Directions are afterwards given that they fhould be taken between the mouth of the *Seyne* and *Harfleur*.

Lampreys are fometimes found fo large as to weigh four or five pounds.

The mouth is round and placed rather obliquely below the end of the nofe: the edges are jagged,

* *Rymer*, VIII. 429.
† Idem. IX. 544.

which

which enables them to adhere the more ſtrongly to the ſtones, as their cuſtom is, and which they do ſo firmly as not to be drawn off without ſome difficulty.

We have heard of one weighing three pounds, which was taken out of the *Eſk*, adhering to a ſtone of twelve pounds weight, ſuſpended at its mouth, from which it was forced with no ſmall pains.

There are in the mouth twenty rows of ſmall teeth, diſpoſed in circular orders, and placed far within.

The color is duſky, irregularly marked with dirty yellow, which gives the fiſh a diſagreeable look.

Not the Muræna.   We believe that the ancients were unacquainted with this fiſh ; ſo far is certain, that which Doctor *Arbuthnot*, and other learned men, render the word *lamprey*, is a ſpecies unknown in our ſeas, being the *muræna* of *Ovid*, *Pliny*, and others, for which we want an *Engliſh* name. This fiſh, the *Lupus* (our Baſſe) and the *Myxo* * (a ſpecies of mullet) formed that pride of *Roman* banquets, the *Tripatinam* †, ſo called according to *Arbuthnot*, from their being ſerved up in a machine with three bottoms.

---

* Perhaps the ſpecies called by *Rondeletius*, *Muge*, and *Maxon*. de Piſc. P. 295.

† Atque ut luxu quoque aliqua contingat auctoritas figlinis, *Tripatinam*, inquit *Feneſtella*, appellabatur, ſumma cænarum lautitia. una erat *Murænarum*, altera *Luporum*, tertia *Myxonis* piſcis. *Plinii* Hiſt. Nat. lib. XXXV. c. 12.

The

The words *Lampetra* and *Petromyzon*, are but of modern date, invented from the nature of the fish; the first a *Lambendo petras*, the other from Πέτρος, and Μυσαω, because they are supposed to lick, or suck the rocks.

La Lamproye d'eaue doulce. *Belon*, 67.	Petromyzon fluviatilis. *Lin. syst.* 394.
Lampredæ alterum genus. *Gesner pisc.* 597.	Nein-oga, natting. *Faun. Suec. No.* 290. Petromyzon pinna dorsali posteriori angulata. *Ibid.*
Lampetra medium genus. *Wil. Icth.* 106. *Raii syn. pisc.* 35.	Gronov. *Zooph. No.* 159.
Neunaugen. *Kram.* 282.	

28. Lesser.

THIS species sometimes grows to the length of ten inches.

Descrip.

The mouth is formed like that of the preceding. On the upper part is a large bifurcated tooth; on each side are three rows of very minute ones: on the lower part are seven teeth, the exterior of which on each side is the largest.

The irides are yellow. As in all the other species, between the eyes, on the top of the head, is a small orifice of great use to clear its mouth of the water that remains on adhering to the stones, for through that orifice it ejects the water in the same manner as cetaceous fish.

On the lower part of the back is a narrow fin, beneath that rises another, which at the beginning

is

is high and angular, then grows narrow, furrounds the tail, and ends near the anus.

Color.　　　The color of the back is brown or dufky, and fometimes mixed with blue; the whole under fide filvery. Thefe are found in the *Thames, Severn,* and *Dee,* are potted with the larger kind, and are by fome preferred to it, as being milder tafted. Vaft quantities are taken about *Mortlake,* and fold to the *Dutch* for bait for their cod fifhery. Above 450,000 have been fold in a feafon at forty fhillings per thoufand. Of late, about 100,000 have been fent to *Harwich* for the fame purpofe. It is faid that the *Dutch* have the fecret of preferving them till the Turbot fifhery.

29. Pride.　　Une Civelle, un Lamproyon. *Belon,* 67.
Lampetra parva et fluviatilis. *Rondel. pifc. fl.* 202.
Lampreda minima. *Gefner pifc.* 598.
Pride. *Plot, Oxf.* 182. *Plate* X.
Lampern, or Pride of the *Ifis. Wil. Ichth.* 104. *Raii fyn.*

pifc. 35.
Petromyzon branchialis. *Lin. fyft.* 394.
Lin-ahl. *Faun. Suec.* No. 291.
Petromyzon pinna dorfali pofteriori lineari, labio oris latere poftico lobato. *Ibid.*
Uhlen. *Kram.* 384.
Gronov. Zooph. No. 160.

WE have feen thefe of the length of eight inches, and about the thicknefs of a fwan's quil, but they are generally much fmaller.

They are frequent in the rivers near *Oxford,* particularly the *Ifis,* but not peculiar to that county, ty, being found in others of the *Englifh* rivers, where, inftead of concealing themfelves under the ftones, they lodge themfelves in the mud, and never are obferved to adhere to any thing like other lampreys.

The body is marked with numbers of tranfverfe lines, that pafs crofs the fides from the back to the bottom of the belly, which is divided from the mouth to the anus by a ftrait line.

The back fin is not angular like that of the former, but of an equal breadth. The tail is lanceolated, and fharp at the end.

V.
RAY.

Body broad, flat, and thin.
Five apertures on each side placed beneath:
Mouth situated quite below.

* With sharp teeth.

30. SKATE.   Βατὶς? *Arist. hist. an. Lib.* | *Ith.* 69. *Raii syn. pisc.* 25.
I. *c.* 5. *Lib.* VI. *c.* 10. | Raia Batis. *Lin. syst.* 395.
Oppian Halieut. I. 103. | Raia varia, dorso medio glá-
Raia undulata five cinerea. | bro, unico aculeorum or-
Rondel. 346. *Gesner pisc.* | dine in cauda. *Arted. synon.*
791. | 102.
The Skate, or Flaire. *Wil.* | Gronov. Zooph. No. 157.

SIZE.   THIS species is the thinnest in proportion to
its bulk of any of the genus, and also the
largest, some weighing near two hundred pounds.

DESCRIP.   The nose, though not long, is sharp pointed;
above the eyes is a set of short spines: the whole
upper part of that we examined was of a pale
brown. Mr. *Ray* says, some he saw were streaked with black: the lower part is white, marked
with great numbers of minute black spots. The
jaws were covered with small granulated but sharp-
pointed teeth.

The tail is of a moderate length: near the
end are two fins: along the top of it is one row of
spines, and on the edges are irregularly dispersed
a few

SKATE.

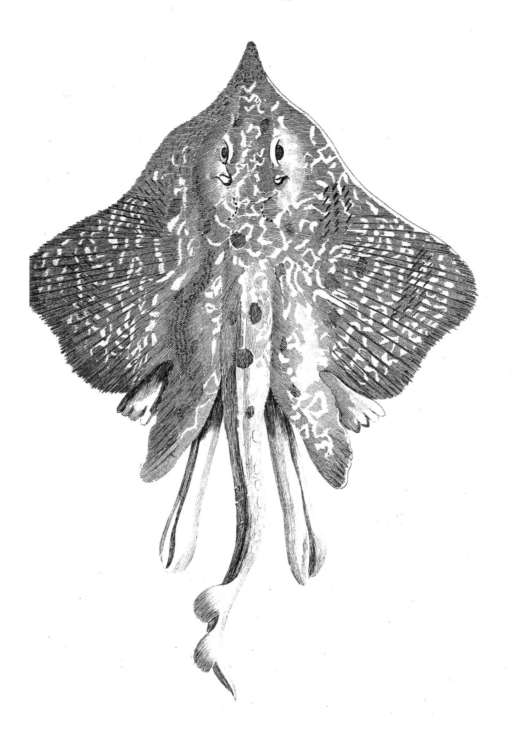

a few others, which makes us imagine with Mr. *Ray*, that in this respect these fish vary, some having one, others more orders of spines on the tail.

It is remarked that in the males of this species the fins are full of spines.

Skates generate in *March* and *April*, at which time they swim near the surface of the water, several of the males pursuing one female. They adhere so fast together in coition, that the fishermen frequently draw up both together, though only one has taken the bait. The females begin to cast their *purses*, as the fishermen call them (the bags in which the young are included) in *May*, and continue doing it till *September*. In *October* they are exceedingly poor and thin; but in *November* they begin to improve, and grow gradually better till *May*, when they are in the highest perfection. The males go sooner out of season than the females.

GENERA-
TION.

Bꞇ? *Arift. hift. an. Lib. V. c. 5. Oppian Halieut.* II. 141.
Bos *Ovidii?* 94. *Plinii, Lib.* IX. *c.* 24.
Raia oxyrhincus. *Rondel.* 347. *Gefner pifc.* 792.

Wil. Ichth. 71. *Raii fyn. pifc.* 26.
Raia oxyrinchus. *Lin. fyft.* 395.
Raia varia tuberculis decem aculeatis in medio dorfo. *Arted. fynon.* 101.

31. SHARP-
NOSED.

IN fishing in the *Menai* (the strait that divides *Anglefea* from *Caernarvonfhire*) *July* 1768, we

G 2

took

took one of this species whose length was near seven feet, and breadth five feet two inches; when just brought on shore, it made a remarkable snorting noise.

The nose was very long, narrow, and sharp-pointed, not unlike the end of a spontoon.

The body was smooth, and very thin in proportion to the size; the upper part ash colored, spotted with numerous white spots, and a few black ones.

The tail was thick; towards the end were two small fins, on each side was a row of small spines, with another row in the middle, which run some way up the back.

The lower part of the fish was quite white.

The mouth very large, and furnished with numbers of small sharp teeth bending inwards.

On its body we found the *hirudo muricata*, which adhered very strongly, and when taken off left a black impression.

This fish has been supposed to be the *Bos* of the antients, which was certainly some enormous species of *Ray*, though we cannot pretend to determine the particular kind: *Oppian* styles it,

Ευρυτατ☺ παντεσσι μετ' ίχθυσιν.

Broadest among fishes.

He adds an account of its fondness of human flesh, and the method it takes of destroying men,

by

by over-laying and keeping them down by its vaft weight till they are drowned. *Phile* gives much the fame relation *.   We are inclined to give them credit, fince a modern writer †, of undoubted authority, gives the very fame account of a fifh found in the *South Seas*, the terror of thofe employed in the pearl fifhery.   It is a fpecies of *ray*, called there *Manta*, or the *Quilt*, from its furrounding and wrapping up the unhappy divers till they are fuffocated; therefore the negroes never go down, without a fharp knife to defend themfelves againft the affaults of this terrible enemy.

Raia alteria afpera. *Rondel.* 352.   32. ROUGH.
Gefner pifc. 794. *Wil. Ichth.* 78.
Raii fyn. pifc. 28.

I TOOK this fpecies in *Loch Broom* in the fhire of

The length from the nofe to the tip of the tail was two feet nine.   The tail was almoft of the fame length with the body.

Nofe very fhort.   Before each eye a large hooked fpine, and behind each another, befet with leffer. The upper part of the body of a cinereous brown

* *De propriet. Anim.* 85.
† *Ulloa's voy.* I. 132. 8vo. edit.

G 3   mixed

mixed with white, and fpotted with black; and entirely covered with fmall fpines. On the tail were three rows of great fpines: all the reft of the tail was irregularly befet with leffer.

The fins, and under fide of the body were equally rough with the upper.

The teeth were flat, and rhomboidal.

53. Fᴜʟʟᴇʀ. Raia fullonica. *Rondel.* 357. *Gefner pifc.* 797. Raia afpera noftras, the white horfe. *Wil. Ichth.* 78. *Raii fyn. pifc.* 26. Raia fullonica. *Lin. fyft.*

Raia dorfo toto aculeato, aculeorum ordine fimplici ad oculos, duplici in cauda. *Arted. fyn.* 101. *Gronov, Zooph.* No. 155.

THIS fpecies derives its *Latin* name from the inftruments fullers make ufe of in fmoothing cloth, the back being rough and fpiney.

The nofe is fhort and fharp. At the corner of each eye a few fpines. The membrane of nictitation is fringed. Teeth fmall, and fharp.

On the upper part of the pectoral fins are three rows of fpines pointing towards the back, crooked, like thofe on a fuller's inftrument.

On the tail are three rows of ftrong fpines: the middle row reaches up part of the back. The tail is flender, and rather longer than the body.

The color of the upper part of the body is cinereous, marked ufually with numerous black

fpots:

fpots: the lower part is white. This, as well as most other species of Rays, vary a little in color, according to age.

This grows to a fize equal to the Skate. It is common at *Scarborough*, where it is called the *White Hans*, or Gullet.

I MET with this fpecies at *Scarborough*, where it is called the *French* Ray.

34. SHAG-
REEN.

It encreafes to the fize of the *Skate*; is fond of *Launces*, or Sandeels, which it takes greedily as a bait.

The form is narrower than that of the common kinds: the nofe long and very fharp: pupil of the eye, fapphirine: on the nofe are two fhort rows of fpines: on the corner of the eyes another of a femicircular form: on the tail are two rows, continued a little up the back, fmall, flender, and very fharp: along the fides of the tail is a row of minute fpines, intermixed with innumerable little *fpiculæ*. The upper part of the body is of a cinereous brown, covered clofely with minute fhagreen-like tubercles, refembling the fkin of the dog-fifh: the under fide of the body is white: from the nofe to the beginning of the pectoral fins is a tuberculated fpace.

The teeth flender, and fharp as needles.

G 4

Iaberete?

35. WHIP.   Iaberete ? BRAZIL : *Marcgrave.* 175.

MR. *Travis*, furgeon at *Scarborough*, had, in the fummer of 1769, the tail of a Ray brought to him by a fifherman of that town : he had taken it in the fea off the coaft, but flung away the body.

It was above three feet long, extremely flender and taper, and deftitute of a fin at the end. I believe it to belong to the fpecies called by the *Brafilians Iaberete* ; and that it is likewife found in the *Sicilian* feas. I once received the tail of one from that ifland, correfponding with the defcription Mr. *Travis* gave : I muft alfo add, that it was entirely covered with hard obtufe tubercles.

Naxn.

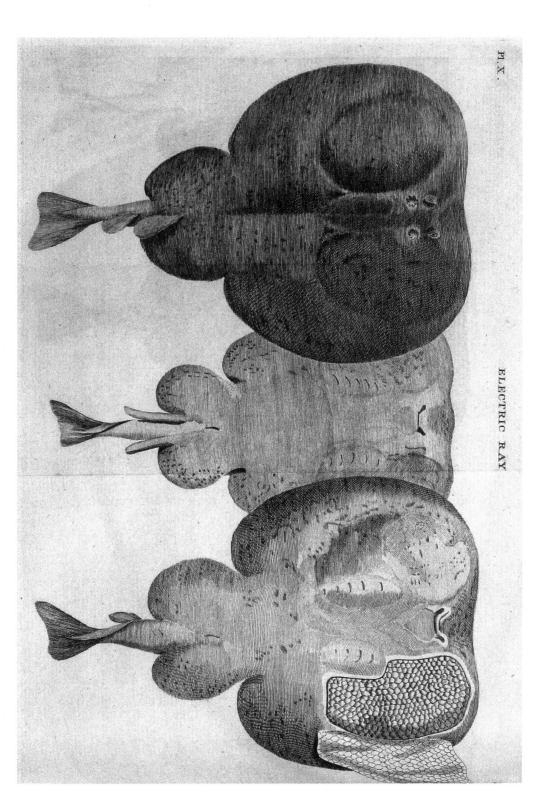

Pl. X.

ELECTRIC RAY

Ναρκη. *Arift. Hift. an. lib.*
V. *c.* 5. IX *c.* 37. *Op-*
pian Halieut. I. 104. II.
56. III. 149.
Torpedo. *Plinii lib.* IX. *c.*
42.
La Tremble ou Torpille.
Belon 78, 81.
Torpedo. *Rondel. Gefner pifc.*

Torpedo. Cramp Fifh. *Wil.*
Icth. 81. *Raii fyn. pifc.* 28.
Smith's Hift. Waterford, 271.
Raia Torpedo. *Lin. fyft.*
395.
Raia tota lævis. *Arted. fynon.*
102. *Gronov. Zooph. No.*
153. *tab.* 9.
Ph. Tr. 1773, 1774.

36. ELEC-
TRIC.

THE narcotic or numbing quality of this fifh has been taken notice of in all ages: it is fo powerful when the fifh is alive, as inftantly to deprive the perfon who touches it of the ufe of his arm, and even to affect him if he touches it with a ftick. *Oppian* goes fo far as to fay, that it will benumb the aftonifhed fifherman, even through the whole length of line and rod.

Ναὶ μὲν κι ΝΑΡΚΗ σφέτερον νόον ὲκ απολείπει,
Πληγῆ ἀνίαζεσα. Τιταινομένη δ᾿ ὀδύνησιν
Ὁρμιῆ λαγόνας προστύσσεται. Αἶψα δε χαίτης
Ἱππείης δόνακος τε δίεδραμεν, ἐς δ᾿ ἁλιῆ-
Δεξιτερην ἔσκηψε φερώνυμον ἰχθυ᾿- ἄλγ᾿.
Πολλάκι δ᾿εκ παλάμης κάλαμ᾿- πέσεν, ὅπλά τε θήρης
Τοῖ᾿- γαρ κρύταλλ᾿- ενιζεται ἀντικα χειρί.

The hook'd *Torpedo* ne'er forgets its art,
But foon as ftruck begins to play its part,
And to the line applies its magic fides,
Without delay the fubtile power glides
Along the pliant rod, and flender hairs,

Then

Then to the fisher's hand as swift repairs :
Amaz'd he stands ; his arm's of sense bereft,
Down drops the idle rod ; his prey is left :
Not less benumb'd, than if he had felt the whole
Of frost's severest rage beneath the *arctic* pole.

But great as its powers are when the fish is in vigor, they are impaired as it declines in strength, and totally cease when it expires. They impart no noxious qualities to it as a food, being commonly eaten by the *French*, who find them more frequently on their coasts than we do on ours.

Galen affirms, that the meat of the *torpedo* is of service to epileptic patients : and that the shock of the living fish applied to the head is efficacious in removing any pains in that part.

We may mention a double use in this strange power the *torpedo* is endued with ; the one, when it is exerted as a means of defence against voracious fish, who are at a touch deprived of all possibility of seizing their prey.

The other is well explained by *Pliny*, who tells us, it attains by the same powers its end in respect to those fish it wishes to ensnare. *Novit* torpedo *vim suam, ipsa non torpens ; mersaque in limo se occultat piscium qui securi supernatantes obtorpuere, corripiens* *.

* " The *torpedo* is well acquainted with its own powers,
" though itself never affected by them. It conceals itself in
" the mud, and benumbing the fish that are carelesly swim-
" ming about, makes a ready prey of them."

But

But the acknowledgements of every naturalist are due to *John Walsh*, Esquire, for his curious and unwearied researches into the nature of this fish; and for the first certainty we had of its being a native of our seas.   To him I am particularly bound, for being enabled to correct my errors in the former account.

IT is frequently taken in *Torbay*; has been once caught off *Pembroke*, and sometimes near *Waterford* in *Ireland*.   It is generally taken, like other flat fish, with the trawl; but there is an instance of its taking a bait, which vindicates the fine account that *Oppian* has left us of this fish. It commonly lies in water of about forty fathoms depth; and in company with the congenerous Rays.

The *torpedo* brings forth its young at the autumnal *equinox* as affirmed by *Aristotle*.   A gentleman of *la Rochelle*, on dissecting certain females of this species, the 10th of *September*, found in the *matrices*, several of the *fœtuses* quite formed, and nine eggs, in no state of forwardness: superfœtation seems therefore to be a property of this fish.

The food of the *torpedo* is fish; a surmullet and a plaise having been found in the stomach of two of them.   The surmullet is a fish of that swiftness, that it was impossible for the torpedo to take it by pursuit.   It is probable, that by their electric stroke, they stupify their prey; yet the crab and sea leech will venture to annoy them.

They

They will live four and twenty hours out of the fea; and but very little longer if placed in freſh water.

They inhabit ſandy places; and will bury themſelves ſuperficially in it, by flinging the ſand over, by a quick flapping of all the extremities. It is in this ſituation that the *torpedo* gives his moſt forcible ſhock, which throws down the aſtoniſhed paſſenger, who inadvertently treads upon him.

In our ſeas it grows to a great ſize, and above eighty pounds weight. My deſcription was taken from a ſmaller, which I had the pleaſure of doing in company with Mr. *Walſh*.

Its length was eighteen inches from the head to the tip of the tail; the greateſt breadth twelve inches. I could not inform myſelf of the weight of this; but that of one, that meaſured four feet in length, and two and a half in breadth, was fifty-three pounds, avoirdupoiſe.

The tail was ſix inches long; was pretty thick and round: the caudal fin broad and abrupt.

The head and body, which were indiſtinct, were nearly round; about two inches thick in the middle, attenuating to extreme thinneſs on the edges: below the body, the ventral fins formed on each ſide a quarter of a circle. The two dorſal fins were placed on the trunk of the tail.

The eyes were ſmall, placed near each other: behind each was a round ſpiracle, with ſix ſmall cutaneous rags on their inner circumference.

Mouth

Pl. XI.

N.º 37.

THORN-BACK.

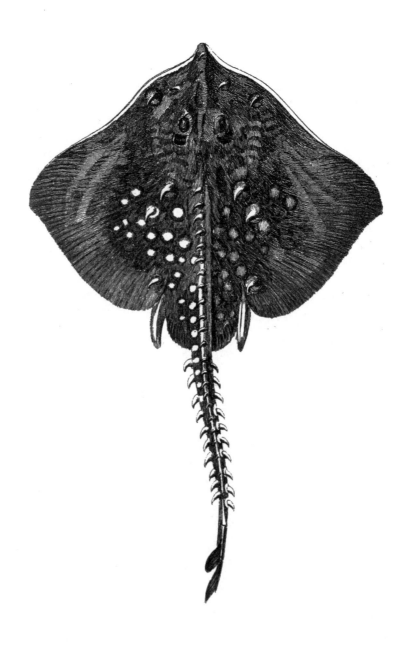

Pl. XII.

THORNBACK.

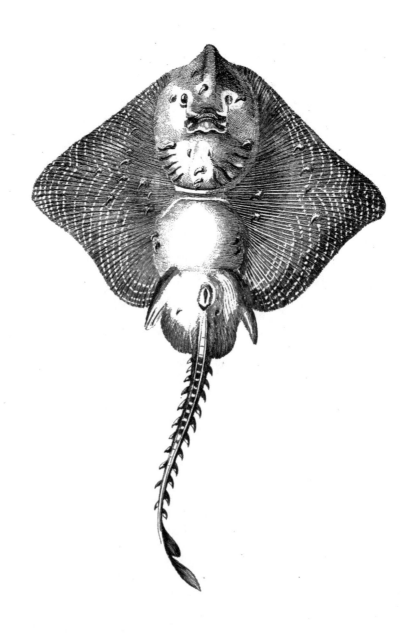

Mouth fmall : teeth minute, fpicular.

Five openings to the gills, as in others of this genus.

The fkin every where fmooth : cinereous brown above; white beneath.

** With blunt Teeth.

La Raye bouclée. *Belon* 70.
Raia clavata. *Rondel.* 353.
  Gefner pifc. 795.
Steinroch. *Shonevelde*, 59.
Thornback. *Wil. Icth.* 74.
  Raii fyn. pifc. 26.

Raia clavata. *Lin. fyft.* 297. 37. Thorn-
Gronov. Zooph. No. 154.   back.
R. aculeata dentibus tubercu-
  lofis, cartilagine tranfverfa
  abdominali. *Arted. fynon.* 94.
Racka. *Faun. Suec.* No. 293.

THIS common fifh is eafily diftinguifhed from the others by the rows of ftrong fharp fpines, difpofed along the back and tail. In a large one we faw, were three rows on the back, and five on the tail, all inclining towards its end.

On the nofe, and on the inner fide of the forehead, near the eyes, were a few fpines, and others were fcattered without any order on the upper part of the pectoral fins.

The mouth was fmall, and filled with granulated teeth.

The upper part of the body was of a pale afh color, marked with fhort ftreaks of black, and the fkin rough, with fmall tubercles like fhagreen.

The

The belly white, croffed with a ftrong femilunar cartilage beneath the fkin: in general the lower part was fmooth, having only a few fpines on each fide.

The young fifh have very few fpines on them, and their backs are often fpotted with white, and each fpot is encircled with black.

FOOD.

This fpecies frequents our fandy fhores, are very voracious, and feed on all forts of flat fifh, and are particularly fond of herrings and fand eels, and fometimes eat cruftaceous animals, fuch as crabs.

Thefe fometimes weigh fourteen or fifteen pounds, but with us feldom exceed that weight.

They begin to generate in *June*, and bring forth their young in *July* and *Auguft*, which (as well as thofe of the fkate) before they are old enough to breed, are called *maids*. The thornback begins to be in feafon in *November*, and continues fo later than fkate, but the young of both are good at all times of the year.

Τρυγων

Τρυγων. *Arift. Hift. an. lib.*
  VIII. *c.* 13. IX. 37. *Op-*
  pian. Halieut. I. 104. II.
  462.
Paftinaca *Plinii lib.* IX. *c.* 42.
  38.
La Paftenade de mer, Tour-
  terelle, ou Tareronde. *Be-*
  lon 83.
Paftinaca. *Rondel.* 331. *Gef-*
  ner pifc. 679.

Steckroche. Grone Tepel.   38. STING.
  Schonevelde, 58.
Paftinaca marina lævis. *Wil.*
  pifc. 67.
Fire Flaire. *Raii fyn. pifc.* 24.
Raia Paftinaca. *Lin. fyft.* 396.
Raia corpore glabro, aculeo
  longo anterius ferrato, cauda
  apterygia. *Arted. fynon.*
  100. *Gronov. Zeoph. No.*
  158.

T HE weapon with which nature has armed this
  fifh, hath fupplied the antients with many
tremendous fables relating to it. *Pliny, Ælian*,*
and *Oppian,* have given it a venom that affects even
the inanimate creation: trees that are ftruck by it
inftantly lofe their verdure and perifh, and rocks
themfelves are incapable of refifting the potent poifon.

The enchantrefs *Circe,* armed her fon with a
fpear headed with the fpine of the *Trygon,* as the
moft irrefiftible weapon fhe could furnifh him with,
and with which he afterwards committed parri-
cide, unintentionally, on his father *Ulyffes.*

That fpears and darts might, in very early
times, have been headed with this bone inftead of
iron, we have no kind of doubt: that of another
fpecies of this fifh being ftill ufed to point the ar-

---

* *Hift. an. lib.* II. *c.* 36.

rows]

rows of some of the *South American Indians*, and is, from its hardness, sharpness, and beards, a most dreadful weapon.

But in respect to its venemous qualities, there is not the left credit to be given to the opinion, though it was believed (as far as it affected the animal world) by *Rondeletius*, *Aldrovand*, and others, and even to this day by the fishermen in several parts of the kingdom. It is in fact the weapon of offence belonging to the fish, capable of giving a very bad wound, and which is attended with dangerous symptoms, when it falls on a tendinous part, or on a person in a bad habit of body. As to any fish having a spine charged with actual poison, we must deny our assent to it, though the report is sanctified by the name of *Linnæus**.

DESCRIP. This species does not grow to the bulk of the others: that which we examined was two feet nine inches from the tip of the nose to the end of the tail; to the origin of the tail one foot three inches; the breadth one foot eight.

The body is quite smooth, of a shape almost round, and is of a much greater thickness, and

---

* *Syst. Nat.* I. 348. He instances the *Pastinaca*, the *Torpedo*, and the *Tetrodon lineatus*. The first is incapable of conveying a greater injury than what results from the meer wound. The second, from its electric effluvia: and the third, by imparting a pungent pain like the sting of nettles, occasioned by the minute spines on its abdomen.

more

more elevated form in the middle than any other *Rays*, but grows very thin towards the edges.

The nose is very sharp pointed, but short; the mouth small, and filled with granulated teeth.

The irides are of a gold color: behind each eye the orifice is very large.

The tail is very thick at the beginning: the spine is placed about a third the length of the former from the body, is about five inches long, flat on the top and bottom, very hard, sharp pointed, and the two sides thin, and closely and sharply bearded the whole way.   The tail extends four inches beyond the end of this spine, and grows very slender at the extremity.

TAIL.

These fish are observed to shed their spine, and to renew them annually; sometimes the new spine appears before the old one drops off, and the *Cornish* call this species *Cardinal Triloft*, or three tailed, when so circumstanced.

The color of the upper part of the body is a dirty yellow, the middle part of an obscure blue: the lower side white, the tail and spine dusky.

Vol. III.   H   Slender

VI.
SHARK.

Slender body growing lefs towards the tail.

Two fins on the back.

Rough fkin.

Five apertures on the fides of the neck.

Mouth generally placed far beneath the end of the nofe.

The upper part of the tail longer than the lower.

* Without the anal fin.

39. ANGEL.   *Plin. Arift. Hift. an. lib.* **V.** *c.* 5, &c. *Athenæus, lib.* **VII.** *p.* 319.
Oppian Halieut. I. 388, 742.
Squatina *Plin. lib.* IX. *c.* 12.
Rhina, fc. Squatus. *lib.* XXXII. *c.* 11.
L'Ange, ou Angelot de mer. *Belon* 69.
Squatina. *Rondel.* 367. *Gefner*

pifc. 899. *Wil. Icth.* 79.
Monk, or Angel Fifh. *Raii fyn. pifc.* 26.
Squalus fquatina. *Lin. fyft.* 398. S. pinna ani nulla, caudæ duabus, ore terminali, naribus cirrofis. *Ibid.*
Sq. pinna ani carens, ore in apice capitis. *Arted. fyn.* 95.
Gronov. Zooph. No. 151.

THIS is the fifh which conneçts the genus of Rays and Sharks, partaking fomething of the charaçter of both; yet in an exception to each in the fituation of the mouth, which is placed at the extremity of the head.

It is a fifh not unfrequent on moft of our coafts, where it prowls about for prey like others of the kind. It is extremely voracious, and, like the

Ray,

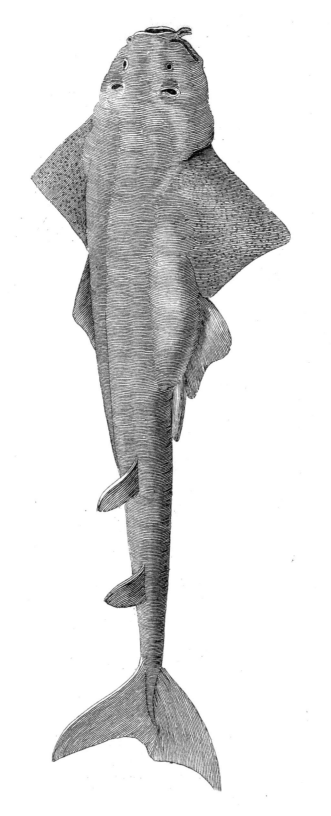

ANGEL SHARK.

Ray, feeds on flounders and flat fish, which keep
at the bottom of the water, as we have often found
on opening them.   It is extremely fierce and dangerous to be approached.   We knew an instance FIERCENESS.
of a fisherman, whose leg was terribly tore by a
large one of this species, which lay within his nets
in shallow water, and which he went to lay hold
of incautiously.

The aspect of these, as well as the rest of the
genus have much malignity in them: their eyes are
oblong, and placed lengthways in their head, sunk
in it, and overhung by the skin, and seem fuller
of malevolence than fire.

Their skin is very rough; the antients made use
of it to polish wood and ivory *, as we do at present that of the greater dog-fish.   The flesh is now
but little esteemed on account of its coarseness and
rankness, yet *Archestratus* (as quoted by *Athenæus*,
p. 319.) speaking of the fish of *Miletus*, gives this
the first place in respect to its delicacy of the whole
cartilaginous tribe.

They grow to a great size; we have seen them DESCRIP.
of near an hundred weight.

The head is large, the teeth broad at their base,
but slender and very sharp above, and disposed in
five rows all round the jaws.   Like those of all
Sharks, they are capable of being raised or depressed by means of muscles uniting them to the

---

* Qua lignum et ebora poliuntur.   *Plinii lib.* IX. c. 12.

H 2   jaws

jaws, not being lodged in fockets as the teeth of cetaceous fifh are.

The tongue is large; the eyes fmall; the pupil of a pale green; the irides white, fpotted with brown: behind each eye is a femilunar orifice.

The back is of a pale afh color, and very rough; along the middle is a prickly tuberculated line: the belly is white and fmooth.

The pectoral fins are very large, and extend horizontally from the body to a great diftance; they have fome refemblance to wings, fo writers have given this the name it bears in this work.

The ventral fins are placed in the fame manner, and the double penis is placed in them, which forms another character of the males in this and the laft genus.

The tail is bifurcated, the upper lobe rather the longeft: not very remote from the end on the back are two fins.

---

40. Picked. 'Ακανθιας γαλεος. *Arift. Hift.*
an. Lib. VI. *c.* 10. *Oppi-*
an Halieut. I. 380.
Επινωτις *Athenæi, Lib.* VII. *p.*
L' Efguillats *Belon,* 61.
Galeus acanthias. *Rondel.* 373.
Gefner pifc. 607.
Sperhaye, Dornhundt. *Scho-*
nevelde, 29.
Galeus acanthias five fpinax.
Wil. Ithb. 56.

The picked dog, or hound fifh. *Raii fyn. pifc.* 21.
Squalus fpinax. *Lin. fyft.* 397.
Sq. pinna ani nulla, dorfa-libus fpinofis, corpore tere-tiufculo. *Ibid.*
Sq. pinna ani nulla, corpore fubrotundo. *Arted. fynon.* 94.
Hai. *Faun. Suec.* No. 295.
Gronov. Zooph. 149.

Name. THE picked dog fifh takes its name from a ftrong and fharp fpine placed juft before each

of

Pl. XIII.

BASKING SHARK.

of the back fins, diftinguifhing it at once from the reft of the *Britifh* fharks.

The nofe is long, and extends greatly beyond the mouth, but is blunt at the end.   DESCRIP.

The teeth are difpofed in two rows, are fmall and fharp, and bend from the middle of each jaw towards the corners of the mouth.

The firft back fin is placed nearer the head than the tail; the other is fituated very near the latter.

The tail is finned for a confiderable fpace beneath, and the upper part is much the longeft.

The back is of a brownifh afh color; the belly white.

It grows to the weight of about twenty pounds.

This fpecies fwarms on the coafts of *Scotland*, where it is taken, fplit and dried: and is a food among the common people.   It forms a fort of internal commerce, being carried on women's backs, fourteen or fixteen miles up the country, and fold; or exchanged for neceffaries.

** With the anal fin.

SQUALUS maximus. SQ. dentibus caninis, pinna dorfali anteriore majore. *Syft. nat.* 400.
Brugden. Squalus maximus.

Gunner Act. Nidros. III. 41.BASKING.
33. *Tab.* II.
Sun-fifh. *Smith's hift. Cork,*
II. 292. *Hift. Waterford,*
271.

THIS fpecies has been long known to the inhabitants of the fouth and weft of *Ireland*

H 3   and

and *Scotland,* and thofe of *Caernarvonfhire* and *Anglefea*; but having never been confidered in any other than a commercial view, has till this time remained undefcribed by any *Englifh* writer; and what is worfe, miftaken for and confounded with the *luna* of *Rondeletius,* the fame that our *Englifh* writers call the *fun-fifh.*

The *Irifh* and *Welch* give it the fame name, from its lying as if to fun itfelf on the furface of the water; and for the fame reafon we have taken the liberty of calling it the *bafking fhark.* It was long taken for a fpecies of whale, till we pointed out the branchial orifices on the fides, and the perpendicular fite of the tail.

Thefe are migratory fifh, or at left it is but in a certain number of years that they are feen in multitudes on the *Welch* feas, though in moft fummers a fingle and perhaps ftrayed fifh appears. They inhabit the Northern feas, even as high as the *arctic* circle.

They vifited the bays of *Caernarvonfhire* and *Anglefea* in vaft fhoals, in the fummers of 1756 *, and a few fucceeding years, continuing there only the hot months, for they quitted the coaft about *Michaelmas,* as if cold weather was difagreeable to them.

They appear in the Firth of *Clyde*; and among

---

* Some old people fay they recollect the fame fort of fifh vifiting thefe feas in vaft numbers about forty years ago.

the

the *Hebrides* in the month of *June*, in small droves of seven or eight; but oftener in pairs. They continue in those seas, till the latter end of *July*, when they disappear.

They had nothing of the fierce and voracious nature of the shark kind, and were so tame as to suffer themselves to be stroked: they generally lay motionless on the surface, commonly on their bellies, but sometimes, like tired swimmers, on their backs.

Their food seemed to consist entirely of sea plants, no remains of fish being ever discovered in the stomachs of numbers that were cut up, except some green stuff, the half digested parts of *algæ*, and the like. *Linnæus* says, it feeds on *medusæ*.

FOOD.

At certain times they were seen sporting on the waves, and leaping with vast agility several feet out of the water. They swim very deliberately, with the dorsal fins above water.

Their length was from three to twelve yards, and sometimes even longer.

Their form was rather slender, like others of the shark kind.

The upper jaw was much longer than the lower, and blunt at the end. The mouth placed beneath, and each jaw furnished with numbers of small teeth: those before were much bent, those more remote in the jaws were conic and sharp pointed.

TEETH.

On the sides of the neck were five large transverse apertures to the gills.

On the back were two fins; the first very large,

H 4

not

not directly in the middle, but rather nearer the head; the other small, and situated near the tail. On the lower part were five others; viz. two pectoral fins; two ventral fins, placed just beneath the hind fin of the back; and a small anal fin. Near these, the male had two genitals, as in other sharks; and between these fins was situated the pudendum of the female.

The tail was very large, and the upper part remarkably longer than the lower.

The color of the upper part of the body was a deep leaden; the belly white.

The skin was rough, like shagreen, but less so on the belly than the back.

Within side the mouth, towards the throat, was a very short sort of whalebone.

LIVER.

O I L.

The liver was of a great size, but that of the female was the largest; some weighed above a thousand pounds, and yielded a great quantity of pure and sweet oil, fit for lamps, and also much used by the people who took them, to cure bruises, burns, and rheumatic complaints. A large fish has afforded to the captors a profit of twenty pounds. They were viviparous, a young one about a foot in length being found in the belly of a fish of this kind.

The measurements of one, I found dead on the shore of *Loch Ranza* in the isle of *Arran*, were as follow. The whole length twenty seven feet, four inches: first dorsal fin, three feet; second,

one

one foot ; pectoral fin, four feet ; ventral, two feet : the upper lobe of the tail, five feet; the lower, three.

They will permit a boat to follow them, without accelerating their motion, till it comes almost within contact ; when a harpooneer strikes his weapon into them, as near to the gills as possible. But they are often so insensible, as not to move till the united strength of two men have forced in the harpoon deeper. As soon as they perceive themselves wounded, they fling up their tail and plunge headlong to the bottom; and frequently coil the rope round them in their agonies, attempting to disengage the harpoon from them by rolling on the ground, for it is often found greatly bent.

As soon as they discover that their efforts are in vain, they swim away with amazing rapidity, and with such violence, that there has been an instance of a vessel of seventy tons having been towed away against a fresh gale. They sometimes run off with two hundred fathoms of line, and with two harpoons in them ; and will employ the fishers for twelve, and sometimes twenty four hours before they are subdued. When killed, they are either hawled on shore, or if at a distance from land, to the vessel's side. The liver (the only useful part) is taken out, and melted into oil in kettles provided for that purpose. A large fish will yield eight barrels of oil ; and two of worthless sedement.

The fishers observed on them a sort of leech of
a reddish

a reddiſh color, and about two feet long, but which fell off when the fiſh was brought to the ſurface of the water, and left a white mark on the ſkin.

42. WHITE.   Λαμια? *Ariſt. Hiſt. an. Lib.* V. *c.* 5. IX. *c.* 37.
Λαμιη. *Oppian Halieut.* I. 370. V. 36.
Καϱχαϱιας Κυων. *Athen. Lib.* VII. *p.* 310.
Lamia? *Plinii, Lib.* IX. *c.* 24.
Le chien carcharien ou Perlz fiſch de Noïvege. *Belon,* 52, 87.

Lamia. Tiburo. *Rondel.* 489, 390.
Canis Carcharias. *Geſner piſc.* 173.
White Shark. *Wil. Iɛ̃h.* 47. *Raii ſyn. piſc.* 18.
Squalus carcharias. Sq. dorſo plano dentibus ſerratis. *Lin. ſyſt.* 400.
Arted. ſynon. 89. *Gronov. Zooph.* No. 143.

SIZE.   THIS grows to a very great bulk, *Gillius* ſays, to the weight of four thouſand pounds; and that in the belly of one was found a human corps entire, which is far from incredible, conſidering their vaſt greedineſs after human fleſh.

They are the dread of the ſailors in all hot climates, where they conſtantly attend the ſhips in expectation of what may drop overboard; a man that has that misfortune periſhes without redemption: they have been ſeen to dart at him, like gudgeons to a worm. A maſter of a *Guinea* ſhip informed me, that a rage of ſuicide prevailed among his new bought ſlaves, from a notion the unhappy creatures had, that after death they ſhould be reſtored

reftored again to their families, friends, and country. To convince them at left that they fhould
not re-animate their bodies, he ordered one of
their corpfes to be tied by the heels to a rope, and
lowered into the fea, and though it was drawn up
again as faft as the united force of the crew could
be exerted, yet in that fhort fpace the fharks had
devoured every part but the feet, which were fecu-
red at the end of the cord.

Swimmers very often perifh by them; fometimes they lofe an arm or leg, and fometimes are
bit quite afunder, ferving but for two morfels for
this ravenous animal: a melancholy tale of this
kind is related in a *Weft India* ballad, preferved in
Doctor *Percy*'s Reliques of ancient *Englifh* Poetry *.

The mouth of this fifh is furnifhed with (fometimes) a fixfold row of teeth, flat, triangular,
exceedingly fharp at their edges, and finely ferrated. We have one that is rather more than an
inch and an half long. *Grew* † fays, that thofe in
the jaws of a fhark two yards in length, are not
half an inch, fo that the fifh to which mine belong-
ed muft have been fix yards long, provided the
teeth and body keep pace in their growth ‡.

This dreadful apparatus, when the fifh is in a

* *Vol.* I. 331.

† *Rarities*, 91.

‡ Foffil teeth of this fifh are very frequent in *Malta*, fome
of which are four inches long.

ftate

ftate of repofe, lie quite flat in the mouth, but when he feizes his prey, he has power of erecting them, by the help of a fet of mufcles that join them to the jaw.

The mouth is placed far beneath, for which reafon thefe, as well as the reft of the kind, are faid to be obliged to turn on their backs to feize their prey, which is an obfervation as antient as the days of *Pliny* [*].

The eyes are large; the back broad, flat, and fhorter than that of other fharks. The tail is of a femilunar form, but the upper part is longer than the lower. It has vaft ftrength in the tail, and can ftrike with great force, fo that the failors inftantly cut it off with an axe as foon as they draw one on board.

The pectoral fins are very large, which enables it to fwim with great fwiftnefs.

The color of the whole body and fins is a light afh.

The antients were acquainted with this fifh; and *Oppian* gives a long and entertaining account of its capture. Their flefh is fometimes eaten, but is efteemed both coarfe and rank.

---

[*] *Omnia autem carnivora funt talia et fupina vefcantur.* Lib. IX. c. 24.

Γλαυϰ‑

Γλαυκ⊙. *Ælian an. Lib.* I.   Squalus foſſula triangulari in   43. BLUE.
   c. 16.   extremo dorſo, foraminibus
Galeus glaucus. *Rondel.* 378.   nullis ad oculos. *Arted. ſyn.*
   Geſner piſc. 609.   98.
Blew ſhark *Wil. Iƈtb.* 49.   Squalus glaucus. *Lin. ſyſt.*
   Raii ſyn. piſc. 20.   401.

ÆLIAN relates ſtrange things of the affection this ſpecies bears to its young: among
others, he ſays, that it will permit the ſmall
brood, when in danger, to ſwim down its mouth,
and take ſhelter in its belly. This fact has been
ſince confirmed by the obſervation of one of our
beſt icthyologiſts *, and is no more incredible, than
that the young of the *Opoſſum* ſhould ſeek an aſy-
lum in the ventral pouch of its parent, a fact too
well known to be conteſted. But this degree of
care is not peculiar to the blue ſhark, but we believe common to the whole genus.

This ſpecies frequents many of our coaſts, but
particularly thoſe of *Cornwall* during the pilchard
ſeaſon, and is at that time taken with great iron
hooks made on purpoſe.

It is of an oblong form: the noſe extends far be-   DESCRIP.
yond the mouth: it wants the orifices behind the
eyes, which are uſual in this genus: the noſtrils
are long, and placely tranſverſely. *Artedi* remarks
a triangular dent in the lower part of the back.

       * *Rondeletius,* 388.

                     The

The skin is smoother than that of other sharks: the back is of a fine blue color; the belly of a silvery white.

Linnæus says, that its teeth are granulated; for our part we must confess it is a fish that has not come under our examination, therefore hope to be favored with an accurate description from some naturalist, who lives on the coast it haunts.

We may add, that *Rondeletius* says he was an eye-witness to its fondness for human flesh: that these fish are less destructive in our seas, is owing to the coolness of the climate, which is well known to abate the fierceness of some, as well as the venom of other animals.

44. LONG-TAILED.

Αλωπεξ? *Arist. Hist. an. Lib.* IX. *c.* 37. *Ælian Var. Hist. Lib.* I. *c.* 5.
Oppian Halieut. I. 381. III. 144.
Vulpes *Plinii Lib.* IX. *c.* 43.
Singe de mer. *Belon,* 88.
Vulpes marina. *Rondel.* 337.
Gesner pisc. 1045.

Cercus *Caii opusc.* 110.
Sea Fox, or Ape. *Wil. Icth.* 54. *Raii syn. pisc.* 20.
Squalus cauda longiore quam ipsum corpus. *Arted. syn.* 96.
Sea Fox. Thresher. *Borlase Cornwall.* 265.

TAIL.

THIS fish is most remarkable for the great length of the tail: the whole measure of that we had an opportunity of examining, was thirteen feet, of which the tail alone was more than six, the

LONG TAILED SHARK.

the upper lobe extending greatly beyond the lower, almoſt in a ſtrait line.

The body was round and ſhort: the noſe ſhort but ſharp pointed: the eyes large, and placed immediately over the corners of the mouth, which was ſmall, and not very diſtant from the end of the noſe.

The teeth are triangular, and ſmall for the ſize of the fiſh, and placed in three rows.

The back aſh color: the belly white: the ſkin univerſally ſmooth.

The antients ſtyled this fiſh Αλωπεξ, and *Vulpes*, from its ſuppoſed cunning. They believed, that when it had the misfortune to have taken a bait, it ſwallowed the hook till it got at the cord, which it bit off, and ſo eſcaped.

They are ſometimes taken in our ſeas, and have been imagined to be the fiſh called the *Threſher*, from its attacking and beating the *Grampus* with its long tail, whenever that ſpecies of whale riſes to the ſurface to breathe.

Κυων? *Ariſt. Hiſt. an. Lib.* VI. *c.* 11.
Canicula? *Plinii Lib.* IX. *c.* 46.
Le chien de mer, ou Canicule. *Belon,* 65.
Canis galeus. *Rondel.* 377. *Geſner piſc.* 167.

The Tope. *Wil. Iæth.* 51. *Raii ſyn. piſc.* 20.
Squalus naribus ori vicinis; foraminibus exiguis ad oculos. *Arted. ſynon.* 97.
Squalus galeus. *Lin. ſyſt.* 399. *Gronov. Zooph.* No. 142.

45. Tope.

ONE that was taken on our coaſt the laſt year weighed twenty ſeven pounds, and its length

Size.

was

was five feet; but they grow to a greater size, some, according to *Artedius*, weighing an hundred pounds.

The color of the upper part of the body and fins was a light cinereous; the belly white.

The nose was very long, flat, and sharp pointed; beyond the nostrils semitransparent. The nostrils were placed very near the mouth.

Behind each eye was a small orifice. The teeth were numerous, disposed in three rows, small, very sharp, triangular, and serrated on their inner edge.

The first back fin was placed about eighteen inches from the head; the other very near the tail.

The tail finned beneath, the upper part ended in a sharp angle.

This species is said by *Rondeletius* to be very fierce and voracious, even to pursue its prey to the edge of the shore.

Its skin and flesh has an offensive rank smell; therefore we suppose Mr. *Dale* gave it ironically the title of *Sweet William* *.

---

* *Hist. Harwich*, 420.

Νεβριας,

Νεβρίας, Σκύλιος, Αςερίας ? Greater Cat Fiſh : the Bounce. 46. Spotted.
Ariſt. Hiſt. an. Lib. V. c.   Raii ſyn. piſc. 22.
10. VI. c. 10, 11.   Squalus ex rufo varius, pinna
Ποικιλος ? Oppian Halieut. I.   ani medio inter anum et
381.   caudem pinnatum. Arted.
La Rouſſete commune. Belon,   ſyn. 97.
65.   Squalus canicula. Lin. ſyſt.
Canicula Ariſtotelis. Rondel.   399. Gronov. Zooph. No.
380. Geſner piſc. 168.   145.
Catulus major vulgaris. Wil.   Greater Cat fiſh. Edw. 289.
Iēth. 62.

THIS ſpecies being remarkably ſpotted, may
be the ſame known to antients by the names
expreſſed in the ſynonyms ; but they ſo frequently
leave ſuch ſlight notices of the animals they mention, that we are often obliged to add a doubtful mark (?) to numbers of them.

The weight of one we took was ſix pounds three   Descrip.
ounces, and yet it meaſured three feet eight inches
in length ; ſo light are the cartilaginous fiſh in
reſpect to their ſize.

The noſe was ſhort, and very blunt, not extending above an inch and an half beyond the
mouth. The noſtrils were large, placed near the
mouth, and covered with a large angular flap :
the head very flat.

The eyes were oblong, behind each a large
orifice opening to the inſide of the mouth.

The teeth ſmall, ſharp, ſmooth at their ſides,
ſtrait, and diſpoſed in four rows.

Vol. III.   I   Both

Both the back fins were placed much behind, and nearer the tail than in common.

The tail was finned, and below extended into a fharp angle.

The color of the whole upper part of the body, and the fins, was brown, marked with numbers of large diftinct black fpots : fome parts of the fkin were tinged with red ; the belly was white.

The whole was moft remarkably round, and had a ftrong fmell.

The tendrils that iffue from each end of the purfe of this fifh, are much more delicate and flender than thofe of any other ; are as fine as *Indian* grafs, and very much refemble it.

The female of this fpecies, and we believe of other fharks, is greatly fuperior in fize to the male ; fo that in this refpect there is an agreement between the fifh and the birds of prey *. They bring about nineteen young at a time : the fifhermen believe that they breed at all times of the year, as they fcarce ever take any but what are with young.

To this kind may be added, as a meer variety, the

Catulus maximus. *Wil. Ith.* 63. *Raii fyn. pifc.* 22.
Squalus cinereus, pinnis ventralibus difcretis. *Arted. fyn.* 97.
Squalus ftellaris. *Lin. fyft.* 399.
No. 145. *Gronov. Zooph.*

* Vide *Britifh Zoology, Vol. I.* 130.

The

Pl. XV.

GREATER & LESSER SPOTTED SHARKS.

The chief difference feeming to be in the color and the fize of the fpots; the former being grey, the latter fewer but larger than in the other.

Le mufcarol? *Belon*, 64.
Catulus minor. *Wil. Ith*. 64.
Leffer Rough Hound, or Morgay *Raii fyn. pifc.* 22.
Squalus dorfo vario, pinnis ventralibus concretis. *Arted. fynon*. 97.
Squalus catulus. *Lin. fyft*. 400.
Gronov. Zooph. No. 144.

THE weight of one that was brought to us by a fifherman was only one pound twelve ounces; the length two feet two inches: it is of a flender make in all parts.

The head was flat: the noftrils covered with a long flap: the nofe blunt, and marked beneath with numerous fmall punctures: behind each eye was a fmall orifice: the back fins, like thofe of the former, placed far behind.

The ventral fins are united, forming as if it were but one, which is a fure mark of this fpecies.

The tail finned like that of the greater dog fifh.

The color is cinereous, ftreaked in fome parts with red, and generally marked with numbers of fmall black fpots; but we have obferved in fome that they are very faint and obfcure.

The belly is white.

This fpecies breeds from nine to thirteen young

I 2

at

at a time, is very numerous on fome of our coafts, and very injurious to the fifheries.

48. SMOOTH. ΓαλεⒼ λεῖος? *Arift. Hift. an. Lib.* VI. *c.* 10. *Oppian, Lib.* I. 380. Galeus lævis. *Rondel.* 375. *Gefner pifc.* 608. Muftelus lævis primus. *Wil. Ifth.* 60.

Smooth or unprickly hound. *Raii fyn. pifc.* 22. Squalus dentibus obtufis feu granulofis. *Arted. fyn.* 93. Squalus muftelus. *Lin. fyft.* 400. *Gronov. Zooph.* No. 142.

THIS fpecies is called fmooth, not that the fkin is really fo, but becaufe it wants the fpines on the back, which are the character of the fecond fpecies, the Picked Dog.

The nofe extends far beyond the mouth, and the end blunt: the holes behind the eyes are fmall; the back is lefs flat than that of others of this genus.

The firft back fin is placed midway above the pectoral and ventral fins: the pectoral fins are fmall.

The tail forked, but the upper part is much the longeft.

The teeth refemble thofe of a Ray, rough and fharp.

The color of the back and fides afh, and free from fpots; the belly filvery.

The

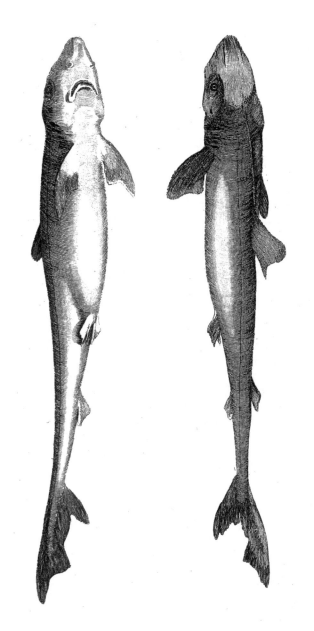

SMOOTH SHARK.

The Porbeagle. *Borlase Cornwall*, 265. *Tab.* 26.

49. PORBEA-
GLE.

THE figure of this fifh, engraved after a drawing by the Rev. Mr. *Jago**, is preferved in
Doctor *Borlase*'s Natural Hiftory of *Cornwall*.

As it is not attended with any account farther
than that it is a *Cornifh* fifh, and a fmall fpecies
of fhark, we are obliged to form the beft defcription we can from the print.

The nofe appears to be very long, flender towards the end, and fharp pointed. The mouth
placed far beneath; the body very thick and
deep, but extremely flender juft at the fetting on
of the tail.

The firft back fin is placed almoft in the middle, the other pretty near the tail.

The belly very deep: the ventral and anal fins
fmall.

The tail bifurcated; the upper fork a little
longer than the lower.

---

* This gentleman was minifter of *Loo*, in *Cornwall*, and
appears to have been well acquainted with the Hiftory of
Fifh. He communicated figures of feveral of the *Cornifh* fifh,
with a brief account of each to *Petiver*, at whofe inftance, as Doctor *Derham* tells us, in the preface to Mr. *Ray's*
Itineraries, p. 69, he added them to the Synopfis *Avium et
pifcium*, p. 162. A few others of his drawings are alfo preferved in the Natural Hiftory of *Cornwall*, and feem to be
executed with fkill and accuracy.

I 3

THIS

THIS species was observed by my friend the Rev. Mr. *Hugh Davies* of *Beaumaris*, who favored me with the description, and an accurate drawing made from the fish taken in a neighboring wear.

The length was seven feet. The snout and body of a cylindrical form. The greatest circumference four feet eight inches.

The nose blunt. The nostrils small. The mouth armed with three rows of slender teeth*, flatted on each side, very sharp, and furnished at the base with two sharp processes. The teeth are fixed to the jaws by certain muscles, and are liable to be raised or depressed at pleasure.

The first dorsal fin was two feet eight inches distant from the snout, of a triangular form: the second very small, and placed near the tail.

The pectoral fins strong and large: the ventral and anal small.

The space between the second dorsal fin and the tail much depressed; the sides forming an acute angle. Above and below was a transverse fossule or dent.

The tail was in the form of a crescent, but the

---

* These teeth are often found fossil, and are styled by *Lhyd Ornithoglossum*, from their resemblance to a bird's tongue.

horns

BEAUMARIS SHARK.

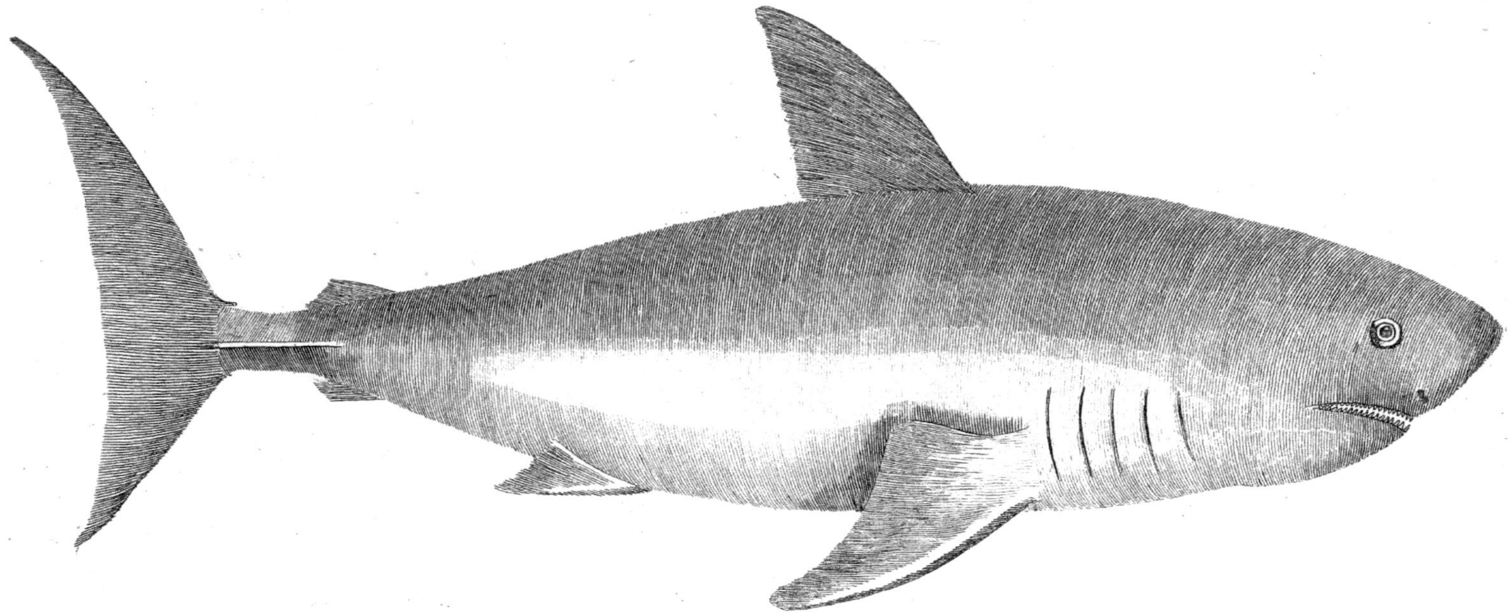

horns of unequal lengths : the upper one foot ten inches; the lower one foot one.

The whole fiſh was a lead color. The ſkin comparatively ſmooth, being far leſs rough than that of the leſſer ſpecies of this genus.

I 4   One

VII.
ANGLER.
One aperture behind each ventral fin.

Large, flat, and circular head and body.

Teeth numerous and small in the jaws, roof of the
mouth, and on the tongue.

Pectoral fins broad and thick.

51. COMMON. Βατραχος. *Arist. Hist. an.
Lib.* IV. *c.* 37. *Oppian
Hali. II.* 86.

Rana piscatrix. *Ovid. Halieut.* 126. *Plinii Lib.* IX.
c. 24.

La Grenouille de mer, ou
pescheuse. Le Diable de
mer, Bauldroy & Pesche-
teau *Belon,* 77.

Rana piscatrix. *Rondel.* 363.
Gesner pisc. 813.

Seheganss, seheteuffel, sehe-
tode. *Schonevelde,* 59.

Toad-fish, Frog-fish, or Sea-
Devil. *Wil. Icth.* 85. *Raii
syn. pisc.* 29.

Lophius ore cirroso: *Arted.
syn.* 87.

Lophius piscatorius. *Lin. syst.*
402.

L. p. depressus capite rotun-
dato. *Faun. Suec.* No. 298.
Gronov. Zooph. No. 207.

NAME. THIS singular fish was known to the antients by the name of ΒατραχⓄ, and *Rana,*
and to us by that of the fishing frog, for it is of a
figure resembling that animal in a tadpole state.
Pliny takes notice of the artifice used by it to
take its prey: *Eminentia sub oculis cornicula turbato
limo exerit, assultantes pisciculos attrahens, donec tam
prope accedant, ut assiliat.* " It puts forth the
" slender horns it has beneath its eyes, enticing
" by that means the little fish to play round, till
" they

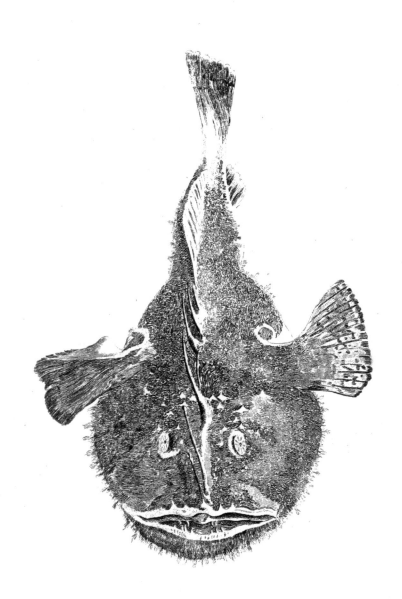

COMMON ANGLER.

" they come within reach, when it fprings on them *."

The fifhing frog grows to a large fize, fome being between four and five feet in length ; and we have heard of one taken near *Scarborough*, whofe mouth was a yard wide. The fifhermen on that coaft have a great regard for this fifh, from a fuppofition that it is a great enemy to the dog fifh †, and whenever they take it with their lines fet it at liberty.

It is a fifh of very great deformity : the head is much bigger than the whole body, is round at the circumference, and flat above : the mouth of a prodigious widenefs.

The under jaw is much longer than the upper : the jaws are full of flender fharp teeth : in the roof of the mouth are two or three rows of the fame : at the root of the tongue, oppofite each other, are two bones of an elliptical form, thick fet, with very ftrong fharp teeth.

The noftrils do not appear externally, but in the upper part of the mouth are two large orifices that ferve inftead of them.

---

* *Cicero*, in his fecond book *De Natura Deorum*, gives much the fame account of this fifh : *Ranæ autem marinæ dicuntur obruere fefe arena folere, et moveri propè aquam, ad quas, quafi ad efcam, pifces cum accefferint, confici a ranis, atque confumi.*

† The bodies of thefe fierce and voracious fifh are often found in the ftomach of the *Fifhing Frog*.

On

On each fide the upper jaw are two fharp fpines, and others are fcattered about the upper part of the head.

Immediately above the nofe are two long tough filaments, and on the back three others; thefe are what *Pliny* calls *cornicula*, and fays it makes ufe of to attract the little fifh. They feem to me like lines flung out for that end: I therefore have changed the old name of FISHING FROG for the more fimple one of ANGLER.

Along the edges of the head and body are a multitude of fhort fringed fkins, placed at equal diftances.

The ventral fins are broad, thick, and flefhy, are jointed like arms, and within fide divided into fingers.

The aperture to the gills is placed behind, each of thefe is very wide, fo that fome writers have imagined it to be a receptacle for the young in time of danger.

The back fin is placed very low near the beginning of the tail: the anal fin is placed beneath, almoft oppofite the former.

The body grows flender near the tail, the end of which is quite even.

The color of the upper part of this fifh is dufky, the lower part white; the fkin fmooth.

Fifhing

Fishing Frog of Mount's-Bay. *Borlase Cornwall*, 266. *Tab.* 52. LONG.
     27. *fig.* 6. *Phil. Transf. Vol.* LIII. 170.

T HIS is a species at present unknown to us, except by description.

It is, says Doctor *Borlase*, of a longer form than
the common kind : the head more bony, rough,
and aculeated. It had no finlike appendages round
the head, but on each side the thinner part of the
body, beginning beneath the dorsal fin, and reaching within two inches of the tail, was a series of
them, each three quarters of an inch in length.

At the end of the pectoral fins were spines an
inch and three quarters in length ; at the end of the
tail others three quarters of an inch long.

One

VIII. STUR-
GEON.

One narrow aperture on each side.

The mouth placed far below, tubular and without teeth.

The body long, and often angular.

53. STUR-
GEON.

Ονισκος. *Athen. Lib.* VIII. 315. A'κκιπησιος ? *Athen. p.* 294.
Acipenser ? *Plinii Lib.* IX. c. 17. *Ovidii Halieut.* ?
L' Esturgeon. *Belon,* 89.
Acipenser. *Rondel.* 410. *Gesner pisc.* 2.
Sturio. *Gesner pisc.*
Stoer. *Schonevelde,* 9.
Sturgeon. *Wil. Ichth.* 239.

Raii *syn. pisc.* 112.
Schirk. *Kram.* 383.
Acipenser corpore tuberculis spinosis exasperato. *Arted. syn.* 91.
Acipenser sturio. *Lin. syst.* 403. *Mus. Ad. Fred.* 54. *Tab.* 18. *fig.* 2.
Stor. *Faun. Suec.* No. 299.
Seb. *Mus.* III. 101. *Tab.* 29. No. 19.

THAT this is the 'Ονισκος of *Dorion,* as quoted by *Athenæus,* is very probable, as well from the account he gives of its form, as of its nature. He says its mouth is always open, with which it agrees with the Sturgeon, and that it conceals itself in the hot months : this shews it to be a fish of a cold nature, which is confirmed by the history of the *European* fish of this species, given by Mr. *Forster**, in his Essay on the *Volga,* who relates that they are scarce ever found in that river

* *Phil. Transf.* LVII. 352.

in

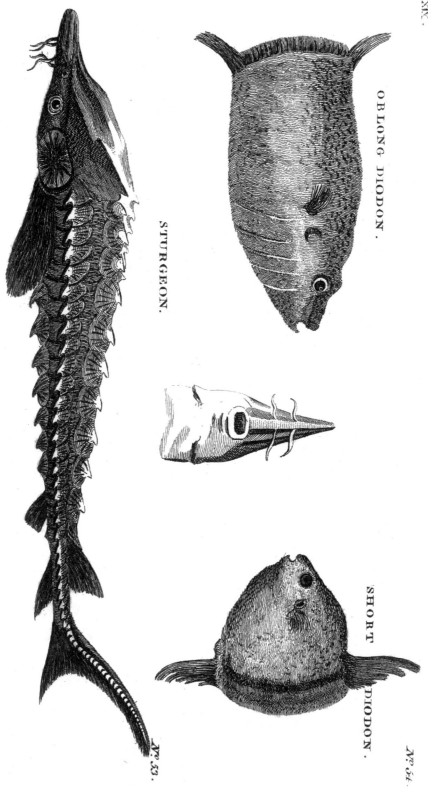

OBLONG DIODON.

STURGEON.

SHORT DIODON.

Nº 53.

Nº 54.

in fpring or fummer, but in vaft quantities in autumn and winter, when they crowd from the fea
under the ice, and are then taken in great numbers.

Whether the *acipenfer* is the fturgeon of the
moderns, may be doubted, otherwife *Ovid* would
never have fpoke of it as a foreign fifh :

> *Tuque peregrinis, Acipenfer, nobilis undis.*
> And, thou, a fifh in foreign feas renowned.

It being well known that it is not uncommon
in the *Mediterranean*, and even in the mouth of
the *Tiber*, at certain feafons; but this paffage leaves
us as much in the dark as to the particular fpecies intended, by the word *acipenfer*, as the defcription *Pliny* has given us ; for that philofopher
relates, that its fcales are placed in a contrary direction to thofe of other fifh, being turned towards the
mouth, which difagrees with the character of all
that are known at prefent.   Whatever fifh it might
be, it was certainly the fame with the *Elops*, or
Helops, as appears from *Pliny*, who makes it fyno-
nimous with the *acipenfer* [*], and from another line
of the poet beforementioned :

> *Et pretiofus Helops noftris incognitus undis.*
> The pretious *Helops* ftranger to our feas.

The fturgeon annually afcends our rivers, but in
no great numbers, and is taken by accident in the

MIGRA-
TORY.

[*] Quidam eum *Elopem* vocant. *Lib.* IX. *c.* 17.

salmon

falmon nets. It feems a fpiritlefs fifh, making no manner of refiftance when entangled, but is drawn out of the water like a lifelefs lump. It is a fifh that is feldom taken far out at fea, but frequents fuch parts as are not remote from the æftuaries of great rivers. It is admired for the delicacy and firmnefs of its flefh, which is white as veal, and extremely good when roafted. It is generally pickled. The moft we receive comes either from the *Baltic* rivers, or *North America:* thofe cured at *Pillau* have been, till of late, in the greateft repute; but through the encouragement given by the fociety inftituted for promoting trade and manufactures, the fturgeon from our colonies begins to rival thofe of the *Baltic*.

Great numbers are taken during fummer in the lakes *Frifchehaff*, and *Curifch-haff* near *Pillau*, in large nets made of fmall cord. The adjacent fhores are formed into diftricts, and farmed out to companies of fifhermen, fome of which are rented for fix thoufand guilders, or near three hundred pounds *per annum*.

They are found in vaft abundance in the *American* rivers in *May*, *June*, and *July*, at which time they leap fome yards out of the water, and falling on their fides, make a noife to be heard in ftill weather at fome miles diftance *.

Caviare is made of the roes of this, and alfo of

* *Catefby Carol. App.* 33.

all

all the other forts of sturgeons, dried, salted, and packed up close. The best is said to be made of those of the *Sterlet* *, a small species frequent in the *Yaik* and *Volga*. *Icthyocolla* †, or ising-glass, is also made of the *sound* of our fish, as well as that of the others, but the *Beluga* affords the best ‡.

The sturgeon grows to a great size, to the length of eighteen feet, and to the weight of five hundred pounds, but it is seldom taken in our rivers of that bulk. The largest we have known caught in those of *Great Britain* weighed four hundred and sixty pounds, which was taken about two years ago in the *Esk*, where they are more frequently found than in our southern waters.

* *Strahlenberg's Hist. Russia,* 337.

† *Phil. Trans.* LVII. 354. A very small quantity is made from this species, and that only designed as presents to great men, as Mr. *Forster* assured me.

‡ The antients were acquainted with the fish that afforded this drug. *Pliny lib.* XXXII. *c.* 7. mentions it under the name of *Icthyocolla,* and says, that the glue that was produced from it had the same title ; and afterwards adds, that it was made out of the belly of the fish. The *Mario,* said by *Pliny lib.* IX. *c.* 15. to be found in the *Danube* and the *Borysthenes,* was certainly of this genus, a cartilaginous fish *(nullis ossibus spinisve interfitis)* resembling a small porpesse *(Porculo marino fimillimus ;)* and very probably may be the same with the *Beluga,* which, according to Mr. *Forster, Phil. Trans.* LVII. 354. has a short blunt nose, agreeing in that respect with the porpesse.

The

The nofe is very long, flender, and ends in a point. The eyes are extremely fmall; the noftrils placed near them: on the lower part of the nofe are four cirri or beards: the mouth is fituated far beneath, is fmall, and unfupported by any jaw bones; neither has it any teeth. The mouth of a dead fifh is always open. When alive it can clofe or open it at pleafure, by means of certain mufcles.

The body is long, pentagonal, and covered with five rows of large bony tubercles: one row of which is placed on the back, and two on each fide. The whole under fide of the fifh, from the end of the nofe to the vent, is flat; on the back, not remote from the tail, is a fingle fin. It has befides two pectoral fins, two ventral, and one anal fin. The tail is bifurcated, but the upper part much longer than the lower.

The upper part of the body is of a dirty olive color; the lower part filvery; the middle of the tubercles white.

In the manner of breeding it is an exception a-mong the cartilaginous fifh, being like the bony fifh oviparous, fpawning in winter.

A very

A very deep body, and as if cut off in the middle.   IX.
Mouth fmall.   DIODON.

Two teeth only in each jaw.

Sun-Fifh from Mount's-Bay.   Oftracion lævis.   *Gronov.* 54. OBLONG.
  Borlafe Cornwall, 268. *tab.*   *Zooph.* No. 185.
  26. *fig.* 7.

*R*ONDELETIUS has given this genus the fy-
    nonym of *Orthragorifcus,* as if it was that
which *Pliny* * intended by the fame name; but
the account left us by that naturalift is fo brief,
that we do not think ourfelves authorized to place
it as a fynonymous creature. He fays no more
than that it was the greateft of fifh, and that it
grunted when it was firft taken, from which probably rofe the name, for according to *Athenæus,*
ὀρθραγορσκ℗ † was that given to a young pig.
We are inclined to believe, that this fifh had
efcaped the notice of *Pliny,* otherwife he muft have
unavoidably made fome remark on its ftriking
figure.

 This fifh grows to a great bulk : that which   SIZE.
was examined by *Salvianus* ‡ was above a hun-

  * *Lib.* XXXII. *c.* 2.
  † *Lib.* IV. *p.* 140.
  ‡ *Hift. Pifc.* 155.

VOL. III.   K   dred

dred pounds in weight: and Doctor *Borlafe* mentions another taken at *Plymouth* in 1734, that weighed five hundred.

DESCRIP.

In form it refembles a bream, or fome deep fifh cut off in the middle. The mouth is very fmall, and contains in each jaw two broad teeth, with fharp edges.

The eyes are little; before each is a fmall femilunar aperture; the pectoral fins very fmall, and placed behind them. The dorfal fin and the anal fin are high, and placed at the extremity of the body: the tail fin is narrow, and fills all the abrupt fpace between thofe two fins.

The color of the back is dufky, and dappled; the belly filvery: between the eyes and the pectoral fins are certain ftreaks pointing downwards. The fkin is free from fcales.

When boiled, it has been obferved to turn into a glutinous jelly, refembling boiled ftarch when cold, and ferved the purpofes of glue, on being tried on paper and leather. The meat of this fifh is uncommonly rank: it feeds on fhell-fifh.

There feems to be no fatisfactory reafon for the old *Englifh* name. Care muft be taken not to confound it with the fun-fifh of the *Irifh*, which differs in all refpects from this.

Orthragorifcus

Orthragorifcus five Luna pifcis. *Rondel.* 424.

Mola *Salviani*, the Sun-fifh. *Wil. Icth.* 151. *Raii fyn. pifc.* 51.

Oftracion cathetoplateus fubrotundus inermis afper, pinnis pectoralibus horizontalibus, foraminibus quatuor in capite. *Arted. fynon.* 83.

Tetraodon mola. T. lævis, compreffus, cauda truncata, pinna breviffima dorfali analique annexa. *Lin. fyft.* 412. *Gronov. Zooph.* No. 186.

Brunnich pifc. Maffil. No. 16.

Sun-fifh, from Loo. *Borlafe Cornwall*, 267. *tab.* 26. *fig.* 6.

55. SHORT.

THIS differs from the former, in being much fhorter and deeper. The back and the anal fins are higher, and the aperture to the gills not femilunar, but oval. The fituation of the fins are the fame in both.

This fpecies was taken off *Penzance*, and is engraved in Doctor *Borlafe*'s Natural Hiftory of *Cornwall*, from one of Mr. *Jago*'s drawings. Both kinds are taken on the weftern coafts of this kingdom, but in much greater numbers in the warmer parts of *Europe*.

Mr. *Brunnich* informs us, that between *Antibes* and *Genoa*, he faw one of this fpecies lie afleep on the furface of the water: a failor jumped overboard and caught it.

K 2   Tetraodon

56. Globe.   Tetraodon lævigatus. *Lin. ſyſt.* 411.

THIS ſpecies is common to *Europe* and *South Carolina*. As yet only a ſingle ſpecimen has been diſcovered in our ſeas; taken at *Penzance* in *Cornwall*.

The length was one foot ſeven: the length of the belly, when diſtended, one foot; the whole circumference in that ſituation two feet ſix.

The form of the body is uſually oblong, but when alarmed it has the power of inflating the belly to a globular ſhape of great ſize. This ſeems deſigned as a means of defence againſt fiſh of prey: as they have leſs means of laying hold of it; and are beſides terrified by the numbers of ſpines with which that part is armed; and which are capable of being erected on every part.

The mouth is ſmall: the irides white, tinged with red: the back from head to tail almoſt ſtrait, or at leaſt very ſlightly elevated; of a rich deep blue color. It has the pectoral, but wants the ventral fins. The dorſal is placed low on the back; the anal is oppoſite: the tail almoſt even divided by an angular projection in the middle: tail and fins brown.

The belly and ſides are white, ſhagreened or wrinkled; and beſet with innumerable ſmall ſharp ſpines, adhering to the ſkin by four proceſſes.

Thick

Pl. XX.

GLOBE DIODON.

Pl. XXI.

I

LUMP SUCKER.

II

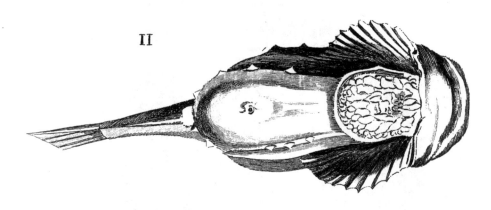

III

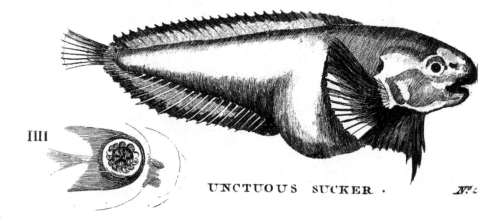

IIII

UNCTUOUS SUCKER.

Thick body, arched back.

Ventral fins united.

Four branchioftegous rays.

X.
SUCKER.

Lumpus anglorum. *Gefner* Paralip. 25.

Seehaefs, Haffpodde. *Scho-nevelde.* 41.

Lump, or Sea-Owl, *Scotis* Cock paddle. *Wil. Icth.* 208. *Raii fyn. pifc.* 77.

Cyclopterus. *Arted. fynon.*

87. *Gronov. Zooph.* No. 197.

Cyclopterus Lumpus. C. cor-pore fquamis offeis angula-to. *Lin. fyft.* 414.

Sjurygg-fifk, Stenbit, Quabb-fu. *Faun. Suec.* No. 320.

57. L U M P.

THIS fingular fifh encreafes to the weight of feven pounds, and the length of nineteen inches: the fhape of the body is like that of the bream, deep and very thick, and it fwims edge-ways. The back is fharp and elevated, the belly flat.

The irides are of a cherry color; lips, mouth, and tongue, of a deeper red: the jaws lined with innumerable fmall teeth; the tongue very thick; along the ridge of the back is a row of large bony tubercles; from above the eye to with-in a fmall fpace of the tail is another row; beneath that a third, commencing at the gills; and on each fide the belly a fourth row, confifting of five tu-bercles like the other: the whole fkin is rough, with fmall tubercles.

DESCRIP.

K 3

On

On the upper part of the back is a thick ridge improperly called a fin, being deftitute of fpines; beneath that is the dorfal fin, of a brownifh hue, reaching within an inch of the tail: on the belly, juft oppofite, is another of the fame form. The belly is of a bright crimfon color: the pectoral fins are large and broad, almoft uniting at their bafe. Beneath thefe is the part by which it adheres to the rocks, &c. It confifts of an oval aperture, furrounded with a flefhy mufcular and obtufe foft fubftance, edged with fmall threaded appendages, which concur as fo many clafpers: tail and vent fins purple.

By means of this part it adheres with vaft force to any thing it pleafes. As a proof of its tenacity we have known, that on flinging a fifh of this fpecies juft caught, into a pail of water, it fixed itfelf fo firmly to the bottom, that on taking the fifh by the tail, the whole pail by that means was lifted, though it held fome gallons, and that without removing the fifh from its hold.

Thefe fifh refort in multitudes during fpring to the coaft of *Sutherland*, near the *Ord of Caithnefs*. The feals which fwarm beneath, prey greatly on them, leaving the fkins; numbers of which thus emptied float at that feafon afhore. It is eafy to diftinguifh the place where feals are devouring this or any unctuous fifh, by a fmoothnefs of the water immediately above the fpot: this fact is now eftablifhed; it being a tried property of oil to

ftill

ſtill the agitation of the waves, and render them ſmooth *.

Great numbers of theſe fiſh are found in the *Greenland* ſeas during the months of *April* and *May*, when they reſort near the ſhore to ſpawn. Their roe is remarkably large, which the *Greenlanders* boil to a pulp, and eat. They are extremely fat, which recommends them the more to the natives, who admire all oily food: they call them *Nipiſets*, or *Cat-fiſh*, and take quantities of them during the ſeaſon †.

This fiſh is ſometimes eaten in *England*, being ſtewed like carp, but is both flabby and inſipid.

Liparis ? *Rondel.* 272. *Geſner* *piſc.* 483.	Cyclopterus Liparis C. cor- pore nudo, pinnis dorſali anali caudalique unitis. *Lin.* *ſyſt.* 414.	58. Unc- tuous.
Liparis noſtras *Dunelm et Ebo- rac.* Sea Snail. *Wil Icth.* *App.* 17. *Raii ſyn. piſc.* 74. *Pet. Gaz. tab.* 51. *fig.* 5.	Cyclogaſter. *Gronov. Zooph.* No. 198.	
Liparis. *Arted. ſynon.* 177.		

THIS fiſh takes the name of ſea ſnail from the ſoft and unctuous texture of its body, reſembling that of the land ſnail. It is almoſt tranſparent, and ſoon diſſolves and melts away.

It is found in the ſea near the mouths of great rivers. We have ſeen it in *January* full of ſpawn.

* *Philoſ. Tranſ.* 1774. *p.* 445.
† *Crantz's Hiſt. Greenland,* I. 96.

K 4   The

DESCRIP.   The length is five inches : the color when fresh taken a pale brown, sometimes finely streaked with a darker; the shape of the body round, but near the tail compressed sideways: the belly is white and very protuberant.

The head is large, thick, and round. There are no teeth in the mouth, but the jaws are very rough: the tongue very large: the eyes very small.

The orifice to the gills is very small. It has six branchiostegous rays.

The pectoral fins are very broad, thin, and transparent, and almost unite under the throat. The first ray next the throat is very long, extends far beyond the rest, and is as fine as a hair. Over the base of each is a sort of operculum, or lid, ending in a point : this is capable of being raised or depressed at pleasure.

Behind the head begins the dorsal fin, which extends quite to the end of the tail : the ventral fin begins at the anus, and unites with the other at the tail.

Beneath the throat is a round depression of a whitish color, like the impression of a seal, surrounded with twelve small pale yellow tubera, by which it is probable it adheres to the stones like the other species.

Lesser

BIMACULATED SUCKER.

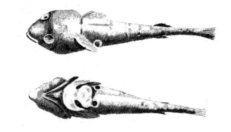

JURA SUCKER. *Nº*

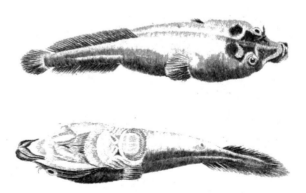

Leffer Sucking Fifh. *Borlafe* Lepadogafter. Le Barbier ou   59. JURA.
Nat. Hift. Cornwal, 269.   Porteecuelle. *Gouan pifc.*
Tab. xxv. *fig. 28.*   177. *Tab.* 1. *fig. 6, 7.*

THIS fpecies is found in *Cornwal.* I alfo difcovered it in the Sound of *Jura.*

Its length is about four inches. The fkin
without fcales, flippery, and of a dufky color.
The body taper. The nofe grows flenderer from
the head, and ends round.

The teeth fmall. Before each eye is a fmall fi-
lament. Behind the eyes are two femilunar marks.

In the middle of the back an oval mark formed by fmall dots, of a whitifh color. The dorfal
fin lies near the tail, and confifts of eleven rays:
the anal is placed oppofite, and has nine rays.
The tail is rounded. The ventral have four rays,
are joined by an intervening membrane with an
oval depreffion in the middle. Beyond that is a-
nother ftrong membrane with a fimilar depreffion.
By means of thefe inftruments it adheres to ftones
or rocks.

Nofe

XI.
PIPE.

Nofe long and tubular.

No orifice to the gills :

The breathing aperture on the hind part of the head.

No ventral fins.

The body covered with a ſtrong cruſt.

60. Longer. Acus *Ariſtotelis* caudâ ſerpentinâ. *Sib. Scot.* 24. *Tab.* 19.

Typhle altera. *Geſner piſc.* 1025.

Syngnathus corpore quadrangulo, pinnâ caudæ carens ? *Arted. Spec.* 3.

Syngnathus barbarus. S. pinnis caudæ anique nullis, corpore ſexangulato ? *Lin. ſyſt.* 417.

THIS ſpecies, deſcribed by Sir *Robert Sibbald*, was two feet in length; that we examined only ſixteen inches.

The noſe was an inch long, compreſſed ſideways, and the end of the lower mandible turned up : the aperture of the mouth was very ſmall.

The irides were red; behind each eye was a deep brown line.

The body, in the thickeſt part, was about equal to a ſwan's quil, hexangular from the end of the dorſal fin; from thence to the tail quadrangular. The belly was ſlightly carinated, and marked along the middle with a duſky line. Under the tail commencing at the *anus* is a ſulcus or groove, ſix

inches

Sp.1.

1.

2.

3.

PIPE FISH.

Nº 60.

Nº 61.

Nº 62.

inches and a half long, covered by two longitudinal valves which concealed a multitude of young
fiſh. On cruſhing this part, hundreds may be
obſerved to crawl out.

The general color of the fiſh was an olive brown :
the ſides marked with numbers of bluiſh lines pointing from the back to the belly, which, in dried fiſh,
ſeemed like the ſigns of ſo many joints.   Thoſe in
a freſh ſubject ceaſed beyond the vent ; all beyond
that was ſpotted with brown.

The dorſal fin was narrow and thin, conſiſting
of forty rays, was two inches long, and placed rather nearer to the head than the tail.

The vent was ſeven inches from the tip of the
noſe ; the body to that orifice was of an equal
thickneſs, but from thence tapered to a very ſmall
point, having no mark of a fin.

The pectoral fins had twelve rays ; the anal three.

When this fiſh and the next ſpecies are dried,
they appear covered with numbers of angular cruſts,
finely radiated from their centre.

As we want a generical name in our language
for this genus, we call it the *Pipe Fiſh*, from its
ſlender body.

L'Orueul

61.Shorter. L'Orueul marin. *Belon*, 446.
Acus fecunda fpecies, five,
acus *Ariftotelis*. *Rondel.* 229.
Typhle. *Gefner pifc.* 1025.
Trummeter, Meherfchlange.
Schonevelde, 11.
Acus *Ariftotelis* feu fecunda.
Wil. Iab. 158. *Raii fyn.
pifc.* 47.
Syngnathus corpore medio

heptagono, caudâ **pinnatâ.**
Arted. fynon. 2.
Syngnathus acus. S. pinnis
caudæ ani pectoralibufque
radiatis, corpore feptem-
angulato. *Lin. fyft.* 416.
Kantnahl. *Faun. Suec.* No. 376.
Syngnathus cauda pinnata.
Gronov. Zooph. No. 172.
Sea-adder. *Borlafe Cornw.* 267.

THIS is fhorter and thicker than the former,
yet I have feen one of the length of fixteen
inches. The middle of the body in fome is hexangular, in others heptangular. *Linnæus* conftitutes
two fpecies of them, his *Syngnathus Typhle*, and
his *Syngnathus Acus*; but we join with Doctor
Gronovius, in thinking them only varieties of the
fame fifh.

The mouth is formed like that of the former:
the irides are yellow: clofe behind the head are the
pectoral fins, which are fmall and fhort.

On the lower part of the back is one narrow
fin; beyond the vent the tail commences, which is
long and quadrangular.

At the extremity is a fin round and radiated.

The body is covered with a ftrong cruft, elegantly divided into fmall compartments.

The belly is white; the other parts brown.

Befides thefe fpecies of hard-fkinned Pipe fifh,
we

we have been informed, that the *Syngathus Hippo-campus* of *Linnæus*, or what the *Engliſh* improperly call the ſea horſe, has been found on the ſouthern ſhores of this kingdom.

Acui *Ariſtotelis* congener piſ-ciculus, pueris Cornubien-ſibus *Sea Adder*, Acus Lum-briciformis, aut Serpenti-num. *Wil. Iĉth.* 160. *Raii ſyn. piſc.*
Syngnathus teres, pinnis pec-toralibus caudaque carens. *Arted. ſynon.* 2.
Syngnathus ophidion. *Lin. ſyſt.* 417.
Haſsnahl, Tangſnipa. *Faun. Suec.* No. 375.

62. LITTLE.

THE little pipe fiſh ſeldom exceeds five inches in length, is very ſlender, and tapers off to a point. It wants both the pectoral and tail fins; is covered with a ſmooth ſkin, not with a cruſt as the two former kinds are.

The noſe is ſhort and turns a little up; the eyes prominent.

On the back is one narrow fin.

This ſpecies is not viviparous: on the belly of the female is a long hollow, to which adhere the eggs, diſpoſed in two or three rows. They are large, and not numerous.

The ſynonym of *Serpent* is uſed in ſeveral lan-guages to expreſs theſe fiſh: the *French* call one ſpecies *Orueul,* from a ſort of ſnake not unlike the blindworm: the *Germans* call it *Meherſchlange*; and the *Corniſh,* the ſea adder.

D i v.

Div. III.   B O N Y   F I S H.

Sect.   I.   A P O D A L.

XII.
E E L.

Body long, flender, and flippery.

Noftrils tubular.

Back, ventral, and tail fins, united.

Aperture to the gills fmall, and placed behind the
  pectoral fins.

Ten branchioftegous rays.

53. COMMON.

Εγχιλυς. *Arift. Hift. an. Lib.*
  IV. *c.* 11. VI. 14. 16.
  Oppian Halieut. I. 516. IV.
  450.
Anguilla *Plinii Lib.* IX. *c.* 21.
L' Anguille. *Belon,* 291. *Obf.*
  55.
Anguilla. *Rondel. fluv.* 198.
  Gefner pifc. 40.
Ael. *Schonevelde,* 14.

The Eel. *Wil. pifc.* 109. *Raii*
  fyn. pifc. 37.
Muræna unicolor maxilla inferiore longiore. *Arted. fyn.*
  39.
Muræna anguilla. *Lin. fyft.*
  426. *Gronov. Zooph.* No.
  166.
Ahl. *Faun. Suec.* No. 301.
Aal. *Kram.* 387.

T HE eel is a very fingular fifh in feveral
      things that relate to its natural hiftory,
and in fome refpects borders on the nature of the
reptile tribe.

It is known to quit its element, and during
night to wander along the meadows, not only for
                                            change

change of habitation, but alfo for the fake of prey, feeding on the fnails it finds in its paffage.

During winter it beds itfelf deep in the mud, and continues in a ftate of reft like the ferpent kind. It is very impatient of cold, and will eagerly take fhelter in a whifp of ftraw flung into a pond in fevere weather, which has fometimes been practifed as a method of taking them. *Albertus** goes fo far as to fay, that he has known eels to fhelter in a hay-rick, yet all perifhed through exeefs of cold.

It has been obferved, that in the river *Nyne*†, there is a variety of fmall eel, with a leffer head and narrower mouth than the common kind, that it is found in clufters in the bottom of the river, and is called the *Bed-eel:* thefe are fometimes roufed up by violent floods, and are never found at that time with meat in their ftomachs. This bears fuch an analogy with the cluftering of blindworms in their quiefcent ftate, that we cannot but confider it as a further proof of a partial agreement in the nature of the two genera.

The ancients adopted a moft wild opinion about the generation of thefe fifh, believing them to be either created from the mud, or that the fcrapings

GENERA-
TION.

---

* *Gefner pifc.* 45.

† *Morton's Hift. Northampt.* 419. *Pliny* obferves, that the eels of the lake *Benacus* collect together in the fame manner in the month of *October,* poffibly to retreat from the winter's cold. *Lib.* ix. *c.* 22.

of

of their bodies which they left on the ftones, were animated and became young eels. Some moderns gave into thefe opinions, and into others that were equally extravagant. They could not account for the appearance of thefe fiſh in ponds that never were ſtocked with them, and that were even ſo remote as to make their being met with in ſuch places a phænomenon that they could not ſolve. But there is much reaſon to believe, that many waters are ſupplied with thefe fiſh by the aquatic fowl of prey, in the ſame manner as vegetation is ſpread by many of the land birds, either by being dropped as they carry them to feed their young, or by paſſing quick thro' their bodies, as is the caſe with herons; and ſuch may be occaſion of the appearance of thefe fiſh in places where they were never ſeen before. As to their immediate genera-

VIVIPA-
ROUS.

tion, it has been ſufficiently proved to be effected in the ordinary courſe of nature, and that they are viviparous.

They are extremely voracious, and very deſtructive to the fry of fiſh.

No fiſh lives ſo long out of water as the eel: it is extremely tenacious of life, as its parts will move a conſiderable time after they are flayed and cut into pieces.

DESCRIP.

The eel is placed by *Linnæus* in the genus of *Muræna,* his firſt of the apodal fiſh, or ſuch which want the ventral fins.

The eyes are placed not remote from the end of
the

the nofe: the irides are tinged with red: the under jaw is longer than the upper: the teeth are fmall, fharp, and numerous: beneath each eye is a minute orifice: at the end of the nofe two others, fmall and tubular.

The fifh is furnifhed with a pair of pectoral fins, rounded at their ends. Another narrow fin on the back, uniting with that of the tail; and the anal fin joins it in the fame manner beneath.

Behind the pectoral fins is the orifice to the gills, which are concealed in the fkin.

Eels vary much in their colors, from a footy hue to a light olive green; and thofe which are called filver eels, have their bellies white, and a remarkable clearnefs throughout.

Silver Eels.

Befides thefe there is another variety of this fifh known in the *Thames* by the name of *Grigs*, and about *Oxford* by that of *Grigs* or *Gluts*. Thefe are fcarce ever feen near *Oxford* in the winter, but appear in fpring, and bite readily at the hook, which common eels in that neighbourhood will not. They have a larger head, a blunter nofe, thicker fkin, and lefs fat than the common fort; neither are they fo much efteemed, nor do they often exceed three or four pounds in weight.

Grigs.

Common eels grow to a large fize, fometimes fo great as to weigh fifteen or twenty pounds, but that is extremely rare. As to inftances brought by *Dale* and others, of thefe fifh encreafing to a fuperior magnitude, we have much reafon to fufpect

Vol. III.   L   them

them to have been congers, fince the enormous fifh they defcribe, have all been taken at the mouths of the *Thames* or *Medway*.

The eel is the moft univerfal of fifh, yet is fcarce ever found in the *Danube*, tho' it is very common in the lakes and rivers of *Upper Auftria*.

The *Romans* held this fifh very cheap, probably from its likenefs to a fnake.

> Vos anguilla manet longæ cognata colubræ*,
> Vernula riparum pinguis torrente cloaca.

> For you, is kept a fink-fed fnake-like eel.

On the contrary, the luxurious *Sybarites* were fo fond of thefe fifh, as to exempt from every kind of tribute the perfons who fold them †.

---

* *Juvenal. Sat.* v. 103.
† *Athenæus. Lib.* xii. *p.* 521.

Κὸγγϛος.

Κόγγρος. *Arist. Hist. an. lib.*
  I. &c.
Γόγγρος *Oppian Halieut.* I.
  113. 521.
Conger. *Ovidii Halieut.* 115.
  Plinii lib. IX. *c.* 16. 20.
Le Congre. *Belon* 159,
Conger. *Rondel.* 394. *Gesner*
  pisc. 290.
The Conger, or Conger Eel.

Wil. Icth. III. *Raii syn.* 64. Conger,
  pisc. 37.
Muræna supremo margine pinnæ dorsalis nigro. *Arted.*
  synon. 40.
Muræna Conger. M. rostro
  tentaculis duobus, linea laterali ex punctis albida,
  Lin. syst. 426.

Size,

THE conger grows to a vast size. Doctor
  Borlase, to whom we are obliged for several
informations relating to this species, assures us,
that they are sometimes taken near *Mount's-Bay* of
one hundred pounds weight *.

Descrip.

They differ from the common eel in the following particulars : 1. Their color in general is more
dark. 2. Their eyes much larger in proportion,
3. The irides of a bright silvery color. 4. The
lower jaw is rather shorter than the upper. 5.
The side line is broad, whitish, and marked with
a row of small spots ; Mr. *Ray* says a double row,
but we did not observe it in the fish we examined.
6. The edges of the dorsal and anal fins are black.
7. They have more bones than the common eel,

* We have heard of some taken near *Scarborough* that were
ten feet and a half long, and eighteen inches in circumference
in the thickest part.

L 2   especially

especially along the back quite to the head. 8. They grow to a much larger size.

As to the distinction that Mr. *Ray*, and other writers, make of the small beards at the end of the nose, we think it not to be depended on, being sometimes found in both kinds, and sometimes entirely wanting.

We believe they generate like the fresh-water species: innumerable quantities, of what are supposed to be their fry, come up the *Severn* about the month of *April*, preceding the *Shads*, which it is conjectured migrate into that river to feed on them: ELVERS. they are called *Elvers*. They quite swarm during their season, and are taken in a kind of sieve made of hair-cloth, fixed to a long pole; the fisherman standing on the edge of the water during the tide, puts in his net as far as he can reach, and drawing it out again takes multitudes at every sweep, and will take as many during one tide as will fill a bushel. They are dressed, and reckoned very delicate.

Congers are extremely voracious, preying on other fish, and on crabs at the time they have lost their shell, and are in a soft state. They and eels in general are also particularly fond of carcasses of any kind, being frequently found lodged in such that are accidentally taken up.

These fish are an article of commerce in *Cornwall*; numbers are taken on that coast, and exported to *Spain* and *Portugal*, particularly to *Barcelona*.

celona. The quantities that were fent from *Mount's-Bay* for five years, were as follow:

	Cwt.	qr.	lb.
1756	46	0	13
1757	164	0	21
1758	164	1	3
1759	213	0	3
1760	71	3	0

Some are taken by a fingle hook and line, but (becaufe that way is tedious, and does not anfwer the expence of time and labour) they are chiefly caught by *Bulters*, which are ftrong lines five hundred feet long, with fixty hooks, each eight feet afunder, baited with pilchards or mackrel: the *Bulters* are funk to the ground by a ftone faftened to them: fometimes fuch a number of thefe are tied together as to reach a mile.

CAPTURE.

We have been told that the fifhermen are very fearful of a large conger, leaft it fhould endanger their legs by clinging round them; they therefore kill them as foon as poffible by ftriking them on the navel.

They are afterwards cured in this manner: they are flit, and hung on a frame till they dry, having a confiderable quantity of fat, which it is neceffary fhould exude before they are fit for ufe. It is remarkable that a conger of a hundred weight will wafte by drying to twenty-four pounds; the

CURE.

L 3

people

people therefore prefer the fmalleft, poffibly becaufe they are fooneft cured.   During the procefs
there is a confiderable ftench; and it is faid that
in the fifhing villages the poultry are fed with the
maggots that drop from the fifh.

The *Portuguefe* and *Spaniards* ufe thofe dried
congers after they have been ground into a powder,
to thicken and give a relifh to their foups.   We
think they are fold for about forty fhillings the
quintal, which weighs one hundred and twenty-fix
pounds.

A fifhery of congers would be of great advantage to the inhabitants of the *Hebrides*.   Perhaps
they would at firft undertake it with repugnancy,
from their abfurd averfion to the eel kind.

Blunt

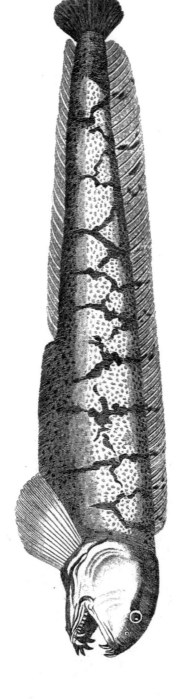

WOLF FISH.

Blunt head : long body.

One dorſal fin reaching almoſt from the head to
the tail.

Fore teeth conic and large.

Grinders flat and round.

Seven branchioſtegous rays.

Anarrhicas. *Gefner Paralip.* 4.
Lupus marinus *Caii opuſc.*
113.
Lupus marinus noſtras, quem
incolæ Wolff. *Schonevelde,*
45. *Tab.* 5.
Cat-Fiſh. *Sib. Scot.* III. 25.
Tab. 16.
Wolf Fiſh, Sea Wolf, or
Woof. *Wil. Icth.* 130. *Raii*

ſyn. piſc. 40.
Steen-bider. *Pontop. Norway,*
II. 151.
Kigutilik *i. e.* dentatus.
Crantz's Greenl. I. 96.
Anarhichas. *Arted. ſynon.* 39.
Anarhichas Lupus. *Lin. ſyſt.*
430.
Zee Wolf. *Gronov. Muſ.* No.
44. *Zooph.* No. 400.

THIS fiſh ſeems to be confined to the
northern parts of the globe. We find it
in the ſeas of *Greenland,* in thoſe of *Iceland* * and
Norway, on the coaſts of *Scotland,* and of *Yorkſhire,* and laſtly, in that part of the *German* ocean,
which waſhes the ſhores of *Holland,* the moſt
ſouthern of its haunts we can with any certainty
mention.

* Where it is called *Steinbeiſſer. Schonevelde,* 45.

L 4   It

It is a moſt ravenous and fierce fiſh, and when taken faſtens on any thing within its reach : the fiſhermen dreading its bite, endeavor as ſoon as poſſible to beat out its fore teeth, and then kill it by ſtriking it behind the head. *Schonevelde* relates, that its bite is ſo hard that it will ſeize on an anchor, and leave the marks of its teeth in it ; and the *Daniſh* and *German* names of *Steenbider* and *Steinbeiſſer*, expreſs the ſenſe of its great ſtrength, as if it was capable of cruſhing even ſtones with its jaws.

FOOD. It feeds almoſt entirely on cruſtaceous animals, and ſhell fiſh, ſuch as crabs, lobſters, prawns, muſcles, ſcollops, large whelks, &c. theſe it grinds to pieces with its teeth, and ſwallows with the leſſer ſhells. It does not appear they are diſſolved in the ſtomach, but are voided with the fœces, for which purpoſe the aperture of the anus is wider than in other fiſh of the ſame ſize.

It is full of roe in *February*, *March*, and *April*, and ſpawns in *May* and *June*.

This fiſh has ſo diſagreeable and horrid an appearance, that nobody at *Scarborough* except the fiſhermen will eat it, and they prefer it to holibut. They always before dreſſing take off the head and ſkin.

SIZE. The ſea wolf grows to a large ſize : thoſe on the *Yorkſhire* coaſt are ſometimes found of the length of four feet, and, according to Doctor *Gronovius*, have been taken near *Shetland* ſeven feet long, and even

even more. That which we examined was three feet two inches and an half from the tip of the nose to the end of the tail : the length of the head was eight inches, from the gills to the vent, ten ; from thence to the tip of the tail, twenty and one half.

The circumference of the head was seventeen inches, at the shoulders twenty, but near the tail only four and a half.

Its weight was twenty pounds and a quarter.

The head is a little flatted on the top : the nose blunt ; the nostrils very small ; the eyes small, and placed near the end of the nose. *Irides* pale yellow.

The teeth are very remarkable, and finely a-   Teeth. dapted to its way of life. The fore teeth are strong, conical, diverging a little from each other, stand far out of the jaws, and are commonly six above, and the same below, though sometimes there are only five in each jaw : these are supported within- side by a row of lesser teeth, which makes the num- ber in the upper jaw seventeen or eighteen, in the lower eleven or twelve.

The sides of the under jaw are convex inwards, which greatly adds to their strength, and at the same time allows room for the large muscles with which the head of this fish is furnished.

The *dentes molares*, or grinding teeth of the under jaw, are higher on the outer than the inner edges, which inclines their surfaces inward : they join to
the

the canine teeth in that jaw, but in the upper are separate from them.

In the centre are two rows of flat strong teeth, fixed on an oblong basis upon the bones of the palate and nose.

BUFONITES.   Thefe and the other grinding teeth are often found foffil, and in that ftate called *Bufonites,* or *Toad-ftones :* they were formerly much efteemed for their imaginary virtues, and were fet in gold, and worn as rings.

The two bones that form the under jaw are united before by a loofe cartilage, which mechanifm admitting of a motion from fide to fide, moft evidently contributes to the defign of the whole, *viz.* a facility of breaking, grinding, and comminuting its teftaceous and cruftaceous food. At the entrance of the gullet, above and below, are two echinated bones : thefe are very fmall, being the lefs neceffary, as the food is in a great meafure comminuted in the mouth by aid of the grinders.

The body is long, and a little compreffed fideways ; the fkin fmooth and flippery : it wants the lateral line.

The pectoral fins confift of eighteen rays, are five inches long, and feven and a quarter broad.

The dorfal fin extends from the hind part of the head almoft to the tail ; the rays in the frefh fifh are not vifible.

The anal fin extends as far as the dorfal fin.

The

The tail is round at its end, and confifts of thirteen rays.

The fides, back, and fins, are of a livid lead color; the two firft marked downwards with irregular obfcure dufky lines: thefe in different fifh have different appearances. The young are of a greenifh caft, refembling the fea wrack, which they refide amongft for fome time after their birth.

We think ourfelves much indebted to Mr. *Travis*, Surgeon, at *Scarborough*, for his ingenious remarks on this fifh, as well as on feveral others that frequent that coaft, being a gentleman much fkilled in icthyology, and extremely liberal in communicating his knowlege.

Head

XIV.
LAUNCE.

Head flender.
Body long and fquare.
Upper lip doubled in.
Dorfal and anal fin reaching almoft to the tail.
Seven branchioftegous rays.

66. SAND.

Ammodytes pifcis, ut nos vocavimus pro *anglico* San-dilz. *Gefner paralip.* 3.
Tobian, vel Tobias Sandtfpir-ing. *Schonevelde*, 76.
Ammodytes *Gefneri, Wil. Icth.* 113.

Sand Eels, or Launces. *Raii fyn. pifc.* 38, 165.
Ammodytes. *Arted. fynon.* 29.
Ammodytes Tobianus. *Lin. fyft.* 430.
Tobis. *Faun. Suec.* 302. *Gronov. Zooph.* No. 404.

THE launce is found on moft of our fandy fhores during fome of the fummer months: it conceals itfelf on the recefs of the tides beneath the fand, in fuch places where the water is left, at the depth of about a foot, and are in fome places dug out, in others drawn up by means of a hook con-trived for that purpofe. They are commonly ufed for baits for other fifh, but they are alfo very delicate eating.

Thefe fifh are found in the ftomachs of the *Por-peffe*, an argument that the laft roots up the fand with its nofe as hogs do the ground.

SIZE.

They grow fometimes the length of nine or ten inches:

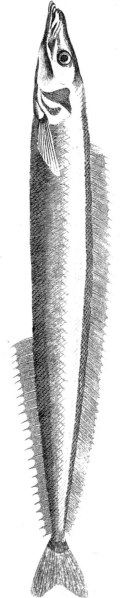

LAUNCE.

MORRIS.

inches: the females are longer and flenderer than the males.

The form of the body is fquare, the fides are rounded, and the angles not fharp: it is neverthelefs long and flender.

The head is fmall and taper; the under jaw much longer than the upper: the upper jaw is moveable, capable of being protruded, fo that when open the gape is very wide.

The irides are filvery.

The dorfal fin runs almoft the whole length of the back, is very narrow, and confifts of fiftyeight rays: the pectoral fins fmall, and have twelve: the anus is placed much nearer the tail than the head, is narrow, and extends almoft to the former.

The tail is forked, but the lobes rounded at their extremities.

The color of the back is blue, varying with green: on each fide the back is a narrow dufky line or two. The fides and belly are filvery; the lateral line ftrait.

Small

XV.
MORRIS.

Small head.

Body extremely thin, compreffed fideways.

No pectoral fins.

67. ANGLE-
SEA.

Leptocephalus. *Gronov. Zooph.* No. 409. *tab.* 13. *fig.* 3.

THIS fpecies was difcovered in the fea near
Holyhead by the late Mr. *William Morris*,
and, in memory of our worthy friend, we have
given it his name. On receiving it from Mr. *Morris*, we communicated it to that accurate Icthyologift, Doctor *Laurence Theodore Gronovius*, of *Ley-
den*, who has defcribed it in his *Zoophylacium*,
under the title of *Leptocephalus*, or fmall head.

DESCRIP.

The length was four inches; the head very fmall;
the body compreffed fideways, extremely thin, and
almoft tranfparent, about the tenth of an inch thick,
and in the deepeft part about one-third of an inch;
towards the tail it grew more flender, and ended
in a point; towards the head it floped down, the
head lying far beneath the level of the back.

The eyes large; the teeth in both jaws very
fmall.

The lateral line ftrait: the fides marked with
oblique ftrokes, that met at the lateral line.

The

The aperture to the gills large.

It wanted the pectoral, ventral, and caudal fins: the dorsal fin was extremely low, and thin, extending the whole length of the back very near the tail. The anal fin was of the same delicacy, and extended to the same distance from the anus.

The

XVI.
SWORD
FISH.

The upper jaw extending to a great length,
   hard, flender, and pointed.

No teeth.

Eight branchioftegous rays.

Slender body.

68. Sicilian

Ξιφιας. *Arift. Hift. an. lib.*
II. *c.* 13. VIII. *c.* 19. *Op-*
pian Halieut. lib. II. 462.
III. 442.
Xiphias. *Ovid Halieut.* 97.
Xiphias, *i. e.* Gladius *Plinii*
lib. XXXII. *c.* 2 *.
L'Heron de mer, ou grand
   Efpadaz *Belon,* 102.
Xiphias. *Rondel.* 251.
Xiphias, *i. e.* Gladius pif-

cis. *Gefner pifc.* 1049.
Caii opufc. 104.
Schwert-fifche. *Schonevelde,*
   35. Sword Fifh. *Wil. Icth.*
   161. *Raii fyn. pifc.* 52.
Xiphias. *Arted. fynon.* 47.
Xiphias Gladius. *Lin. fyft.*
   432.
Swerd-fifk. *Faun. Suec.* No.
   303.

Place.

THIS fifh fometimes frequents our coafts, but
   is much more common in the *Mediterra-*
nean fea, efpecially in the part that feparates *Italy*
from *Sicily,* which has been long celebrated for
it: the promontory *Pelorus**, now *Capo di Faro,*
was a place noted for the refort of the *Xiphias,* and
poffibly the ftation of the *fpeculatores,* or the perfons
who watched and gave notice of the approach of
the fifh.

* *Athenæus,* 314.

The

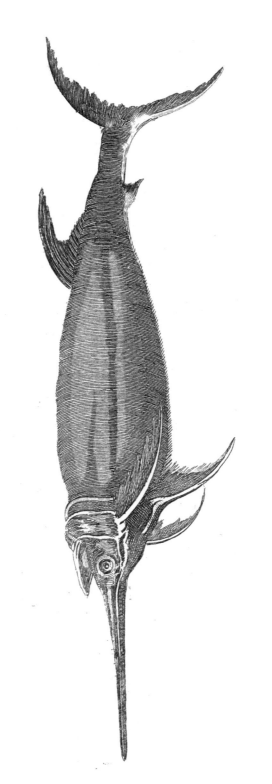

SWORD - FISH.

The antient method of taking them is particularly defcribed by *Strabo**, and agrees exactly with that practifed by the moderns.

A man afcends one of the cliffs that overhangs the fea: as foon as he fpies the fifh, he gives notice either by his voice, or by figns, of the courfe it takes.   Another, that is ftationed in a boat, climbs up the maft, and on feeing the fword fifh, directs the rowers towards it.   As foon as he thinks they are got within reach, he defcends, and taking a fpear in his hand, ftrikes it into the fifh, which, after wearying itfelf with its agitation, is feized and drawn into the boat.   It is much efteemed by the *Sicilians*, who buy it up eagerly, and at its firft coming into feafon give about fix-pence *Englifh* per pound.   The feafon lafts from *May* till *Auguft* †.   The antients ufed to cut this fifh into pieces, and falt it, whence it was called *Tomus Thurianus* ‡, from *Thurii*, a town in the bay of *Tarentum*, where it was taken and cured.

Kircher, in his *Mufurgia*, has preferved a ftrange incantation ufed by the *Sicilian* fifhermen, at the capture of the *Pefce Spada*, as they call it, which is expreffed in the following unintelligible jargon:

* *Lib.* I. *p.* 16.

† *Ray's Travels*, I. 271.

‡ *Tomus Thurianus, quem alii* Xiphiam *vocant.* Plinii *lib.* XXXII. *c.* 11.

Vol. III.   M   Mamaffu

Mamaſſu di pajanu,
Paletta di pajanu,
Majuſſu di ſtignela,
Palettu di pænu pale,
Pale la ſtagnetta,
Mancuta ſtigneta.
Pro naſtu, vardu, preſſu da
Viſu & da terra.

But this uſe of charmed words is not confined to *Sicily*; the *Iriſh* have their ſong at the taking of the razor ſhell; and the *Corniſh* theirs, at the taking of the whiſtle fiſh.

The ſword fiſh is ſaid to be very voracious, and that it is a great enemy to the Tunny, who (according to *Belon*) are as much terrified with it as ſheep are at the ſight of a wolf.

Ac durus Xiphias, *iƈtu non mitior enſis*;
Et pavidi magno fugientes agmine Thunni.
Ovid. Halieut. 97.

Sharp as a ſword the *Xiphias* does appear;
And crowds of flying *Tunnies* ſtruck with fear.

SIZE.   It grows to a very large ſize; the head of one, with the pectoral fins, found on the ſhore near *Laugharn*, in *Caermarthenſhire*, alone weighing ſeventy-five pounds: the ſnout was three feet long, rough, and hard, but not hard enough to penetrate ſhips and ſink them, as *Pliny* pretends *.

* Xiphiam, *id eſt*, Gladium, *roſtro mucronato eſſe, ab boc naves perfoſſas mergi in oceano.* Plin. *Lib.* XXXII. *c.* 11.

The

The fnout is the upper jaw, produced to a SNOUT,
great length, and has fome refemblance to a fword,
from whence the name. It is ccmpreffed at the
top and bottom, and fharp at the point. The
under jaw is four times as fhort as the upper, but
likewife fharp pointed. The mouth is deftitute of
teeth.

The body is flender, thickeft near the head,
and growing lefs and lefs as it approaches the
tail.

The fkin is rough, but very thin : the color
of the back is dufky, of the belly filvery.

The dorfal fin begins a little above the gills, and
extends almoft to the tail : it is higheft at the beginning and the end, but very low in the middle :
a little above the tail, on each fide, the fkin rifes
and forms two triangular protuberances, not unlike
the fpurious fins of the tunny.

The pectoral fins are long, and of a fcythe-like
form, and their firft rays the longeft.

The anus is placed at the diftance of one-third
part of the body from the tail; beneath are two
anal fins.

The tail is exactly of the fhape of a crefcent.

M 2   S E C T

Sect. II. JUGULAR.

XVII.
DRAGO-
NET.

Upper lip doubled.

Eyes near each other.

Two breathing apertures on the hind part of the head.

Firſt rays of the dorſal fin very long.

69. Gemme-
ous.

La tierce eſpece de Exocetus ? *Belon*, 218.

Dracunculus. *Rondel.* 304.

Dracunculus, aranei ſpecies altera. *Geſner piſc.* 80.

Dragon fiſh. *Marten's Spitzberg*, 123.

Yellow Gurnard. *Phil. Tranſ.* No. 293.

Lyra Harvicenſis. *Pet. Gaz. Tab.* 22. *Dale Harwick*, 431.

Callionymus Lyra. C. dorſalis prioris radiis longitudine corporis. *Lin. ſyſt.* 433. *Faun. Suec.* No. 110.

Uranoſcopus. *Gronov. Zooph.* No. 206.

Floy-fiſke. *Pontop. Norway*, II. III.

Dracunculus marinus. *Borlaſe Cornwall*, 270. *Seb. Muſ.* III. 92. *Tab.* 30. *fig.* 7.

Name.

*L*INNÆUS has given this genus the name of *Callionymus*, a fiſh mentioned by ſeveral of the antients; but the notices they have left of it are ſo very ſlight, as to render it difficult to determine what ſpecies they intended. * *Pliny* makes it a ſynonym to the *Uranoſcopus*, a fiſh frequent in the *Italian* ſeas, but very different from our *Dragonet*, a

* *Lib.* XXXII. *c.* 11.

name

Pl. XXVII.

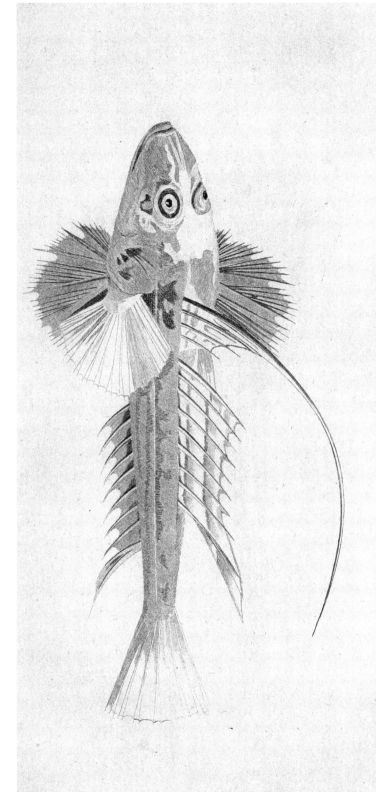

name we have taken the liberty of forming, from the diminutive *Dracunculus*, a title given it by *Rondeletius*, and other authors. The *English* writers have called it the Yellow Gurnard, which having no one character of the *Gurnard* genus, we think ourselves obliged to drop that name.

PLACE.

It is found as far north as *Norway** and *Spitzbergen*, and as far south as the *Mediterranean* sea, and is not unfrequent on the *Scarborough* coasts, where it is taken by the hook in thirty or forty fathoms water. It is often found in the stomach of the Cod-fish.

DESCRIP.

This species grows to the length of ten or twelve inches: the body is slender, round, and smooth.

The head is large, and flat at the top; in the hind part are two orifices, thro' which it breathes, and also forces out the water it takes in at the mouth, in the same manner as the cetaceous fish.

The apertures to the gills are closed: on the end of the bones that cover them is a very singular trifurcated spine.

The eyes are large, and placed very near each other on the upper part of the head, so that they look upwards; for which reason it has been ranked

---

* We have received it, with other curiosities, from that well-meaning prelate, *Erich Pontoppidan*, Bishop of *Bergen*. He was also Vice-Chancellor of the University of *Copenhagen*, in which station he died, *December* 20th. 1764, aged 66, much respected by his countrymen.

M 3 among

among the *Uranofcopi :* the pupils are of a rich fappharine blue, the irides of a fine fiery carbuncle.

The upper jaw projects much farther than the
lower : the mouth is very wide : the teeth are fmall.

The pectoral fins are round, and of a light-
brown color ; the ventral placed before them, are
very broad, and confift of five branched rays.

The firft dorfal fin is very fingular, the firft ray
being fetaceous, and fo long as to extend almoft
to the tail : thofe of the fecond dorfal fins are of a
moderate length, except the laft, which is produced far beyond the others.

The anus is placed about the middle of the belly ;
the anal fin is broad, and the laft ray the fongeft.
Pontoppidan calls this fpecies the flying fifh : whether it makes ufe of any of its fins to raife itfelf out
of the water, as he was informed they did, we cannot pretend to fay.

The tail is rounded and long, and confifts of ten
rays.

Colors.   The fide line is ftrait: the colors are yellow,
blue, and white, and make a beautiful appearance
when the fifh has been juft taken. The blue is of
an inexpreffible fplendor, the richeft cærulean
glowing with a gemmeous brilliancy. The throat
is black. The membranes of all the fins extremely thin and delicate.

Dracunculous

Pl. XXVIII.

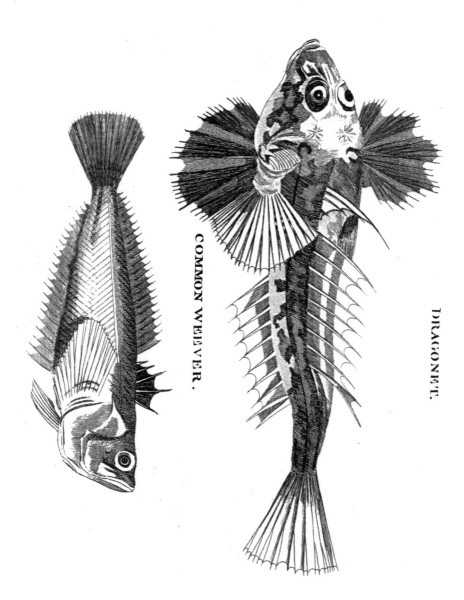

COMMON WEEVER.

DRAGONET.

N.º 72.

N.º 70.

Dracunculus. *Wil. Icth.* 136.   Callionymus Dracunculus. C. 70. SORDID.
  Raii syn. pisc. 79.   dorsalis prioris radiis cor-
Cottus pinna secunda dorsi al-   pore brevioribus. *Lin. syst.*
  ba *Arted. synon.* 77.   434.

THIS species we received from Mr. *Travis.* Its length was only six inches and an half.

The head was compressed; the forehead sloped down to the nose, being not so level as that of the preceding.

The eyes large, and almost contiguous.

The mouth small; the teeth very minute.

Over the gills was a strong trifurcated broad spine.

The first dorsal fin had four rays; the first setaceous, extending a little higher than the others, the last very short: the two first rays and webs were yellow, the others black.

The second had ten soft rays, their ends extending beyond the webs, which were pellucid.

The pectoral fins consisted of twenty rays, and were ferruginous, spotted with a deeper cast of the same: the ventral fins consisted of five broad and much branched rays, like those of the first species.

The anal fin was white, and had ten rays; the tail had ten rays. In both species they are bifurcated at their ends, and the ray next the anal fin in both is very short.

M 4

In

In colors this is far inferior to the former, being of a dirty yellow, mixed with white and duſky ſpots ; the belly is entirely white.

Lower

Lower jaw floping down.
Gill covers aculeated.
Six branchioftegous rays.
Two dorfal fins.
Anus near the breaft.

Δρακων? *Arift. Hift. an. Lib.*
VIII. *c.* 13 *Ælian. Hift.
an. Lib.* II. *c.* 50. *Oppian
Halieut.* II. 459.
Draco marinus *Plinii Lib.* IX.
c. 27. Draco, Dracunculus.
Lib. XXXII. *c.* 11. Ara-
neus. *Lib.* IX. *c.* 48.
La vive. *Belon.* 209.
Draco. *Rondel.* 300. *Gefner
pifc.* 77, 78.

Peter-manniken,   Schwertfif-
che. *Schonevelde* 16.
The Weever. *Wil. Icth.* 238.
Raii fyn. pifc. 91.
Trachinus maxilla inferiore
longiore,   cirris deftituta.
Arted. fyn. 71.
Trachinus Draco.   *Lin. fyft.*
453. *Gronov. Zooph.* No.
274.
Farfing, Fiaffing. *Faun. Suec.*
No. 305.

71. COMMON.

THE qualities of this fifh were well known
to the antients, who take notice of them
without any exaggeration : the wounds inflicted
by its fpines are exceedingly painful, attended
with a violent burning, and moft pungent fhooting,
and fometimes with an inflammation that will extend from the arm to the fhoulder. *

It is a common notion that thefe fymptoms pro-

* It is probable that the malignity of the fymptoms arifes
from the habit of body the perfon is in, or the part in which
the wound is given.

ceed

ceed from fomething more than the fmall wound this fifh is capable of inflicting; and that there is a venom infufed into it, at leaft fuch as is made by the fpines that form the firft dorfal fin, which is dyed with black, and has a moft fufpicious afpect. The remedy ufed by a fifherman in our neighbour-hood is the fea fand, with which he rubs the place affected for a confiderable time. * At *Scarborough*, ftale urine, warmed, is ufed with fuccefs.

This fifh buries itfelf in the fands, leaving only its nofe out, and if trod on immediately ftrikes with great force; and we have feen them direct their blows with as much judgment as fighting cocks. Notwithftanding this noxious property of the fpines, it is exceeding good meat.

NAME.   The *Englifh* name feems to have no meaning, being corrupted from the *French*, *la vive*, fo called as being capable of living long out of the water, according to the interpretation of *Belon*.

DESCRIP.   It grows to the length of twelve inches, but is commonly found much lefs.

The irides are yellow: the under jaw is longer than the upper, and flopes very much towards the belly: the teeth are fmall.

The back is ftrait, the fides flat, the belly pro-

* In the *Univerfal Mufeum* for *November* 1765, is an inftance of a perfon who was reduced to great danger by a wound from this fifh, and who was cured by the application of fweet oil, and taking *opium* and *venice treacle*.

menent

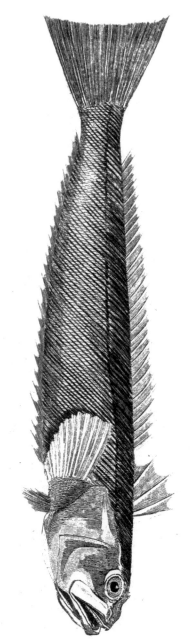

GREATER WEEVER.

minent, the lateral line ftrait: the covers of the gills are armed with a very ftrong fpine.

The firft dorfal fin confifts of five very ftrong fpines, which, as well as the intervening membranes are tinged with black ; this fin, when quiefcent, is lodged in a fmall hollow.

The fecond confifts of feveral foft rays, commences juft at the end of the firft, and continues almoft to the tail. The pectoral fins are broad and angular; the ventral fins fmall.

The vent is placed remarkably forward, very near the throat: the anal fin extends to a fmall diftance from the tail, is a little hollowed in the middle, but not fo much as to be called forked.

The fides are marked lengthways with two or three dirty yellow lines, and tranfverfely by numbers of fmall ones: the belly filvery.

Draco major feu araneus. *Salvian.* 70.
Greater Weever. *Tour Scotland,* 1769, octavo.

72. GREAT.

THE length eleven inches: greateft depth one and three quarters: head flat: eyes large: edges of the jaws rough with minute teeth; lower jaw the longeft: head covered with minute tubercles: cheeks and gills with minute fcales: on the gills is a fharp fpine.

Firft dorfal fin black, with five fpines: the fecond reaches almoft to the tail: in the pectoral fins are thirteen branched rays: in the ventral, fix: the anal extends oppofite to the fecond dorfal fin: tail large, triangular, even at the end.

The fcales run in oblique lines from the back to the belly, with a divifion between each row.

Inhabits the fea near *Scarborough.*

Head

XIX.
COD FISH.

Head fmooth.

Seven flender branchioftegous rays.

Body oblong; fcales deciduous.

All the fins covered with a common fkin.

Ventral fins flender, and ending in a point.

Teeth in the jaws; and in the palate, a feries of minute teeth clofely fet together.

* With three dorfal fins; the chin bearded.

73. COMMON.

La Morue. *Belon,* 121.
Molva. *Rondel.* 280.
Molva five morhua altera. *Gefner pifc.* 88.
Kablauw. *Schonevelde,* 18.
Afellus major vulgaris. *Wil. Icth.* 165.
Cod-fifh, or Keeling. *Raii*

fyn. pifc. 53.
Gadus dorfo tripterygio, ore cirrato, cauda æquali fere cum radio primo fpinofo. *Arted. fynon.* 35.
Gadus morhua. *Lin. fyft.* 436.
Gronov. Zooph. No. 319.
Cabblia. *Faun. Suec.* No. 398.

THIS fifh is found only in the northern part of the world; it is, as *Rondeletius* calls it, an ocean fifh, and never met with in the *Mediterranean* fea*. It affects cold climates, and feems confined between the latitudes 66 and 50: what are caught north and fouth of thofe degrees being

* None (fays Captain *Armftrong* in his hiftory of *Minorca)* of the *Afelli* or cod fifh kind, frequent our fhores. *p.* 163.

either

either few in quantity, or bad in quality. The *Greenland* fiſh are ſmall and emaciated through want of food, being very voracious, and having in thoſe ſeas a dearth of proviſion.

This locality of ſituation is common to many other ſpecies of this genus, moſt of them being inhabitants of the cold ſeas, or ſuch that lie within zones that can juſt clame the title of temperate. There are neverthelefs certain ſpecies found near the *Canary Iſlands*, called *Cherny* *, of which we know no more than the name; but according to the unfortunate Captain *Glaſs*, are better taſted than the *Newfoundland* kind.

The great rendezvouz of the cod fiſh is on the Banks of *Newfoundland*, and the other ſand banks that lie off the coaſts of *Cape Breton*, *Nova Scotia*, and *New England*. They prefer thoſe ſituations, by reaſon of the quantity of worms produced in thoſe ſandy bottoms, which tempt them to reſort there for food: but another cauſe of the particular attachment the fiſh have to theſe ſpots, is their vicinity to the polar ſeas, where they return to ſpawn; there they depoſe their roes in full ſecurity, but want of food forces them, as ſoon as the firſt more ſouthern ſeas are open, to repair thither for ſubſiſtence.

Few are taken north of *Iceland*, but on the ſouth and weſt coaſts they abound: they are again found

* *Hiſt. Canary Iſlands,* 198.

to

to fwarm on the coafts of *Norway*, in the *Baltic*, off the *Orkney* and the *Weftern Ifles*; after which their numbers decreafe, in proportion as they advance towards the fouth, when they feem quite to ceafe before they reach the mouth of the Straits of *Gibraltar*.

Before the difcovery of *Newfoundland*, the greater fifheries of cod were on the feas of *Iceland*, and off our *Weftern Ifles*, which were the grand refort of fhips of all the commercial nations; but it feems that the greateft plenty was met with near *Iceland*. The *Englifh* reforted thither before the year 1415: for we find that *Henry* V. was difpofed to give the King of *Denmark* fatisfaction for certain irregularities committed on thofe feas by his fubjects. In the reign of *Edward* the IV. the *Englifh* were excluded from the fifhery by treaty; and forbidden to refort there under pain of forfeiture of life and goods. Notwithftanding this, our monarch afterwards gave licence to a fhip of *Hull* to fail to *Iceland*, and there relade fifh and other goods, without regard to any reftrictions to the contrary. Our right in later times was far from being confirmed, for we find Queen *Elizabeth* condefcending to afk permiffion to fifh in thofe feas from *Chriftian* the IV. of *Denmark*, yet afterwards fhe fo far repented her requeft, as to inftruct her embaffadors to that court, to infift on the right of a free and univerfal fifhery *. How far fhe fucceeded, I do

---

* *Rymer's Fœd.* XVI. 275, 425.

not

not know: but it appears, that in the reign of her fucceffor, our countrymen had not fewer than a hundred and fifty fhips employed in the *Iceland* fifhery. I fuppofe this indulgence might arife from the marriage of *James* with a Princefs of *Denmark*.

But the *Spanifh*, the *French*, and the *Bretons*, had much the advantage of us in all fifheries at the beginning, as appears by the ftate of that in the feas of *Newfoundland* in the year 1578*, when the number of fhips belonging to each nation ftood thus :

Spaniards, 100, befides 20 or 30 that came from *Bifcaie*, to take whale for train, being about five or fix thoufand tons.

Portuguefe, 50, or three thoufand tons.

French and *Bretons*, 150, or feven thoufand tons.

Englifh, from 30 to 50.

But Mr. *Anderfon*, in his Dictionary of Commerce, I. 363, fays, that the *French* began to fifh there fo early as 1536; and we think we have fomewhere read, that their firft pretence for fifhing for cod in thofe feas, was only to fupply an *Englifh* convent with that article.

The encreafe of fhipping that refort to thofe fertile banks, are now unfpeakable : our own country ftill enjoys the greateft fhare, which ought to be efteemed our chiefeft treafure, as it brings wealth to individuals, and ftrength to the ftate.

* *Hackluyt's Coll. Voy.* III. 132.

All

All this immenfe fifhery is carried on by the hook and line only *; the bait is herring, a fmall fifh called a *Capelin*, a fhell fifh called *Clams*, and bits of fea fowl; and with thefe are caught fifh fuffi-cient to find employ for near fifteen thoufand *Bri-tifh* feamen, and to afford fubfiftence to a much more numerous body of people at home, who are engaged in the various manufactures which fo vaft a fifhery demands.

FOOD.

The food of the cod is either fmall fifh, worms, teftaceous, or cruftaceous animals, fuch as crabs, large whelks, &c. and their digeftion is fo power-ful, as to diffolve the greateft part of the fhells they fwallow. They are very voracious, and catch at any fmall body they perceive moved by the water, even ftones and pebbles, which are often found in their ftomachs.

THE SOUNDS.

Fifhermen are well acquainted with the ufe of the air-bladder or *found* of the cod, and are very dexterous in perforating this part of a live fifh with a needle, in order to difengage the inclofed air; for without this operation it could not be kept un-der water in the well-boats, and brought frefh to market. The *founds* of the cod falted is a delica-

* We have been informed that they fifh from the depth of fifteen to fixty fathoms, according to the inequality of the *Bank*, which is reprefented as a vaft mountain, under water, above five hundred miles long, and near three hundred broad, and that feamen knew when they approach it by the great fwell of the fea, and the thick mifts that impend over it.

cy

cy often brought from *Newfoundland*. *Isinglass* is alfo made of this part by the *Iceland* fifhermen: as the procefs may be of fervice to inftruct the natives of the North of *Scotland* where thefe' fifh are plentiful, I beg leave to give it in the Appendix, extracted from a ufeful paper on the fubject, in the *Ph. Tr.* of 1773, by *Humphrey Jackfon*, Efq.

Providence hath kindly ordained, that this fifh, fo ufeful to mankind, fhould be fo very prolific as to fupply more than the deficiencies of the multitudes annually taken.   *Leuwenhoek* counted nine millions three hundred and eighty-four thoufand eggs in a cod fifh of a middling fize, a number fure that will baffle all the efforts of man, or the voracity of the inhabitants of the ocean to exterminate, and which will fecure to all ages an inexhauftible fupply of grateful provifion.

In our feas they begin to fpawn in *January*, and depofite their eggs in rough ground, among rocks. Some continue in roe till the beginning of *April.* The cod fifh in general recover quicker after fpawning than any other fifh, therefore it is common to take fome good ones all the fummer.   When they are out of feafon they are thin tailed and loufy, and the lice chiefly fix themfelves on the infide of their mouths.

The fifh of a middling fize are moft efteemed for the table, and are chofen by their plumpnefs and roundnefs, efpecially near the tail, by the depth of the fulcus or pit behind the head, and by

Vol. III   N   the

the regular undulated appearance of the sides, as
if they were ribbed. The glutinous parts about
the head lose their delicate flavor after it has been
twenty-four hours out of the water, even in winter,
in which these and other fish of this genus are in
highest season.

Size. The largest that we ever heard of taken on our
coasts, weighed seventy-eight pounds, the length
was five feet eight inches; and the girth round
the shoulders five feet. It was taken at *Scarborough*
in 1755, and was sold for one shilling. But the
general weight of these fish in the *Yorkshire* seas,
is from fourteen to forty pounds.

Descrip. This species is short in proportion to its bulk,
the belly being very large and prominent.

The jaws are of an equal length, at the end of
the lower is a small beard; the teeth are disposed
in the palate as well as jaws.

The eyes are large.

On the back are three soft fins; the first has
fourteen, the two last nineteen rays a-piece. The
ventral fins are very slender, and consist but of six
rays; the two first extending far beyond the
other. It has two anal fins; the first consisting of
twenty, the last of sixteen rays.

The tail is almost even at the end : the first ray
on each side is short, and composed of a strong
bone.

The color of this fish is cinereous on the back
and sides, and commonly spotted with yellow: the

belly

belly is white, but they vary much, not only in color* but in fhape, particularly that of the head.

The fide line is white and broad, ftrait, till it reaches oppofite the vent, when it bends towards the tail.

Side Line.

Aigrefin, ou aiglefin. *Belon.* 118.
Tertia afellorum fpecies. *Rondel.* 277.
Tertia afel. Sp. Eglefinus. *Gefner pifc.* 86.
Onos five afinus veterum. *Turner epift. ad. Gefner.*
Afellus minor, Schelfifch. *Schonevelde.* 18.
Hadock. *Wil. Icth.* 170.

Raii fyn. pifc. 55.
Gadus dorfo tripterygio, ore cirrato, max. fup. longiore, corpore albicante, cauda parum bifurca. *Arted. fynon.* 36.
Gadus Æglefinus. G. tripterygius cirratus albicans, cauda biloba. *Lin. fyft.* 435.
Kolja. *Faun. Suec.* No. 306.
Gronov. Zooph. No. 321.

74. Hadock.

OUR countryman *Turner* conjectured this fpecies to have been the Ονℭ, or *Afinus*, of the antients, and *Belon* that it was the Κριὸς, and the Πρόϭατος of *Oppian*. We have carefully confulted moft of the antient naturalifts, but cannot difcover any marks by which we can determine the fpecies they intended. The words† Ονℭ, ‡ *Afinus,*

Name,

* Codlings are often taken of a yellow, orange, and even red color, while they remain among the rocks, but on changing their place affume the color of other cod fifh.

† *Arift. Hift. an. Lib.* VIII. *c.* 15. *Oppian Halieut.* I. 151. III. 191.

‡ *Ovidii Halieut. Lin.* 131. *Plinii Lib.* IX. *c.* 16. 17.

N 2 *Afellus,*

Afellus, * *Callarias,* and *Bacchus,* are familiarly applied to feveral of our fpecies of cod fifh by the more modern writers ; yet the antients from whom they are borrowed, have not authorized the application to any particular kind, either by defcription or any other method.

Different reafons have been affigned for giving the name of Ονⓔ, or *Afinus* to this genus, fome imagining it to be from the color of the fifh, others becaufe it ufed to be carried on the backs of affes to market; but we fhall drop this uncertain fubject, and proceed to what we have fuller affurance of.

SEASON. Large hadocks begin to be in roe the middle of *November,* and continue fo till the end of *January* ; from that time till *May* they are very thin tailed, and much out of feafon. In *May* they begin to recover, and fome of the middling-fized fifh are then very good, and continue improving till the time of their greateft perfection. The fmall ones are extremely good from *May* till *February,* and fome even in *February, March,* and *April,* viz. thofe which are not old enough to breed.

The fifhermen affert, that in rough weather hadocks fink down into the fand and ooze in the bottom of the fea, and fhelter themfelves there till the ftorm is over, becaufe in ftormy weather they take none, and thofe that are taken immediately

* *Lib. c.* 17.

fmall

after a ftorm are covered with mud on their backs.

In fummer they live on young herrings and other fmall fifh; in winter on the ftone-coated worms *, which the fifhermen call *hadock meat*.

The grand fhoal of hadocks comes periodically on the *Yorkfhire* coafts. It is remarkable that they appeared in 1766 on the 10th of *December*, and exactly on the fame day in 1767 : thefe fhoals extended from the fhore near three miles in breadth, and in length from *Flamborough* head to *Tinmouth* caftle, and perhaps much farther northwards. An idea may be given of their numbers by the following fact : three fifhermen, within the diftance of a mile from *Scarborough* harbour, frequently loaded their *coble* or boat with them twice a-day, taking each time about a ton of fifh : when they put down their lines beyond the diftance of three miles from the fhore, they caught nothing but dog fifh, which fhows how exactly thefe fifh keep their limits.

The beft hadocks were fold from eightpence to a fhilling per fcore, and the poor had the fmaller fort at a penny, and fometimes a halfpenny per fcore †.

The large hadocks quit the coaft as foon as they

---

* A fpecies of *Serpula*.

† Here Mr. *Travis*, to whom I am much obliged for a moft accurate account of the *Yorkfhire* fifh, with great humanity projects an inland navigation, to convey at a cheap and eafy method, thofe gifts of Providence to the thoufands of poor manufacturers who inhabit the diftant parts of that vaft county.

N 3   go

go óut of feafon, and leave behind great plenty of fmall ones. It is faid that the large ones vifit the coafts of *Hamburgh* and *Jutland* in the fummer.

It is no lefs remarkable than providential, that all kinds of fifh (except mackrel) which frequent the *Yorkfhire* coaft, approach the fhore, and as if it were offer themfelves to us, generally remaining there as long as they are in high feafon, and retire from us when they become unfit for ufe.

It is the commoneft fpecies in the *London* markets.

They do not grow to a great bulk, one of four-teen pounds being of an uncommon fize, but thofe are extremely coarfe; the beft for the table weigh-ing from two to three pounds.

The body is long, and rather more flender than thofe of the preceding kinds : the head flopes down to the nofe : the fpace between the hind part of the firft dorfal fin is ridged : on the chin is a fhort beard.

On the back are three fins refembling thofe of the common cod-fifh : on each fide beyond the gills is a large black fpot. Superftition affigns this mark to the impreffion St. *Peter* left with his finger and thumb when he took the tribute out of the mouth of a fifh of this fpecies, which has been continued to the whole race of hadocks ever fince that miracle.

The lateral line is black : the tail is forked.

The color of the upper part of this fpecies is dufky

duſky or brown; the belly and lower part of the ſides ſilvery.

Irides ſilvery: pupil large and black.

Aſellus mollis latus. Mr. *Liſter apud Wil. Icth. App.* 22.

Whiting Pout, *Londinenſibus. Raii ſyn. piſc.* 55.

Gadus dorſo tripterygio, ore cirrato, longitudine ad latitudinem tripla, pinna ani prima oſſiculorum triginta. *Arted. ſynon.* 37.

Gadus barbatus. G. tripterygius cirratus maxilla inferiore punctis utrinque ſeptem. *Lin. ſyſt.* 437. *Gronov. Zooph.* No. 320.

Sma-Torſk. *Faun. Suec.* No. 311.

75. Pout.

THIS ſpecies never grows to a large ſize, ſeldom exceeding a foot in length.

It is diſtinguiſhed from all others by its great depth; one of the ſize abovementioned being near four inches deep in the broadeſt part.

The back is very much arched, and carinated. The ſcales larger than thoſe of the cod fiſh. The mouth ſmall; the beard ſhort. On each ſide of the lower jaw are ſeven or eight punctures.

The firſt dorſal fin is triangular, and terminates in a long fibre: the color of the fins and tail black: at the bottom of the pectoral fins is a black ſpot.

The lateral line is white, broad, and crooked.

The tail is even at the end, and of a duſky color.

N 4   The

The color of the body is white, but more obscure on the back than the belly, and tinged with
yellow.

It is called at *Scarborough* a *Kleg*. It is a very
delicate fish.

76. BIB.   Afellus nanus, Dwergdorfch,   Gadus dorfo tripterygio, ore
                Krumftert? *Schonevelde*, 20.   cirrato, officulo pinnarum
            Bib & Blinds *Cornubienfibus.*   ventralium primo in longam
              Wil. Icth. 169.   fetam producto. *Arted. fynon.*
            Afellus lufcus. *Raii fyn. pifc.*   35.
              54.   Gadus lufcus. *Lin. fyft.* 437.

THIS species grows to the length of one foot.
The greatest depth three inches and a half.
The scales are large, and so far from adhering to
the skin, as is afferted by naturalists, are extremely deciduous.

The body is deep, the sides compreffed. The
eyes covered with a loofe membrane, which it can
blow up at pleafure, like a bladder. The mouth
is small : beneath the chin a beard, an inch long.

In the first dorfal fin twelve rays : in the second,
which is longeft, twenty-three : in the third, twenty.
The pectoral fins about sixteen : the ventral six
or feven, of which the first ray is long, and feta-

ceous :

Pl. XXX.

G. Wilkinson del

P. Mazell Sculp

POOR.

N.º 77.

BIB.

N.º 76.

ceous: the firſt anal fin has twenty-ſeven; the laſt twenty-one rays.

The back is of a light olive: the ſides finely COLOR.
tinged· with gold: the belly white: the anal fins
duſky, edged with pure white; the tail with black.

Le Merlan? *Belon*, 120.
Anthiæ ſecunda ſpecies. *Ron-*
del. 191. *Geſner piſc.* 56.
Aſellus mollis minor, ſeu
  aſellus omnium minimus.
MOLLO *Venetiis*. CAPELAN
Maſſiliæ. *Wil. Icth.* 171.
Poor or   Power *Cornub.*

Mr. *Jago. Raii ſyn. piſc.*   **77. POOR.**
  163. *fig.* 6.
Gadus dorſo tripterygio, ore
  cirrato, corpore ſeſcunciali,
  ano in medio corporis.
Arted. ſynon. 36.
Gadus minutus. *Lin. ſyſt.* 438.

THIS is the only ſpecies of cod fiſh with three
dorſal fins that we (at this time) are aſſured
is found in the *Mediterranean* ſea. It is taken near
Marſeilles, and ſometimes in ſuch quantities as to
become a nuſance; for no other kinds of fiſh are
taken during their ſeaſon *. It is eſteemed good,
but incapable of being ſalted or dried: *Belon* ſays,
that when it is dried in the ſun, it grows as hard
as horn; *C'eſt dela que les* ANGLOIS *l'ont nommé*
Bouclzs horn.

It is the ſmalleſt ſpecies yet diſcovered, being DESCRIP.
little more than ſix inches long.

On the chin is a ſmall beard: the eyes are co-

* *Rondel.* 191.

              vered

vered with a loofe membrane: on the gill-covers, and the jaws are on each fide, nine punctures.

The firft dorfal fin has twelve rays; the fecond nineteen; the third feventeen.

The pectoral fins thirteen; the ventral fins fix: the firft anal fin twenty-feven; the fecond feventeen.

The color on the back is a light brown; on the belly a dirty white.

We owe the difcovery of this kind in our feas to the Rev. Mr. *Jago.*

** Three dorfal fins : chin beardlefs.

78. COAL.

Colfifch. *Belon,* 128.
Colfifch *Anglorum. Gefner pifc.* 89.
Afellus niger. Kolfifch. Koler. *Schonevelde,* 19.
Cole fifh *Septentrionalium anglorum.* Rawlin Pollack *Cornubienfium. Wil. pifc.*

168. *Raii fyn. pifc.* 54.
Gadus dorfo tripterygio, ore imberbi, maxilla inferiore longiore et linea laterali recta. *Arted. fynon.* 34.
Gadus carbonarius. *Lin. fyft.* 438. *Gronov. Zooph.* No. 317.

THE coal fifh takes its name from the black color that it fometimes affumes. *Belon* calls it the *Colfifch,* imagining it was fo named by the *Englifh,* from its producing the *Icthyocolla,* but *Gefner* gives the true etymology.

Thefe

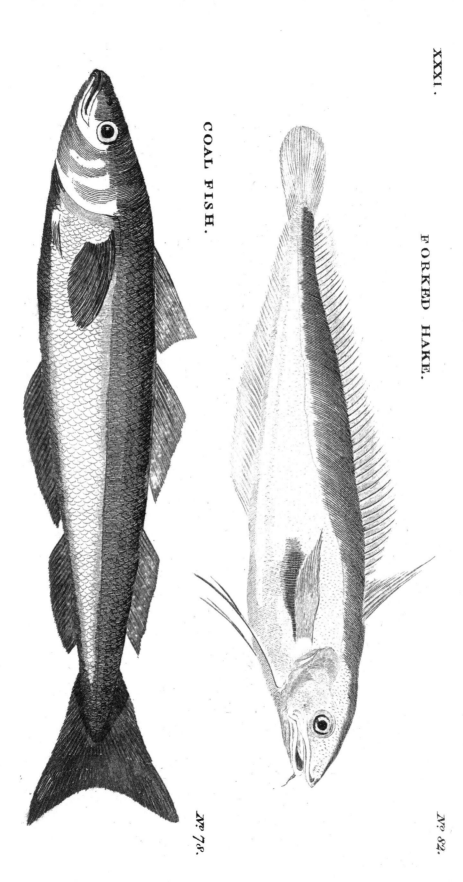

FORKED HAKE.

COAL FISH.

N.º 82.

N.º 78.

Thefe fifh are common on moft of our rocky and deep coafts, but particularly thofe of the north of *Great Britain*. They fwarm about the *Orknies*, where the fry are the great fupport of the poor.

The young begin to appear on the *Yorkfhire* coaft the beginning of *July* in vaft fhoals, and are at that time about an inch and an half long. In *Auguft* they are from three to five inches in length, and are taken in great numbers with the angling rod, and are then efteemed a very delicate fifh, but grow fo coarfe when they are a year old that few people will eat them. Fifh of that age are from eight to fifteen inches long, and begin to have a little blacknefs near the gills, and on the back, and the blacknefs encreafes as they grow older.

Young.

The fry is known by different names in different places: they are called at *Scarborough Parrs*, and when a year old, *Billets*. About nine or ten years ago fuch a glut of *Parrs* vifited that part, that for feveral weeks it was impoffible to dip a pail into the fea without taking fome.

Tho' this fifh is fo little efteemed when frefh, yet it is falted and dried for fale; a perfon laft year having cured above a thoufand at *Scarborough*.

The coal fifh is of more elegant form than the cod fifh: they generally grow to the length of two feet and an half, and weigh about twenty-eight or thirty pounds at moft. The head is fmall; the under jaw a little longer than the upper:

Descrip.

the

the irides filvery, marked on one fide with a black fpot.

It has three dorfal fins, the firft confifts of four-teen, the next of twenty, the laft of twenty-two rays.

The pectoral fins of eighteen; the ventral of fix: the firft anal fin of twenty-two, the fecond of nineteen.

The tail is broad and forked.

Thefe fifh vary in color. We have feen fome whofe back, nofe, dorfal fins and tail were of a deep black: the gill covers filver and black: the ventral and anal fins white; the belly of the fame color.

We have feen others dufky, others brown, but in all the lateral line was ftrait and white, and the lower part of the ventral and anal fins white.

79. POLLACK. Afellus virefcens, Schwartres Kolmulen. *Schonevelde*, 20. Afellus flavefcens; Gelbe Kolmulen. *Ibid.* Afellus Huitingo-Pollachius. *Wil. Icth.* 167. Whiting Pollack. *Raii fyn. pifc.* 53. Gadus dorfo tripterygio, ore imberbi, max. inf. longiore, linea laterali curva. *Arted. fynon.* 35. Gadus Pollachius. *Lin. fyft.* 439. *Gronov. Zooph.* No. 318. *Norwegis* Scy. *Bahufiis* Grafik? *Faun. Suec.* No. 309.

THIS fpecies is common on many of our rocky coafts: during fummer they are feen

in

in great fhoals frolicking on the furface of the water, and flinging themfelves into a thoufand forms. They are at that time fo wanton as to bite at any thing that appears on the top of the waves, and are often taken with a goofe's feather fixed to the hook. They are a very ftrong fifh, being ob-ferved to keep their ftation at the feet of the rocks in the moft turbulent and rapid fea.

They are a good eating fifh: they do not grow to a very large fize; at left the biggeft we have feen did not exceed fix or feven pounds: but we have heard of fome that were taken in the fea near *Scarborough*, which they frequent during winter, that weighed near twenty-eight pounds. They are there called *Leets*.

The under jaw is longer than the upper; the head and body rifes pretty high, as far as the firft dorfal fin.

The fide line is incurvated, rifing towards the middle of the back, then finking and running ftrait to the tail; it is broad, and of a brown color.

The firft dorfal fin has eleven rays, the mid-dle nineteen, the laft fixteen: the tail is a little forked.

The color of the back is dufky, of fome in-clining to green: the fides beneath the lateral line marked with lines of yellow; the belly white.

Secunda

80. Whi-
ting.

Secunda afellorum fpecies.
Rondel. 276.
Merlanus. *Rondel. Gefner pifc.*
85.
Afellus candidus primus,
Witling. *Schonevelde,* 17.
Afellus mollis major, feu al-
bus. *Wil. Ićth.* 170.
Whiting. *Raii fyn. pifc.* 55.

Gadus dorfo tripterygio, ore
imberbi corpore albo, maxilla fuperiore longiore. *Ar-
ced. fynon* 34.
Gadus merlangus. *Lin. fyft.*
438. *Gronov. Zooph.* No.
315.
Hwitling, Widding. *Faun.
Suec.* No. 310.

WHITINGS appear in vaft fhoals in our feas in the fpring, keeping at the diftance of about half a mile to that of three from the fhore. They are caught in vaft numbers by the line, and afford excellent diverfion.

They are the moft delicate, as well as the moft wholefome of any of the genus, but do not grow to a large fize; the biggeft we ever faw * not exceeding twenty inches, but that is very uncommon, the ufual length being ten or twelve.

It is a fifh of an elegant make : the upper jaw is the longeft; the eyes large, the nofe fharp, the teeth of the upper jaw long, and appear above the lower when clofed.

The firft dorfal fin has fifteen rays, the fecond eighteen, the laft twenty.

* We have been informed that whitings, from four to eight pounds in weight, have been taken in the deep water at the edge of the *Dogger-Bank.*

The

The color of the head and back is a pale brown; the lateral line white, and is crooked; the belly and fides filvery; the laft ftreaked lengthways with yellow.

** With only two dorfal fins.

81. HAKE.

Le Merluz. *Belon*, 115.
Afellus, ὄνος, ὄνισκος. *Rondel.* 272.
Merlucius. *Gefner pifc.* 84.
Afellus primus five Merlucius. *Wil. Icth.* 174.
The Hake. *Raii fyn. pifc.*

Gadus dorfo dipterygio, maxilla inferiore longiore. *Arted. fynon.* 36.
Gadus Merlucius. *Lin. fyft.* 439. *Faun. Suec.* No. 314. *Gronov. Zooph.* No. 315.

A FISH that is found in vaft abundance on many of our coafts, and of thofe of *Ireland.* There was formerly a vaft ftationary fifhery of *Hake* on the *Nymph Bank* off the coaft of *Waterford,* immenfe quantities appearing there twice a year; the firft fhoal coming in *June,* during the *Mackrel* feafon, the other in *September,* at the beginning of the *Herring* feafon, probably in purfuit of thofe fifh: it was no unufual thing for fix men with hooks and lines to take a thoufand *Hake* in one night, befides a confiderable quantity of other fifh. Thefe were falted and fent to *Spain,* particularly

to

to *Bilboa.* * We are at this time uninformed of the ftate of this fifhery, but find that Mr. *Smith,* who wrote the hiftory of the county of *Waterford,* complains even in his time (1746) of its decline. Many of the gregarious fifh are fubject to change their fituations, and defert their haunts for numbers of years, and then return again. We fee, p. 102, how unfettled the *Bafking Shark* appears to be: Mr. *Smith* inftances the lofs of the *Hadock* on the *Waterford* fhores, where they ufed to fwarm; and to our knowlege we can bring the capricioufnefs of the herrings, which fo frequently quit their ftations, as another example.

Sometimes the irregular migration of fifh is owing to their being followed and haraffed by an unufual number of fifh of prey, fuch as the fhark kind.

Sometimes to deficiency of the fmaller fifh, which ferved them as food.

And laftly, in many places to the cuftom of trawling, which not only demolifhes a quantity of their fpawn, which is depofited in the fand, but alfo deftroys or drives into deeper waters numberlefs worms and infects, the repaft of many fifh.

The hake is in *England* efteemed a very coarfe fifh, and is feldom admitted to table either frefh or falted †.

* *Smith's Hift. Waterford,* 261.

† When cured it is known by the name of *Poor John.*

The

Thefe fifh are from a foot and an half to near twice that length: they are of a flender make, of a pale afh color on their backs, and of a dirty white on their bellies.

Their head is flat and broad; the mouth very wide; the teeth very long and fharp, particularly thofe of the lower jaw.

The firft dorfal fin is fmall, confifting of nine rays; the fecond reaches from the bafe of the former almoft to the tail, and is compofed of forty rays, of which the laft are the higheft: the pectoral fins have about twelve, the ventral feven: the anal thirty-nine.

The tail is almoft even at the end.

Galee, claria marina. *Belon,* 126.
Phycis. *Rondel.* 186. *Gefner pifc.* 718.
Tinca marina. *Aldr. Wil.*

Icth. 205. *Raii fyn. pifc.* 75.
Phycis. *Arted. fynon. App.* III.
Blennius Phycis. *Lin. fyft.* 442.

82. Forked.

THIS is the fifh to which *Rondeletius* gives the name of *Phycis,* borrowing it from *Ariftotle* and *Pliny,* who have not fo fufficiently characterized it, as to enable us to judge what fpecies they intended. It is found in the *Mediterranean* more frequently than in our feas, and we believe is the fifh mentioned by Mr. *Armftrong,* and Doctor *Cleghorn*,

* *Armftrong,* 161. *Cleghorn,* 43.

in their hiftories of *Minorca*, under the name of *Molio*, *Mollera*, and *Molle*. It is known on the coaft of *Cornwall* by the name of the great forked beard*, where it was firft difcovered by Mr. *Jago*. We place it in this genus, as it has more the appearance of the cod fifh kind, the hake efpecially, than of the *Blenny*, into which genus *Linnæus* has flung it; we therefore have given this fpecies the name of the *Forked Hake*.

The length of one that was taken on the *Flintfhire* fhores was eleven inches and an half, its greateft depth three inches; but according to Doctor *Borlafe*, fome grow to be above eighteen inches long.

The head floped down to the nofe in the fame eafy manner with others of this genus: the mouth large: befides the teeth in the jaws was a triangular congeries of fmall teeth in the roof of the mouth.

At the end of the lower jaw was a fmall beard. The firft dorfal fin was triangular; the firft ray extended far beyond the reft, and was very flender: the fecond fin began juft behind the firft, and extended almoft to the tail: the ventral fins were three inches long, and confifted of only two rays, joined at the bottom, and feparated or bifurcated towards the end: the vent was in the middle of the body: the anal fin extended from thence juft

---

* *Barbus major* Cornubienfis *cirris bifurcatis:* the great forked beard. Mr. *Jago*. *Raii fyn. pifc.* 163. *fig.* 7.

to

to the tail: the lateral line was incurvated: the tail was rounded.

The color was a cinereous brown.

Barbus minor *Cornubienfis* cirris bifurcis.   The Leffer Forked   83. LEST.
  Beard. Mr. *Jago*.   *Raii fyn. pifc*. 164. *fig*. 8.

WE never faw this fpecies, and having but very imperfect defcriptions of it, cannot with any certainty pronounce it to be of this genus, but are unwilling to feparate them, as we found them united by that judicious Icthyologift Mr. *Jago*.

It is faid not to exceed five inches in length: the firft dorfal fin (in the print) is fhorter than that of the preceding; the fecond refembles that of the other kind: the ventral fins bifurcated. It has a fmall beard, and a rounded tail, but the head is fhorter and more fteep; the color black, the fkin fmooth, and the appearance difagreeable.

O 2   THIS

84. TRIFUR-
CATED.

THIS new fpecies was communicated to me by the Reverend Mr. *Hugh Davies* of *Beaumaris*, and was taken near that place.

Its length was twelve inches: the color a deep brown; excepting the folding of the lips, which were fnow white, giving it a ftrange appearance.

The head depreffed and very broad: eyes large: irides yellowifh: mouth very wide, with irregular rows of incurvated teeth. In the roof of the mouth a femilunar congeries of teeth. No tongue.

From the fetting on of the pectoral fins the body was compreffed, but remarkably fo, as it approached the tail, growing very flender near that part. On the beginning of the back was a *fulcus*, in which was the rudiment of a firft dorfal fin; the fecond reached almoft to the tail, and the anal correfponded. Above the pectoral fins, on each fide, was a row of tubercles from which commenced the lateral line, which was (midway) incurvated. The ventral fins were trifurcated: the tail rounded.

In a prone fituation this fifh made a ftrange appearance, fo is reprefented in that as well as another attitude.

Ling

Pl. XXXII.

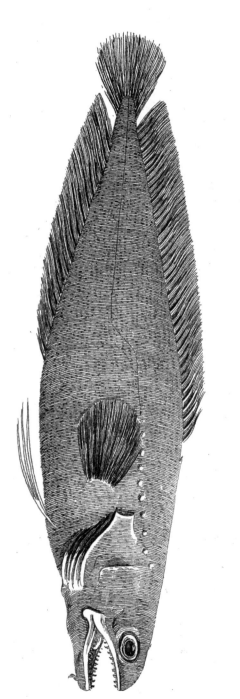
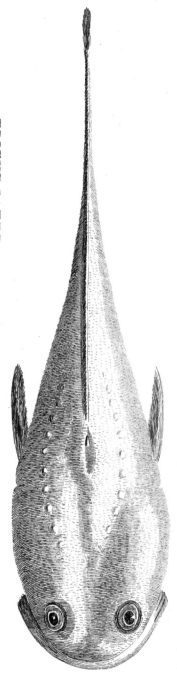

TRIFURCATED HAKE.

Ling, Lingfifche. *Belon*, 130.
  Gefner pifc. 95.
Molva major *Charleton ex. pifc.*
  3.
Afellus longus, eine Lenge.
  Schonevelde, 18.
Ling. *Wil. Icth.* 175. *Raii*

fyn. pifc. 56.
Gadus dorfo dipterygo, ore
  ferrato, maxilla fuperiore
  longiore. *Arted. fynon.* 36.
Gadus molva. *Lin. fyft.* 439.
Langa. *Faun. Suec.* No. 313.

THE ling takes its name from its length, being corrupted from the word *long*. It abounds about the *Scilly Ifles*, on the coafts of *Scarborough*, and thofe of *Scotland* and *Ireland*, and forms a confiderable article of commerce *.

In the *Yorkfhire* feas they are in perfection from the beginning of *February* to the beginning of *May*, and fome till the end of that month. In *June* they fpawn, depofiting their eggs in the foft oozy ground of the mouth of the *Tees :* at that time the males feparate from the females, and refort to fome rocky ground near *Flamborough Head*, where the fifhermen take great numbers without ever finding any of the female or roed fifh among them.

While a ling is in feafon its liver is very white, and abounds with a fine flavored oil ; but as foon

Oil.

* This branch of trade was confiderable fo long ago as the reign of *Edward* III. an act for regulating the price of *Lob, Ling*, and *Cod*, being made in his 31ft year.

O 3   as

as the fish goes out of feafon, the liver becomes red as that of a bullock, and affords no oil. The fame happens to the cod and other fish in a certain degree, but not fo remarkably as in the ling. When the fish is in perfection, a very large quantity of oil may be melted out of the liver by a flow fire, but if a violent fudden heat be ufed for that purpofe, they yield very little. This oil, which nature hoards up in the cellular membranes of fifhes, returns into their blood, and fupports them in the engendring feafon, when they purfue the bufinefs of generation with fo much eagernefs as to neglect their food.

Vaft quantities of ling are falted for exportation, as well as for home confumption. When it is cut or fplit for curing, it muft meafure twenty-fix inches or upwards from the fhoulder to the tail; if lefs than that it is not reckoned a fizeable fish, and confequently not entitled to the bounty on exportation; fuch are called *Drizzles*, and are in feafon all fummer.

DESCRIP.   The ufual fize of a ling is from three to four feet; but we have heard of one that was feven feet long.

The body is very flender; the head flat; the upper jaw the longeft; the teeth in that jaw fmall and very numerous; in the lower, few, flender, and fharp: on the chin is a fmall beard.

The firft dorfal fin is fmall, placed near the head, and confifts of fifteen rays: the fecond is very long,

reaching

reaching almoft to the tail, and confifts of fixty-five rays: the pectoral fins have fifteen radiated rays; the ventral fins fix; the anal fixty-two: the tail is rounded at the end.

Thefe fifh vary in color, fome being of an olive hue on the fides and back, others cinereous; the belly white. The ventral fins white: the dorfal and anal edged, with white. The tail marked near the end with a tranfverfe black bar, and tipt with white.

Strinfias, ou Botatriffa. *Belon*, 300.
Lota. *Rondel. fluviat.* 165. *Gefner pifc.* 599.
Quappen, Elff-quappen, Tider-quappen, Trufchen ? *Schonevelde*, 49.
Burbot, or Bird-bolt. *Plot Staff.* 241. *Tab.* 22. *fig.* 4.
Muftela fluviatilis noftratibus

Eel-pout. *Wil. Icth.* 125. 86. Burbot. *Raii fyn. pifc.* 67.
Aal-rutte, Rutte. *Kram.* 388.
Gadus dorfo dipterygio, ore cirrato, maxillis æqualibus. *Arted. fynon.* 38.
Gadus Lota. *Lin. fyft.* 440. *Gronov Zooph.* No. 97.
Lake. *Faun. Suec.* No. 113.

THIS fifh is found in the *Trent*, but in greater plenty in the river *Witham*, and in the great *Eaft Fen* in *Lincolnfhire*. It is a very delicate fifh for the table, though of a difgufting appearance when alive. It is very voracious, and preys on the fry and leffer fifh. It does not often take a bait, but is generally caught in weels.

It abounds in the lake of *Geneva*, where it is called

Place.

O 4.   ed

ed *Lota*, and it is alſo met with in the *Lago Mag-giore*, and *Lugano*.

The largeſt that we ever heard was taken in our waters weighed between two and three pounds, but abroad they are ſometimes found of double that weight.

Their body has ſome reſemblance to that of an eel, only ſhorter and thicker, and its motions alſo reſemble thoſe of that fiſh : they are beſides very ſmooth, ſlippery, and ſlimy.

The head is very ugly, being flat, and ſhaped like that of a toad : the teeth are very ſmall, but nume-rous : the irides yellow.

On the end of the noſe are two ſmall beards ; on the chin another : the number of its branchi-oſtegous rays are ſeven.

The firſt dorſal fin is ſhort : the ſecond is placed immediately behind it, and extends almoſt to the tail : the pectoral fins are rounded : the ventral fins conſiſt of ſix rays, of which the two firſt are divided near their ends from each other : the vent is placed in the middle of the belly, and the anal fin reaches almoſt to the tail : the tail is rounded at the end.

The color of this ſpecies varies ; ſome are duſky, others of a dirty green, ſpotted with black, and oftentimes with yellow, and the belly in ſome is white ; but the real colors are frequently concealed by the ſlime.

Muſtella

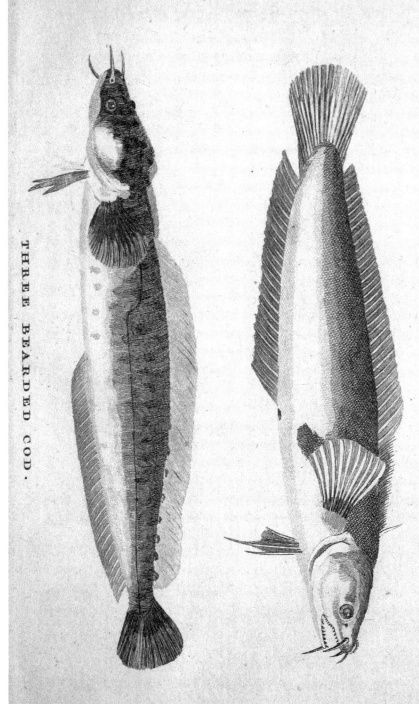

FIVE BEARDED COD.

THREE BEARDED COD.

Muftella vulgaris. *Rondel.* 281 *Gefner pifc.* 89. Sea Loche *Ceftriæ*, Whiftle fifh *Cornubiæ*. *Wil. Icth.*

121. *Raii fyn. pifc.* 67. Rockling, Mr. *Jago. Raii fyn. pifc.* 164. *fig.* 9.

87. THREE BEARDED.

THIS fpecies commonly frequents the rocky fhores of thefe iflands, and is fometimes taken with a bait.

It grows to the length of nineteen inches; the weight two pounds two ounces: the head is large and flat: the eyes not remote from the end of the nofe: the body is long, flender, and compref-fed fideways, efpecially towards the tail: at the end of the upper jaw are two beards; on the chin one.

The teeth are numerous and fmall, difpofed along the jaws in form of a broad plate: in the roof of the mouth is a fet of fmall teeth, difpofed in a triangular form.

The number of branchioftegous rays is feven.

The firft dorfal fin is lodged in a deep furrow juft beyond the head, and confifts of a number of fhort unconnected rays: the fecond rifes juft be-hind it, and reaches very near the tail: the pecto-ral fins are broad and round: the ventral fins fmall; the fecond ray the longeft: the anal fin reaches al-moft to the tail: the tail rounded at the end.

The fcales are very fmall: the color of the body and head a reddifh yellow, marked above the lateral

line

line with large black fpots : the back fin and tail
are darker; the vent fin of a brighter red, but all
are fpotted. The lateral line bends in the middle,
then paffes ftrait to the tail.

| 88. FIVE BEARDED. | Gadus dorfo dipterygio, ful-co magno ad pinnam dorfi primam, ore cirrato? *Arted. fynon.* 37. Gadus muftela. G. diptery- | gius cirris 5, pinna dorfali priore exoleta. *Lin. fyft.* 440. *Gronov. Zooph.* No. 314. |

M R. *Willughby* makes this fpecies with five
beards, a variety only of the former; but
having opportunity of examining feveral fpecimens, we muft diffent from his opinion, having
always obferved the number of the beards in the
fpotted kind not to exceed three, nor the number
in the brown kind to be lefs than five. The
firft ray of the dorfal fin is very long. There is
alfo fome difference in the form as well as color, this
fpecies being rather thicker in proportion than the
former.

Excepting thefe particulars, and the number of
beards, there is a general agreement in the parts of
both. The beards on the upper jaw are four, viz.
two at the very end of the nofe, and two a little
above them : on the end of the lower jaw is a fingle
one.

Thefe fifh are of a deep olive brown, their belly
whitifh.

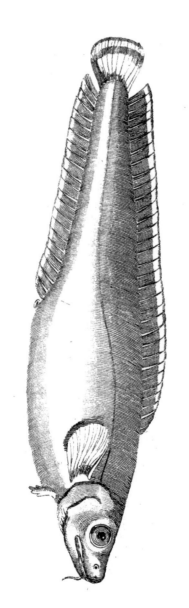

TORSK.

whitifh. They grow to the fame fize as the former.

The *Cornifh* fifhermen are faid to whiftle, and
make ufe of the words *Bod, Bod, vean,* when they
are defirous of taking this fifh, as if by that they
facilitated the capture. In the fame manner the
Sicilian fifhermen repeat their *Mamaffu di pajanu,*
&c. when they are in purfuit of the *Sword Fifh* *.

..* With only one dorfal fin.

THIS fifh has been hitherto fuppofed to be of
the fection of this genus, which has three
dorfal fins. The fpecies known in *Sweden* by that
name is included in that divifion; and as fuch I
defcribed it in the former edition from the account
Linnæus has given us. But from the information of
the Rev. Mr. *Low,* minifter of *Birfa, Orkney,* who
in 1774, made (at my requeft) the voyage of the
Shetland iflands, I find the *Britifh Torfk* to be totally different; and will occafion the addition of a
fourth divifion in this genus.

The *Torfk* is defcribed and engraven in Mr.
Strom's hiftory of *Sondmoer,* under the fame name. †

89. Torsk.

* *Vide p.* 162.

† *Eller Torfk,* p. 272. tab. I. fig. 19. and when dried,
Klip-fifh.

The

The figure agrees with that Mr. *Low* favoured me with.

The *Torfk*, or as it is called in the *Shetlands*, *Tufk* and *Brifmak* is a northern fifh; and as yet undifcovered lower than about the *Orknies*, and even there it is rather fcarce. In the feas about *Shetland*, it fwarms, and forms (barrelled or dried) a confiderable article of commerce.

The length of the fpecimen, Mr. *Low* defcribed for me, was twenty inches, the greateft depth four and a half.

The head fmall, the upper jaw a little longer than the lower: both jaws furnifhed with multitudes of fmall teeth: on the chin was a fmall fingle beard: from the head to the dorfal fin was a deep furrow. The dorfal fin began within fix inches from the tip of the nofe, and extended almoft to the tail.

The pectoral fins fmall, and rounded; the ventral fhort, thick and flefhy, ending in four *cirrhi*.

The belly from the throat grows very prominent: the anal fin was long, and reached almoft clofe to the tail, which is fmall and circular. The number of rays could not be counted with accuracy by reafon of their foftnefs, and the thicknefs of the fkin: the fide line fcarcely difcernible.

The color of the head dufky: the back and fides yellow: belly white: edges of the dorfal, anal, and caudal fins white: the other parts dufky: the pectoral fins brown.

I flatter myfelf, that in a fmall time, the public

lic will receive from Mr. *Low,* a fuller account of this important fiſh, in a comprehenſive hiſtory of the iſlands of *Orkney,* and *Shetland.*

Head

XX.
BLENNY *.

Head blunt at the end, and very fteep.

Body fmooth and flippery.

Teeth flender.

Body compreffed fideways.

Ventral fins confifting generally of only two
    united rays.

One dorfal fin.

Six branchioftegous rays.

* With a crefted head.

90. CRESTED. Adonis, ou exocetus. *Belon,* *fyn. pifc.* 73.
219.
Galerita. *Rondel.* 204. *Gef-*   Blennius crifta capitis tranf-
ner pifc. 14, 17, 18.   verfa cutacea. *Arted. fynon.*
Alauda criftata, five Galeri-   44.
ta. *Wil. Icth.* 134. *Raii*   Blennius Galerita. *Lin. fyſt.*
                                    441.

THIS fpecies is found, though not frequently,
    on our rocky fhores, and is commonly about
four or five inches long.

On the head is a fmall creft-like fin, which it

* There being no *Englifh* name for this genus, *Blenny* is
given it, derived from the word *Blennius,* the generical term
ufed by *Artedius,* who forms it from Βλὲννα *mucus,* it being of
a fliмy nature.

                                                        can

Pl. XXXV.

N.º 91.

GATTORUGINE.

CRESTED BLENNY

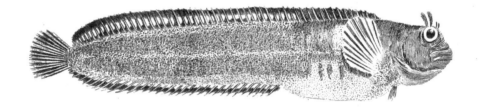

SPOTTED BLENNY.

can erect or depress at pleasure. On the top of the head, between the eyes, is a triangular lump pointing backwards, and red about its edges.

The skin at the corner of the upper jaw is loose, and projects.

From the hind part of the head, almost to the tail, extends the dorsal fin: the ventral fin is small: the vent is placed under the ends of the pectoral fins.

The body is smooth and flippery: the color brown, and spotted.

Scorpioides. *Rondel.* 204.
 Gesner pisc. 847.
Gattorugine *Venetiis. Wil.*
 Icth. 132. *Raii syn. pisc.* 72.
Blennius pinnulis duabus ad

oculos, pinna ani officulorum viginti trium. *Arted. synon.* 44.
Blennius Gattorugine. *Lin. syst.* 442.

91. GATTO-RUGIN.

THIS curious kind was discovered to be a *British* fish on the *Anglesea* coast.

PLACE.

Its length was seven inches and an half: the body was smooth, and compressed on the sides: the belly a little prominent: the vent situated as in the preceding fish.

DESCRIP.

The teeth slender, almost setaceous, and very close set: between the eyes was a small hollow, and above each, just on the summit, was a narrow loose membrane, trifurcated at the top, which distinguishes this from all other species.

The

The pectoral fins broad and rounded, confisting of fourteen rays, which extend beyond the webs, making the edges appear fcalloped.

The ventral fins like thofe of others of the genus: the dorfal fin confifted of fourteen ftrong fpiny rays, and nineteen foft rays; the laft of which were higher than the fpiny rays.

The anal fin had twenty-one rays: the ends in every fin extending beyond their webs.

The tail was rounded at the end, and confifted of twelve rays, divided towards their extremities.

This fifh in general was of a dufky hue, marked acrofs with wavy lines: the belly of a light afh color.

The lower part of the pectoral fins, and the ends of the ventral fins, of an orange color.

** With a fmooth head.

92. Smooth.

La tierce efpece de Exocetus? *Belon*, 219.
Alauda non criftata. *Rondel.* 205. *Gefner pifc.* 18.
Mulgranoc, & Bulcard *Cornubiæ. Wil. Icth.* 133. *Raii fyn. pifc.* 73.
Cataphractus lævis *Cornubien-fis*, Smooth Shan. *Mr. Jago apud Raii fyn. pifc.* 164. *fig.* 10.
Blennius maxilla fuperiore longiore, capite fummo acuminato. *Arted. fynon.* 45.
Blennius Pholis. *Lin. fyft.* 443. *Gronov. Zooph. No.* 259.

Place.

WE difcovered this fpecies in plenty lying under the ftones among the tang on the rocky

PL. XXXVI.

SMOOTH BLENNY.

N.º 92.

rocky coasts of *Anglesea*, at the lower water-mark. It was very active and vivacious, and would by the help of its ventral fins creep up between the stones with great facility. It bit extremely hard, and would hang at ones finger for a confiderable time. It was very tenacious of life, and would live for near a day out of water.

It feeds on shells and small crabs, whose remains we found in its stomach.

The length in general was five inches: the head large, and sloping suddenly to the mouth: the irides red.

DESCRIP.

The teeth slender, very sharp, and close set: there were twenty-four in the upper, and nineteen in the lower jaw.

The pectoral fins broad and rounded, confifting of thirteen rays: the ventral fins of only two thick rays, feparated near their ends.

The dorfal fin confifted of thirty-two foft rays, and reached from the hind part of the head almoft to the tail.

The vent was in the middle of the body: the anal fin extended almoft to the tail, and confifted of nineteen rays, tipt with white.

The tail rounded at the end, and compofed of twelve branched rays.

The color varied, fome were quite black, but generally they were of a deep olive, prettily marbled with a deeper color; others fpotted with white:

VOL. III. P the

the laft often difpofed in rows above and beneath the lateral line.

93.Spotted. Gunnellus *Cornubienfium*, non-nullis *Butter-fifh*, q. d. Li-paris. *Wil. Icth.* 115. *Raii fyn. pifc.* 144.
Blennius maculis circiter de-cem nigris limbo albicante utrinque ad pinnam dorfa-lem. *Arted. fynon.* 45.
Blennius Gunnellus. B. pinna
dorfali ocellis X nigris. *Lin. fyft.* 443. *Faun. Suec.* No. 318.
Seb. Muf. III. p. 91. *Tab.* 30. *fig.* 6.
Pholis maculis annulatis ad pinnam dorfalem, pinnis ventralibus obfoletis. *Gronov. Zooph.* No. 267.

THIS fpecies is found in the fame place with the preceding, lurking like it under ftones, is equally vivacious, and is ufed as a bait for larger fifh.

Its length is fix inches: the depth only half an inch: the fides very much compreffed, and ex-tremely thin.

The head and mouth is fmall; the laft points upwards, and the lower jaw flopes confiderably to-wards the throat.

The teeth are very fmall; the irides whitifh.

The pectoral fins rounded, and of a yellow color: inftead of the ventral fins are two minute fpines.

The dorfal fin confifts of feventy-eight fhort fpiny rays, and runs the length of the back almoft to the tail: on the top of the back are eleven round fpots, which

N.º 94.

N.º 96.

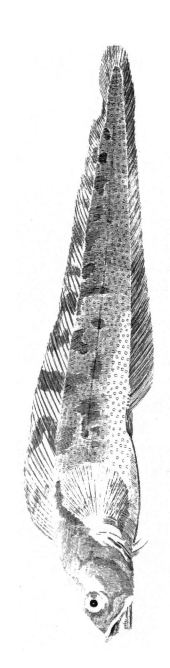

SPOTTED GOBY.

VIVIPAROUS BLENNY.

PL. XXXVII.

which reach the lower half of the dorfal fin; they are black, half encircled with white.

The vent is in the middle of the body; the anal fin extends from it almoft to the tail.

The tail is rounded, and of a yellow color.

The back and fides are of a deep olive: the belly whitifh.

Muftela marina vivipara, Ael-quappe, Ael-puet, Ael-moder. *Schonevelde*, 50. *Tab.* 4. Guffer, Eelpout. *Sib. Scot.* III. 25. Muftela vivipara *Schoneveldii. Wil. Ichth.* 122. *Raii fyn. pifc.* 69. Blennius capite dorfoque fufco flavefcente lituris nigris, pinna ani flava. *Arted. fynon.* 45.	Blennius viviparus, B. ore tentaculis duobus. *Lin. fyft.* 443. Tanglake. *Faun. Suec.* No. 317. *Muf. Ad. Fr.* I. 69. *Tab.* 32. Enchelyopus corpore lituris variegato; pinna dorfi ad caudam finuata. *Gronov. Zooph.* No. 265.

94. VIVIPA-ROUS.

*S*CHONEVELDE firft difcovered this fpecies; Sir *Robert Sibbald* afterwards found it on the *Scotch* coafts; and *Linnæus* has defcribed it in his account of his *Swedifh* Majefty's *Mufeum.*

They are viviparous, bringing forth two or three hundred young at a time. Their feafon of parturition is a little after the depth of winter. Before *Midfummer* they quit the bays and fhores, and retire into the deep, where they are commonly tak-

P 2 en.

en. They are a very coarfe fifh, and eat only by the poor.

They are common in the mouth of the river *Efk*, at *Whitby, Yorkfhire*; where they are taken frequently from off the bridge.

They fometimes grow to the length of a foot. Their form flender: their fkin fmooth and flippery. The teeth very minute and fharp: the upper lip thin and fkinny.

The dorfal fin commences juft behind the head, and joins with that of the tail; but near the tail, the reft are fhort, fo as to form the appearance of a divifion. The pectoral fins rounded: the ventral confift of only four fhort rays: the anal extends far, and unites with the tail. The tail round.

The dorfal fin, back, and fides are of a yellowifh brown, ftained with dufky fpots and lines. The end of the tongue, the chin, throat, and anal fin of a fine yellow.

The back-bone is green, as that of a fea-needle.

SECT.

Pl. XXXVIII.

BLACK GOBY.

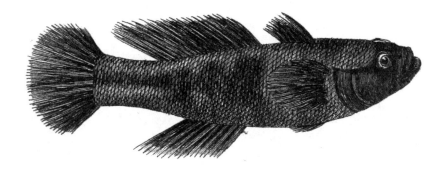

Sect. III.   T H O R A C I C   F I S H.

Eyes placed near each other.
Four branchioſtegous rays,
Ventral fins united.

XXI.
G O B Y *.

Gobio niger. *Rondel,* 200. *Geſner piſc.* 395. Schwartzer Goeb. *Schonevelde,* 36. Sea Gudgeon. Rock-fiſh. *Wil. Iɛth.* 206. *Raii ſyn. piſc.* 76. Gobius ex nigricante varius, pinna dorſi ſecunda oſſiculorum quatuordecim. *Arted. ſynon.* 46. Gobius niger. *Lin. ſyſt.* 449. Eleotris capite cathetoplateo, pinnis ventralibus concretis, *Gronov. Zooph.* No. 281.

95. Black.

IT is to this fiſh that Naturaliſts have given the ſynonym of Κωϐιος, and *Gobio,* names of certain ſpecies mentioned by *Ariſtotle, Pliny,* and *Oppian.* The two firſt have not left any characters for us to diſtinguiſh them by; and *Oppian* at once ſhews that he never intended this kind, as he has placed it among thoſe which are armed with a poiſonous ſpine. *Ariſtotle* was acquainted with two ſpecies; one a ſea fiſh that frequented the rocks, another that was gregarious, and an inha-

---

* Formed from *Gobius,* the generic name beſtowed by Naturaliſts on theſe fiſh.

P 3

bitant

bitant of rivers, which laft feems to have been our common gudgeon.

This fpecies grows to the length of fix inches: the body is foft, flippery, and of a flender form: the head is rather large; the cheeks inflated; the teeth fmall, and difpofed in two rows: from the head to the firft dorfal fin is a fmall fulcus.

The firft dorfal fin confifts of fix rays; the fecond of fourteen; the pectoral fins of fixteen or feventeen, clofely fet together, and the middlemoft the longeft; the others on each fide gradually fhorter.

The ventral fins coalefce and form a fort of funnel, by which thefe fifh affix themfelves immoveably to the rocks, for which reafon they are called *Rock-fifh*.

The tail is rounded at the end.

The color is brown, or deep olive, mixed with dark ftreaks, and fpotted with black: the dorfal and anal fins are of a pale blue, the rays marked with minute black fpots.

Aφυα?.

Αφυα ? *Athen. Lib.* VII. *p.*
284.
Aphia. *Belon,* 207.
Aphya cobites. *Rondel.* 210.
Gefner pifc. 67. *Wil. pifc.*
207. *Raii fyn. pifc.* 76.

Gobius Aphya et Marfio
dictus. *Arted. fynon.* 47.
Gobius Aphya. G. fafciis
etiam pinnarum fufcis. *Lin.
fyft.* 450.

96. SPOT-
TED.

WE faw feveral of this fpecies taken laft fummer on our fandy fhores in the fhrimp nets.

The length of the largeft was not three inches:
the nofe was blunt: the eyes large and prominent,
ftanding far out of the head: the irides fappharine; the head flat; the tongue large; teeth in
both jaws.

The firft dorfal fin confifted of fix rays; the
fecond of eleven, and placed at fome diftance from
the other.

The ventral fins are united: the anal confift of
eleven rays: the tail is even at the end.

The body is of a whitifh color, obfcurely fpotted with ferruginous: the rays of the dorfal fins,
and the tail, barred with the fame color.

P 4   Large

XXII. BULL-
HEAD.

Large flat head, armed with fharp fpines.
Six branchioftegous rays.

97. River. Boῖτος. *Arift. Hift. an. Lib.*
IV. *c.* 8.
Chabot. *Belon,* 213.
Cottus. *Rondel. Fluviat.* 202.
Gobio capitatus. *Gefner pifc.*
401.
Een Miiller. *Schwenckfelt*
Siles. 431.
Bull-head, Miller's Thumb.
Wil. Icth. 137. *Raii fyn.*
pifc. 76.

Cottus alepedotus glaber, ca-
pite diacantho. *Arted. fynon.*
76.
Cottus Gobio. C. lævis, ca-
pite fpinis duabus. *Lin.*
fyft. 452.
Sten - fimpa, Slagg - fimpa.
Faun. Suec. No. 323.
Koppe. *Kram.* 384. *Gronov.*
Zooph. No. 270.

THIS fpecies is very common in all our clear
brooks ; it lies almoft always at the bottom,
either on the gravel or under a ftone : it depofits
its fpawn in a hole it forms in the gravel, and
quits it with great reluctance. It feeds on water
infects ; and we found in the ftomach of one the
remains of the frefh water fhrimp, the *pulex aqua-*
tilis of *Ray.*

This fifh feldom exceeds the length of three
inches and an half: the head large, broad, flat,
and thin at its circumference, being well adapted
for infinuating itfelf under ftones : on the middle
part of the covers of the gills is a fmall crooked
fpine turning inwards.

The

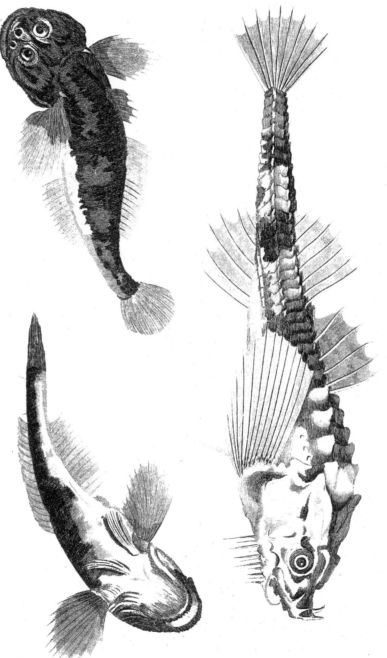

ARMED BULL-HEAD.

RIVER BULL-HEAD.

N.º 98.

N.º 97.

The eyes are very fmall: the irides yellow: the teeth very minute, placed in the jaws and the roof of the mouth.

The body grows flender towards the tail, and is very fmooth.

The firft dorfal fin confifts of fix rays, the fecond of feventeen: the pectoral fins are round, and prettily fcalloped at their edges, and are compofed of thirteen rays; the ventral of only four; the anal of thirteen; the tail of twelve, and is rounded at the end.

The color of this fifh is as difagreeable as its form, being dufky, mixed with a dirty yellow: the belly whitifh.

Cataphractus, Stein-bicker, Miiller, Turfs-bull. *Schonevelde*, 30. *Tab*. 3.
Cataphractus *Schoneveldii* Septentr. *Anglis* a Pogge. *Wil. Icth*. 211. *Raii fyn. pifc.* 77.
Cottus cirris plurimis corpore octagono. *Arted. fynon*. 77.

Cottus Cataphractus. C. loricatus, roftro verrucofo 2 bifidis, capite fubtus cirrofo. *Lin. fyft*. 451.
Botn-mus. *Faun. Suec*. No. 324.
Seb. Muf. III. *Tab*. 28. Gronov. *Zooph*. No. 271.

98. ARMED.

THE pogge is very common on moft of the *Britifh* coafts.

It feldom exceeds five inches and an half in length, and even feldom arrives at that fize.

The head is large, bony, and very rugged: the end of the nofe is armed with four fhort upright

spines:

fpines : on the throat are a number of fhort white beards.

The teeth are very minute, fituated in the jaws.

The body is octagonal, and covered with a number of ftrong bony crufts, divided into feveral compartments, the ends of which project into a fharp point, and form feveral echinated lines along the back and fides from the head to the tail.

The firft dorfal fin confifts of fix fpiny rays : the fecond is placed juft behind the firft, and confifts of feven foft rays.

The pectoral fins are broad and rounded, and are compofed of fifteen rays.

99. FATHER-LASHER.

Scorpios. *Ovid. Halieut.* 116. La Scorpene. *Belon,* 242. Scorpius marinus, Waelkuke, Buloffe, Schorp-fifche. *Schonevelde,* 67. tab. 6. Scorpænæ *Belonii* fimilis *Cornub.* Father-lafher. *Wil. Icth.* 138. *Raii fyn pifc.* 145. Scorpius virginianus. *Idem.* 142. *Wil. Icth.*

App. 25. Cottus fcorpius. C. capite fpinis pluribus, maxilla fuperiore paulo longiore. *Lin. fyft.* 452. Rot-fimpa, Skrabba, Skialryta. *Faun. Suec.* No. 323. Ulke. *Crantz. Greenl.* 1. 95. *Gronov. Zooph.* No. 268. Sea Scorpion. *Edw.* 284.

THIS fifh is not uncommon on the rocky coafts of this ifland : it lurks under ftones, and will take a bait.

DESCRIP.   It does not grow to a large fize, feldom exceeding (as far as we have feen in the fpecimens that are taken on our fhores) eight or nine inches.

The

Pl. XI.

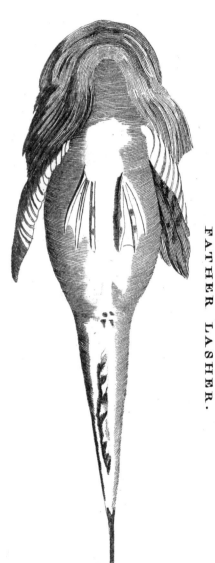

FATHER LASHER.

The head is very large, and has a moft formi-
dable appearance, being armed with vaft fpines,
which it can oppofe to any enemy that attacks
it, by fwelling out its cheeks and gill covers to a
large fize.

> *Et capitis duro nociturus* Scorpios *ictu.*
> The hurtful *Scorpion* wounding with its head.

The nofe, and fpace contiguous to the eyes, are   SPINES.
furnifhed with fhort fharp fpines: the covers of the
gills are terminated by exceeding long ones, which
are both ftrong and very fharp pointed.

The mouth is large: the jaws covered with rows
of very fmall teeth: the roof of the mouth is furnifhed with a triangular fpot of minute teeth.

The back is more elevated than that of others
of this genus: the belly prominent: the fide-line
rough, the reft of the body very fmooth, and
grows flender towards the tail.

The firft dorfal fin confifts of eight fpiny rays;
the fecond of eleven high foft rays: the pectoral
fins are large, and have fixteen; the ventral three;
the anal eight: the tail is rounded at the end,
and is compofed of twelve bifurcated rays.

The color of the body is brown, or dufky and
white marbled, and fometimes is found alfo ftained
with red: the fins and tail are tranfparent, fometimes clouded, but the rays barred regularly with
brown: the belly is of a filvery white.

This kind is very frequent in the *Newfoundland* AMERICAN.
                                              feas,

seas, where it is called *Scolping* : it is also as common on the coast of *Greenland* in deep water near shore. It is a principal food of the natives, and the soup made of it is said to be agreeable as well as wholesome.

Body

Pl. XLI.

SMEAR DAB.

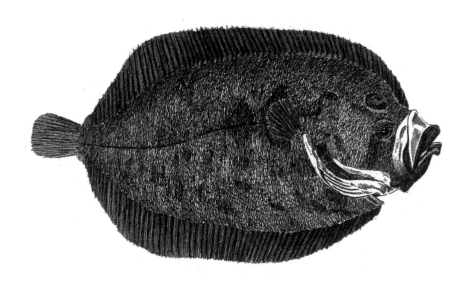

DOREE

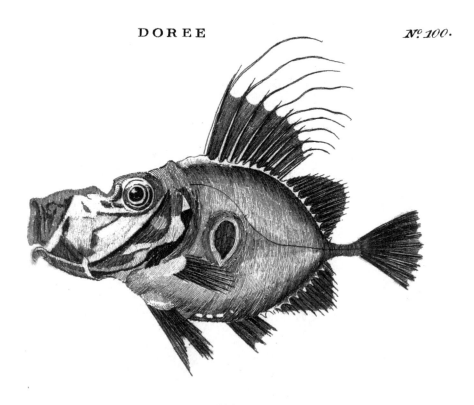

Body very deep, and compreſſed ſideways.
Very long filaments iſſuing from the firſt dorſal fin.
Seven branchioſtegous rays.

Χαλκὺς. *Athen. lib.* VII. 328.
 Oppian Halieut. I. 133.
Faber ? *Ovid Halieut.* 110.
 Zeus idem Faber *Gadibus.*
 Plin. lib. IX. *c.* 18.
La Dorèe. *Belon,* 146.
Faber ſive Gallus marinus.
Rondel. 328. *Geſner piſc.* 369.
A Doree. *Wil. Icth.* 294.
 Raii ſyn. piſc. 99.

Zeus ventre aculeato, cauda
 in extremo circinato. *Arted.*
 ſynon. 78.
Zeus Faber. Z. cauda rotun-
 data, lateribus mediis ocello
 fuſco, pinnis analibus duabus. *Lin. ſyſt.* 454. *Gro-*
 nov. Zooph. No. 311.
Zeus ſpinoſus. *Muſ. Fred.*
 Ad. 67. *tab.* XXXI.

100. DOREE.

SUPERSTITION hath made the *Doree* rival
to the *Hadock,* for the honor of having been
the fiſh out of whoſe mouth St. *Peter* took the tri-
bute-money, leaving on its ſides thoſe inconteſti-
ble proofs of the identity of the fiſh, the marks of
his finger and thumb.

It is rather difficult at this time to determine on
which part to decide the diſpute; for the *Doree*
likewiſe aſſerts an origin of its ſpots of a ſimilar
nature, but of a much earlier date than the former. St. *Chriſtopher* *, in wading through an arm
            of

---

* *Belon, Rondel,* alſo *Aldrovand de piſc.* 40. St. *Chriſtopher*
was of a *Coloſſal* ſtature, as is evident from his image in the
               church

of the fea, having caught a fifh of this kind *en paf-
fant*, as an eternal memorial of the fact, left the
impreffions on its fides to be tranfmitted to all
pofterity.

In our own country it was very long before this
fifh attracted our notice, at left as an edible one.
We are indebted to that judicious actor and *bon
vivant* the late Mr. *Quin*, for adding a moft delicious fifh to our table, who overcoming all the
vulgar prejudices on account of its deformity, has
effectually eftablifhed its reputation.

PLACE.

This fifh was fuppofed to be found only in the
fouthern feas of this kingdom, but it has been difcovered laft year on the coaft of *Anglefea*. Thofe
of the greateft fize are taken in the *Bay of Bifcay*,
off the *French* coafts : they are alfo very common
in the *Mediterranean*; *Ovid* muft therefore have
ftyled it *rarus Faber*, on account of its excellency,
not its fcarcity.

DESCRIP.

The form of this fifh is hideous : its body is
oval, and greatly compreffed on the fides : the head
large : the fnout vaftly projecting : the mouth very
wide : the teeth very fmall.

The eyes great : the irides yellow.

The lateral line oddly diftorted, finking at each
end, and rifing near the back in the middle : beneath it on each fide is a round black fpot.

church of *Notre Dame* at *Paris*, and a ftill larger at *Auxerre :*
the laft we think is near feventy feet high. His hiftory is in
his name, χριστοφορος, being faid to have carried our Saviour,
when a child, over an arm of the fea.

The

Pl. XLII.   Nº 112

LUNULATED
GILT HEAD.

OPAH.   Nº 10.

The firſt dorſal fin conſiſts of ten ſtrong ſpiny rays, with long filaments, reaching far beyond their ends: the ſecond is placed near the tail, and conſiſts of twenty-four ſoft rays, the middlemoſt of which are the longeſt.

The pectoral fins have fourteen rays, the ventral ſeven; the firſt ſpiny, the others ſoft: it has two anal fins; the firſt conſiſts of four ſharp ſpines, the ſecond of twenty-two ſoft ones, and reaches very near the tail.

The tail is round at the end, and conſiſts of fifteen branched rays.

The color of the ſides is olive, varied with light blue and white, and while living is very reſplendent, and as if gilt, for which reaſon it is called the *Doree*.

The largeſt fiſh we have heard of, weighed twelve pounds.

Opah, or King-fiſh. *Ph. Trans. abr.* XI. 879. *Tab.* V. Zeus cauda bifurca, colore   argenteo purpureo ſplendens. *Strom. Sondmor.* 323, 325. *Tab.* 1. *fig.* 20.   101. Opah.

WE have only five inſtances of this fiſh being taken in our ſeas, four of them in the *North*, viz. twice off *Scotland**, once off *Northumberland*,

* The fiſh engraved by Sir *Robert Sibbald, Hiſt. Scot. Tab.* 6. and thus deſcribed, is of this kind. *Piſcis maculis aureis aſperſus non ſcriptus, pollices* 42 *longus*.

one

one in *Filey-Bay*, *Yorkſhire*; and a fifth was caught at *Brixham*, in *Torbay*, in 1772.

The laſt weighed a hundred and forty pounds. The length was four feet and an half: the breadth two feet and a quarter: the greateſt thickneſs, only four inches. Its general color was a vivid tranſparent ſcarlet varniſh, over burniſhed gold, beſpangled with oval ſilver ſpots of various ſizes: the breaſt was an hard bone, reſembling the keel of a ſhip: the fleſh looked, and taſted like beef *.

I find a more ample deſcription of another, by Mr. *Robert Harriſon*, of *Newcaſtle*.

Newcaſtle, *Sept.* 12. 1769; On *Saturday* laſt was thrown upon the ſands at *Blyth*, a very rare and beautiful fiſh, weighing between ſeventy and eighty pounds, ſhaped like the ſea bream. The length was three feet and an half; the breadth from back to belly almoſt two feet; but the thickneſs from ſide to ſide not above ſix inches.

The mouth ſmall for the ſize of the fiſh, forming a ſquare opening, and without any teeth in the jaws. The tongue thick, reſembling that of a man, but rough and thick ſet with beards or prickles, pointing backwards, ſo that any thing might eaſily paſs down, but could not eaſily return back, therefore theſe might ſerve inſtead of teeth to retain its prey. The eyes remarkably large, covered with a membrane, and ſhining with a glare of gold. The cover of the gills like the ſalmon.

† This deſcription was ſent to me by a gentleman, who ſaw the fiſh ſoon after it was taken.

The

The body diminiſhes very ſmall to the tail, which is forked, and expands twelve inches: the gill fins are broad, about eight inches long, and play horizontally: a little behind their inſertion the back fin takes its original, where it is about ſeven inches high, but ſlopes away very ſuddenly, running down very near the tail, and at its termination becomes a little broader: the belly fins are very ſtrong, and placed near the middle of the body: a narrow fin alſo runs from the anus to the tail.

All the fins, and alſo the tail, are of a fine ſcarlet; but the colors and beauty of the reſt of the body, which is ſmooth and covered with almoſt imperceptible ſcales, beggars all deſcription; the upper part being a kind of bright green, variegated with whitiſh ſpots, and enriched with a ſhining golden hue, like the ſplendor of a peacock's feather. This by degrees, vaniſhes in a bright ſilvery, and near the belly the gold again predominates in a lighter ground than on the back.

XXIV.
FLOUN-
DER.

Body quite flat, and very thin.

Eyes, both on the fame fide the head.

Branchioftegous rays from four to feven.

* With the eyes on the right fide.

102. HOLI-
BUT.

Hippogloffus. *Rondel.* 325. *Gefner pifc.* 669.	totus glaber. *Arted. fynon.* 31.
Heglbutte, Hilligbutte. *Schonevelde,* 62.	Pleuronectes Hippogloffus. *Lin. fyft.* 456.
Holibut, *Septentr. Anglis* Turbot. *Wil. Ith.* 99. *Raii fyn. pifc.* 33.	Halg-flundra. *Faun. Saec.* No. 329. *Gronov. Zooph.* No. 247.
Pleuronectes oculis a dextris,	

SIZE.

THIS is the largeft of the genus; fome have been taken in our feas weighing from one to three hundred pounds; but much larger are found in thofe of *Newfoundland, Greenland,* and *Iceland,* where they are taken with a hook and line in very deep water. They are part of the food of the *Greenlanders* *, who cut them into large flips, and dry them in the fun.

They are common in the *London* markets, where they are expofed to fale cut into large pieces. They are very coarfe eating, excepting the part

* *Crantz. Hift. Greenl.* I. 98.

which

which adheres to the fide fins, which is extremely fat and delicious, but furfeiting.

They are the moſt voracious of all flat fiſh.
The laſt year there were two inſtances of their fwal-
lowing the lead weight at the end of a line, with
which the feamen were founding the bottom from
on board a ſhip, one off *Flamborough Head*, the
other going into *Tinmouth Haven* : the latter was
taken, the other difengaged itfelf.

The holibut, in refpect to its length, is the nar-   Descrip.
roweſt of any of this genus except the fole.

It is perfectly fmooth, and free from fpines either above or below. The color of the upper
part is dufky ; beneath of a pure white. We do
not count the rays of the fins in this genus, not
only becaufe they are fo numerous, but becaufe
nature hath given to each fpecies characters fuffi-
cient to diftinguiſh them by.

Thefe flat fiſh fwim fideways ; for which reafon
Linnæus hath ſtyled them *Pleuronectes*.

Q 2   Plateſſa ?

103. Plaise. Platessa? *Aufonii Epist. ad.*   Pleuronectes oculis et tuber-
     Theon. 62.   culis fex a dextra capitis,
     Le Quarlet. *Belon*, 139. lateribus glabris, fpina ad
     Quadratulus. *Rondel.* 318. anum. *Arted. fynon.* 30.
      Gefner pifc. 665.   Pleuronectes Platessa. *Lin.*
     Scholle, Pladife. *Schonevelde,* *fyst.* 456. *Gronov. Zooph.*
     61.   *No.* 246.
     Plaife. *Wil. Icth.* 96. *Raii* Skalla, Rodfputta. *Faun. Suec.*
      fyn. pifc. 31.   No. 328.

THESE fifh are very common on moft of
  our coafts, and fometimes taken of the weight
of fifteen pounds; but they feldom reach that fize,
one of eight or nine pounds being reckoned a large
fifh.

  The beft and largeft are taken off *Rye,* on the
coaft of *Suffex,* and alfo off the *Dutch* coafts. They
fpawn on the beginning of *February.*

  They are very flat, and much more fquare than the
preceding. Behind the left eye is a row of fix tubercles, that reaches to the commencement of the
lateral line.

  The upper part of the body and fins is of a clear
brown, marked with large bright orange-colored
fpots : the belly is white.

Le

Le Flez.   *Belon,* 141.
Paſſeris tertia ſpecies. *Rondel,*
    319. *Geſner piſc.* 666, 670.
Struff-butte. *Schonevelde,* 62.
Flounder, Fluke, or But.
    Wil. Icth. 980. *Raii ſyn.*
    piſc. 32.
Pleuronectes oculis a dextris,
    linea laterali aſpera, ſpi-
nulis ſupiné ad radices pin-
    narum,   dentibus obtuſis.
    Arted. ſynon. 31.
Plueronectes Fleſus.   *Lin.*
    ſyſt. 457.   *Gronov. Zooph.*
    No. 248.
Flundra, Slatt-ſkadda. *Faun.*
    Suec. No. 327.

THE flounder inhabits every part of the *Britiſh* ſea, and even frequents our rivers at a great diſtance from the ſalt waters; and for this reaſon ſome writers call it the *Paſſer fluviatilis.* It never grows large in our rivers, but is reckoned ſweeter than thoſe that live in the ſea. It is inferior in ſize to the plaiſe, for we never heard of any that weighed more than ſix pounds.

It may very eaſily be diſtinguiſhed from the plaiſe, or any other fiſh of this genus, by a row of ſharp ſmall ſpines that ſurround its upper ſides, and are placed juſt at the junction of the fins with the body. Another row marks the ſide-line, and runs half way down the back.

The color of the upper part of the body is a pale brown, ſometimes marked with a few obſcure ſpots of dirty yellow, the belly is white.

We have met with a variety of this fiſh with the eyes and lateral line on the left ſide. *Linnæus* makes a diſtinct ſpecies of it under the name of

Descrip.

Q 3

Pleuro-

Pleuronectes Passer, p. 459; but since it differs in no other respect from the common kind, we agree with Doctor *Gronovius* in not separating them.

105. DAB.   La Limande. *Belon*, 142. Passer asper, sive squamosus. *Rondel.* 319. *Gesner pisc.* 665. Dab. *Wil. Icth.* 79. *Raii syn. pisc.* 32. Pleuronectes oculis a dextra, squamis asperis, spina ad anum, dentibus obtusis. *Arted. synon.* 33. Pleuronectes Limanda. Pl. oculis dextris, squamis ciliatis, spinulis ad radicem pinnarum dorsi, anique. *Lin. syst.* 457.

THE dab is found with the other species, but is less common. It is in best season during *February*, *March*, and *April*: they spawn in *May* and *June*, and become flabby and watery the rest of summer. They are superior in goodness to the plaise and flounder, but far inferior in size.

DESCRIP.   It is generally of an uniform brown color on the upper side, tho' sometimes clouded with a darker. The scales are small and rough, which is a character of this species. The lateral line is extremely incurvated at the beginning, then goes quite strait to the tail. The lower part of the body is white.

106. SMEAR- DAB.   Rhombus lævis *Cornubiensis* maculis nigris, a Kit. Mr. *Jago. Raii syn. pisc.* 162. *fig.* 1.

WE found one of this species at a fishmonger's in *London*, where it is known by the name of the *Smear-dab*.

It

It was a foot and a half long, and eleven inches broad between fin and fin on the wideſt part.

The head appeared very ſmall, as the dorſal fin began very near its mouth, and extended very near to the tail.   It conſiſted of ſeventy nine rays.

The eyes were pretty near each other.   The mouth full of ſmall teeth.

The lateral line was much incurvated for the firſt two inches from its origin, then continued ſtrait to the tail.

The back was covered with ſmall ſmooth ſcales, was of a light brown color, ſpotted obſcurely with yellow.   The belly white, and marked with five large duſky ſpots.

It was a fiſh of goodneſs equal to the common dab.

Βɣγλωσσος. *Athen. lib.* viii. *p.* 288. *Oppian Halieut.* I. 99. La Sole. *Belon,* 142.
Bugloſſus. *Rondel.* 320. *Geſner piſc.* 666.
Tungen. *Schonevelde,* 63.
Pleuronectes oculis a ſiniſtra corpore oblongo, maxilla ſuperiore longiore, ſquamis utrinque aſperis. *Arted. ſyn.* 32.
Pleuronectes Solea *Lin. ſyſt.* 457. *Gronov. Zooph. No.* 251. Tunga, Sola. *Faun. Suec. No.* 326.

107. SOLE.

THE ſole is found on all our coaſts, but thoſe on the weſtern ſhores are much ſuperior in ſize to thoſe of the north.   On the former they are ſometimes taken of the weight of ſix or ſeven

Q 4   pounds,

pounds, but towards *Scarborough* they rarely exceed one pound; if they reach two, it is extremely uncommon.

They are ufually taken in the trawlnet: they keep much at the bottom, and feed on fmall fhell fifh.

Descrip.   It is of a form much more narrow and oblong than any other of the genus. The irides are yellow; the pupils of a bright fappharine color: the fcales are fmall, and very rough: the upper part of the body is of a deep brown: the tip of one of the pectoral fins black: the under part of the body is white: the lateral line ftrait: the tail rounded at the end.

It is a fifh of a very delicate flavour; but the fmall foles are much fuperior in goodnefs to large ones*. The chief fifhery for them is at *Brixham* in *Torbay*.

---

108. Smooth Sole.   Solea? *Ovid. Halieut.* 124.
              Arnogloffus feu Solea lævis. *Wil. Icth.* 102. *Raii fyn. pifc.* 34.

THIS, as defcribed by Mr. *Ray*, (for we have not feen it) is extremely thin, pellucid, and

* By the antient laws of the *Cinque* ports, no one was to take foles from the 1ft of *November* to the 15th of *March*; neither was any body to fifh from fun fetting to fun-rifing, that the fifh might enjoy their night-food.

white,

white, and covered with fuch minute fcales, and thofe inftantly deciduous, as to merit the epithet fmooth.

It is a fcarce fpecies, but is found in *Cornwall*, where, from its tranfparency, it is called the *Lantern Fifh*.

It is probable that *Ovid* intended this fpecies, by his *Solea*; for the common kind does by no means merit his defcription.

Fulgentes SOLEÆ *candore*.

And *Soles* with white refplendent.

** With the eyes on the left fide.

Rhombus. *Ovid Halieut.*
Le Turbot. *Belon*, 134.
Rhombus aculeatus. *Rondel.* 310. *Gefner pifc.* 661.
Steinbutt, Torbutt, Treenbutt, Dornbutt. *Schonevelde*, 60.
Turbot, in the *north* a Bret. *Wil. Icth.* 94.
Rhombus maximus afper non fquamofus. *Raii fyn. pifc.* 31.
Pleuroneĉtes oculis a finiftra, corpore afpero. *Arted. fynon.* 32.
Pleuroneĉtes maximus. *Lin. fyft.* 459. *Gronov. Zooph.* No. 254.
Butta. *Faun. Suec. No.* 325.

109. TURBOT.

TUREOTS grow to a very large fize; we have feen them of three and twenty pounds weight, but have heard of fome that weighed thirty. They are taken chiefly off the north coaft of

SIZE.

England,

England, and others off the *Dutch* coast; but we believe the last has, in many instances, more credit than it deserves for the abundance of its fish.

FISHERY.   The large Turbots, and several other kinds of flat fish, are taken by the hook and line, for they lye in deep water: the method of taking them in wares, or staked nets, is too precarious to be depended on for the supply of our great markets, because it is by meer accident that the great fish stray into them.

It is a misfortune to the inhabitants of many of our fishing coasts, especially those of the north part of *North Wales*, that they are unacquainted with the most successful means of capture: for their benefit, and perhaps that of other parts of our island, we shall lay before them the method practised by the fishermen of *Scarborough*, as it was communicated to us by Mr. *Travis*.

LINES.   When they go out to fish, each person is provided with three lines. Each man's lines are fairly coiled upon a flat oblong piece of wicker-work; the hooks being baited, and placed very regularly in the centre of the coil. Each line is furnished with 14 score of hooks, at the distance of six feet two inches from each other. The hooks are fastened to the lines upon sneads of twisted horse-hair, 27 inches in length.

When fishing there are always three men in each coble, and consequently nine of these lines are fastened together, and used as one line, extending

ing

ing in length near three miles, and furnifhed with
2520 hooks.   An anchor and a buoy are fixed at
the firft end of the line, and one more of each at
the end of each man's lines; in all four anchors,
which are commonly perforated ftones, and four
buoys made of leather or cork.   The line is always laid acrofs the current.   The tides of flood and
ebb continue an equal time upon our coaft, and
when undifturbed by winds run each way about fix
hours.   They are fo rapid that the fifhermen can
only fhoot and haul their lines at the turn of tide;
and therefore the lines always remain upon the
ground about fix hours *.   The fame rapidity of
tide prevents their ufing hand-lines; and therefore
two of the people commonly wrap themfelves in the
fail, and fleep while the other keeps a ftrict look-
out, for fear of being run down by fhips, and
to obferve the weather.   For ftorms often rife fo
fuddenly, that it is with extreme difficulty they can
fometimes efcape to the fhore, leaving their lines
behind.

The coble is 20 feet 6 inches long, and 5 feet   COBLE,
extreme breadth.   It is about one ton burthen,
rowed with three pair of oars, and admirably conftructed for the purpofe of encountering a mountanous fea : they hoift fail when the wind fuits.

---

* In this fpace the *myxine glutinofa* of *Linnæus*, will frequently penetrate the fifh that are on the hooks, and entirely devour them, leaving only the fkin and bones.

The

The five-men boat is 40 feet long and 15 broad, and of 25 tons burthen: it is fo called, tho' navigated by fix men and a boy, becaufe one of the men is commonly hired to cook, &c. and does not fhare in the profits with the other five. All our able fifhermen go in thefe boats to the herring fifhery at *Yarmouth* the latter end of *September*, and return about the middle of *November*. The boats are then laid up until the beginning of *Lent*, at which time they go off in them to the edge of the *Dogger*, and other places, to fifh for turbot, cod, ling, fkates, &c. They always take two cobles on board, and when they come upon their ground, anchor the boat, throw out the cobles, and fifh in the fame manner as thofe do who go from the fhore in a coble; with this difference only, that here each man is provided with double the quantity of lines, and inftead of waiting the return of tide in the coble, return to the boat and bait their other lines; thus hawling one fet, and fhooting another every turn of tide. They commonly run into harbour twice a week to deliver their fifh. The five-men boat is decked at each end, but open in the middle, and has two large lug-fails.

BAIT.   The beft bait for all kinds of fifh is frefh herring cut in pieces of a proper fize; and notwithftanding what has been faid to the contrary, they are taken here at any time in the winter, and all the fpring, whenever the fifhermen put down their nets for that purpofe. The five-men boats always take
some

fome nets for that end.   Next to herrings are the leffer lampreys *, which come all winter by land-carriage from *Tadcafter*.   The next baits in efteem are fmall hadocks cut in pieces, fand worms, muf-cles, and limpets (called here *Flidders*;) and laftly, when none of thefe can be had they ufe bullock's liver.   The hooks ufed here are much fmaller than thofe employed at *Iceland* and *Newfoundland*. Experience has fhewn that the larger fifh will take a living fmall one upon the hook, fooner than any bait that can be put on; therefore they ufe fuch as the fmall fifh can fwallow.   The hooks are two inches and an half long in the fhank, near an inch wide between the fhank and the point.   The line is made of fmall cording, and is always tanned before it is ufed.

Turbots, and all the rays, are extremely delicate in their choice of baits.   If a piece of herring or hadock has been twelve hours out of the fea, and then ufed as bait, they will not touch it.

This and the pearl are of a remarkable fquare form : the color of the upper part of the body is cinereous, marked with numbers of black fpots of different fizes: the belly is white : the fkin is without fcales, but greatly wrinkled, and mixed with fmall fhort fpines, difperfed without any order.

* The *Dutch* alfo ufe thefe fifh as baits in the turbot fifhery, and purchafe annually from the *Thames* fifhermen as much as amounts to 700*l.* worth, for that purpofe.

Paffer

110. Pearl. La Barbue. *Belon*, 137.
Rhombus lævis. *Rondel.* 312.
   Gefner pifc. 662.
Schlichtbutt. *Schonevelde*, 60.
Rhombus non aculeatus fqu-
   amofus *the Pearl. Londinens.*
   Cornub. Lug-aleaf. Wil.
   Icth. 95. *Raii fyn. pifc.* 31.

Pleuronectes oculis a finiftris,
   corpore glabro. *Arted. fyn.*
   31.
Pleuronectes Rhombus. *Lin.*
   fyft 458. *Gronov. Zooph.*
   No. 149.
Pigghvarf. *It. W. Goth.* 178.

IT is frequently found in the *London* markets, but is inferior to the turbot in goodnefs as well as fize.

The irides are yellow : the fkin is covered with fmall fcales, but is quite free from any fpines or inequalities.

The upper fide of the body is of a deep brown, marked with fpots of dirty yellow : the under fide is of a pure white.

111. Whiff. Paffer *Cornubienfis* afper, magno oris hiatu. Mr. *Jago. Raii fyn. pifc.* 163. *fig.* 2.

THIS bears fome refemblance to the *Holibut.* One was brought to me by my fifherman, *October* 31, 1775. Its length was eighteen inches: the greateft breadth not feven, **exclufive of the** fins.

The

The mouth extremely large : teeth very fmall : the under jaw hooks over the upper : the eyes large ; and placed on the   fide.

The fcales great, and rough : the fide-line uncommonly incurvated at the beginning. After making a fharp angle, goes ftrait to the tail, and is tuberculated : the tail is rounded.

The color of the upper part of the body is cinereous brown, clouded in parts, and obfcurely fpotted : the under fide white, tinged with red.

Covers

XXV.
GILT-
HEAD.

Covers of the gills fcaly.

Five branchioftegous rays.

Fore teeth fharp.

Grinders flat.

One dorfal fin, reaching the whole length of the back.

Forked tail.

112. LUNU-
LATED.

Χρυσοφρυς.　*Oppian Halieut.*
I. 169.
Chryfophrys.　*Ovid Halieut.*
III.
Aurata *Plinii, Lib.*IX. *c.* 16.
La Dorade.　*Belon,*　186.
Chryfophry *Caii opufc.* 112.
Aurata. *Rondel.* 115. *Gefner pifc.* 110. 112.

Gilt-head or Gilt-poll. *Wil.
Icth.* 307. *Raii fyn. pifc.* 131.
Sparus dorfo acutiffimo, linea
arcuata inter oculos. *Arted.
fynon.* 63.
Sparus lunula aurea inter oculos. *Lin. fyft.* 467. *Gronov. Zooph.* No. 220.

THIS is one of the *pifces faxatiles,* or fifh that haunt deep waters on bold rocky fhores: thofe that form this genus, as well as the following, feed chiefly on fhell fifh, which they comminute with their teeth before they fwallow; the teeth of this genus in particular being extreme-ly well adapted for that purpofe, the grinders be-ing flat and ftrong, like thofe of certain quadru-peds: befides thofe are certain bones in the low-er part of the mouth, which affift in grinding their food.

They

They are but a coarſe fiſh; nor did the *Ro-mans* hold them in any eſteem, except they had fed on the *Lucrine* oyſter.

> *Non omnis laudem pretiumque* AURATA *meretur,*
> *Sed cui ſolus erit concha* LUCRINA *cibus* *.

> No praiſe, no price a *Gilt-head* e'er will take,
> Unfed with oyſters of the *Lucrine* lake.

They grow to the weight of ten pounds: the form of the body is deep, not unlike that of a bream: the back is very ſharp, and of a duſky green color: the irides of a ſilvery hue: between the eyes is a ſemilunar gold colored ſpot, the horns of which point towards the head: on the upper part of the gills is a black ſpot, beneath that another of purple. DESCRIP.

The dorſal fin extends almoſt the whole length of the back, and conſiſts of twenty-four rays, the eleven firſt ſpiny, the others ſoft: the pectoral fins conſiſt of ſeventeen ſoft rays; the ventral of ſix rays, the firſt of which is very ſtrong and ſpiny: the anal fin of fourteen; the three firſt ſpiny.

The tail is much forked.

It takes its name from its predominant color; that of the forehead and ſides being as if gilt, but the laſt is tinged with brown. COLOR.

---

* *Martial. Lib.* XIII. *Ep.* 90.

R Pagur ?

113. RED.　Pagur ? *Ovid Halieut.* 107.　　Sparus rubefcens, cute ad ra-
　　　　Le Pagrus　*Belon,* 245.　　　　dicem pinnarum dorfi et ani
　　　　Pagrus. *Rondel.* 142. *Gefner*　　in finum producta. *Arted.*
　　　　pifc 656.　　　　　　　　　　*fynon.* 64.
　　　　Sea Bream. *Wil. Icth.* 312.　　Sparus Pagrus. *Lin. fyft.* 469.
　　　　Raii fyn. pifc. 131.

DESCRIP.　　THIS fpecies grows to a fize equal with that
　　　　　　　　of the former: its fhape and the figure of the
teeth are much the fame.

　　The irides are filvery: the infide of the covers
of the gills, the mouth, and the tongue, are of a
fine red.

　　At the bafe of the pectoral fins is a ferruginous
fpot.

　　What is peculiar to this fpecies is, that the fkin
at the end of the dorfal and anal fins is gathered
up, and hides the laft rays.

　　The fcales are large: the tail forked.

COLOR.　　The color of the whole body is red.

Brama

Pl. XLIII

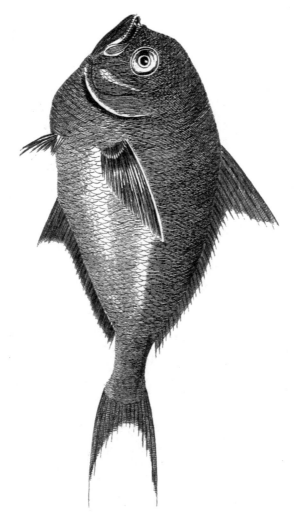

TOOTHED GILT-HEAD.

Brama marina cauda forcipata D. *Jonfton. Raii fyn. pifc.* 115.   114. Tooth-ed.

THIS fpecies was communicated to Mr. *Ray* by his friend Mr. *Jonfton*, a *Yorkfhire* gentleman, who informed him it was found on the fands near the mouth of the *Tees, Sept.* 18, 1681.

It was a deep fifh, formed like a roch, twenty-fix inches long, ten broad, and grew very flender towards the tail.

The eyes large, like thofe of quadrupeds. In the lower jaws were two rows of teeth, flender and fharp as needles; and on each fide a flender canine tooth: in the upper only a fingle row of teeth. The aperture of the gills very large. The body fcaly.

In the middle of the back was one fin extending almoft to the tail; the feven firft rays high, the reft low: behind the vent is another, correfponding: both are entirely covered with fcales flated over each other.

The back black; the fides of a brighter color: the belly quite of a filvery brightnefs.

R 2   Covers

XXVI.
WRASSE.

Covers of the gills fcaly.

Branchioftegous rays unequal in number *.

Teeth conic, long and blunt at their ends. One tuberculated bone in the bottom of the throat: two above oppofite to the other.

One dorfal fin reaching the whole length of the back: a flender fkin extending beyond the end of each ray.

Rounded tail.

115. An-
TIENT.

Vieille, Poule de mer, Gallot, une Roffe. *Belon,* 248.
Turdorum undecimum genus. *Rondel.* 179. *Gefner pifc.* 1019.
Turdus vulgatiffimus. *Wil. Icth.* 319.

Wraffe, or Old Wife. *Raii fyn. pifc.* 136.
Labrus roftro furfum reflexo cauda in extremo circulari. *Arted. fynon.* 56.
Labrus Tinca. *Lin. fyft.* 477.

THIS fpecies is found in deep water adjacent to the rocks. It will take a bait, though its ufual food is fhell-fifh, and fmall cruftacea.

* *Linnæus* fays fix: this fpecies had only four; the fecond, fix; the third and fourth, five. We alfo find the fame variation in the rays of the fins, the numbers being different in fifh of the fame fpecies, not only of this but of other *genera.*

It

It grows to the weight of four or five pounds: it bears fome refemblance to a carp in the form of the body, and is covered with large fcales.

The nofe projects; the lips are large and flefhy, and the one turns up, the other hangs down: the mouth is capable of being drawn in or protruded.

TEETH.

The irides are red: the teeth are difpofed in two rows; the firft are conic, the fecond very minute, and as if fupporters to the others: in the throat juft before the gullet are three bones, two above of an oblong form, and one below of a triangular fhape; the furface of each rifing into roundifh protuberances: thefe are of fingular ufe to the fifh, to grind its fhelly food before it arrives at the ftomach.

The dorfal fin confifts of fixteen fharp and fpiny rays, and nine foft ones, which are much longer than the others.

The pectoral fins large and round, and are compofed of fifteen rays.

The ventral of fix; the firft fharp and ftrong: the anal of three fharp fpines, and nine flexible.

The tail is rounded at the end, and is formed of fourteen foft branching rays.

The lateral line much incurvated near the tail.

Thefe fifh vary infinitely in color: we have feen them of a dirty red, mixed with a certain dufkinefs; others moft beautifully ftriped, efpecially about

COLOR.

R 3   the

the head, with the richeſt colors, ſuch as blue, red, and yellow. Moſt of this genus are ſubjeƈt to vary; therefore care muſt be taken not to multiply the ſpecies from theſe accidental teints, but to attend to the form which never alters.

The *Welch* call this fiſh *Gwrach*, or the old woman; the *Fernch, la Vieille*; and the *Engliſh* give it the name of *Old Wife.*

16 Ballan. THIS is a kind of *Wraſſe*, ſent from *Scarborough* by Mr. *Travis*, differing from the other ſpecies. They appear during ſummer in great ſhoals off *Filey-Bridge:* the largeſt weigh about five pounds.

It was of the form of the common *wraſſe*, only between the dorſal fin and the tail was a conſiderable ſinking: above the noſe was a deep ſulcus: on the fartheſt cover of the gills was a depreſſion radiated from the center.

It had only four branchioſtegous rays.

The dorſal fin had thirty-one rays, twenty ſpiny, eleven ſoft; the laſt branched, and much longer than the ſpiny rays.

The peƈtoral fins had fourteen; the ventral ſix; the firſt of which was ſhort and ſpiny: the anal twelve; the three firſt ſpiny, the nine others branched and ſoft.

The

Pl. XLIV.

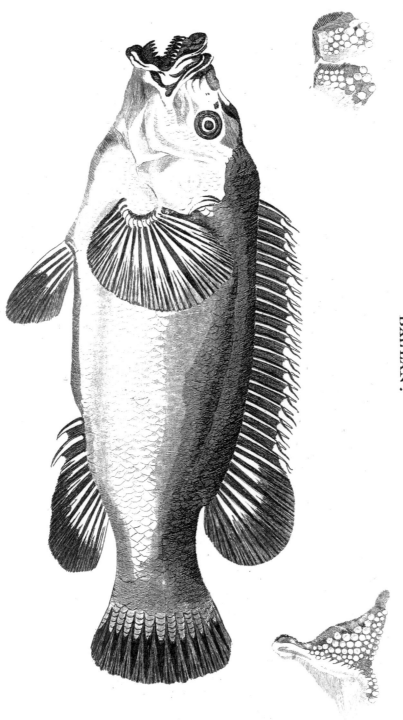

The tail was rounded at the end; at the bottom, for about a third part of the way, between each ray was a row of scales.

The color in general was yellow, spotted with orange.

Labrus bimaculata. L. pinna dorsali ramentacea, macula fusca in latere medio, et ad caudam. *Lin. syst.* 477.	Sciæna bimaculata. *Mus. Ad. Fred.* I. 66. *tab.* XXXI. *fig.* 66.	117. BIMA-CULATED.

M R. *Brunnich* observed this species at *Penzance*, and referred me to *Linnæus*'s description of it in the *Museum Ad. Fred.* where it is described under the name of *Sciæna Bimaculata*.

The body is pretty deep, and of a light color, marked in the middle on each side with a round brown spot; on the upper part of the base of the tail is another: the lateral line is incurvated.

Descrie.

The branchiostegous rays are six in number*: the first fifteen rays of the dorsal fin are spiny; the

---

* *Linnæus*, in his last edition, has removed this species from the genus of *Sciæna*, to that of *Labrus*, though it does not agree with the last in *his* number of branchiostegous rays,

R 4   other

other eleven foft, and lengthened by a fkinny appendage : the pectoral fins confift of fifteen rays; the ventral of fix; the firft fpiny; the fecond and third ending in a flender briftle : the anal fin is pointed; the four firft rays being fhort and fpiny; the reft long and foft.

118. TRIMA-
CULATED.

THE fpecies we examined was taken on the coaft of *Anglefea*; its length was eight inches.

It was of an oblong form; the nofe long; the teeth flender; the fore teeth much longer than the others.

The eyes large : branchioftegous rays, five.

The back fin confifting of feventeen fpiny rays, and thirteen foft ones; beyond each extended a long nerve.

The pectoral fins were round, and confifted of fifteen branched rays.

The ventral fins confifted of fix rays; the firft fpiny.

The anal fin of twelve; the three firft fhort, very ftrong, and fpiny; the others foft and branched.

The tail was rounded.

The lateral line was ftrait at the beginning of the back, but grew incurvated towards the tail.

The

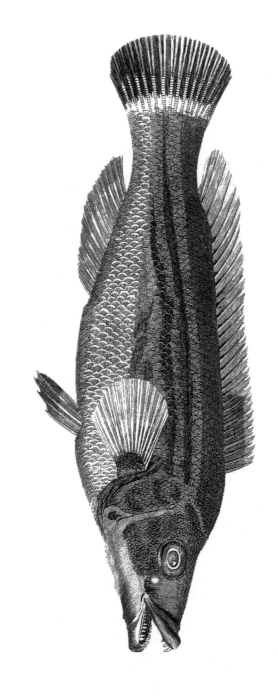

The body covered with large red fcales; the covers of the gills with fmall ones.

On each fide of the lower part of the back fin
were two large fpots, and between the fin and the
tail another.

THIS was taken off [the *Skerry Ifles*, on the   119. STRIP-
coaft of *Anglefea*; its length was ten inches.   ED.

The form was oblong, but the beginning of   DESCRIP.
the back a little arched: the lips large, double,
and much turned up: the teeth like thofe of the
preceding: branchioftegous rays, five.

The number of rays in the back, pectoral, and
ventral fins, the fame as in thofe of the former.

In the anal fin were fifteen rays; the three firft
ftrong and fpiny.

The tail almoft even at the end, being very little
rounded: the covers of the gills cinereous, ftriped
with fine yellow.

The fides marked with four parallel lines of   COLOR.
greenifh olive, and the fame of moft elegant blue.

The back and belly red; but the laft of a much
paler hue, and under the throat almoft yellow.

Along the beginning of the back fin was a
broad bed of rich blue; the middle part white;
the reft red.

At

At the bafe of the pectoral fins was a dark olive fpot.

The ends of the anal fin, and ventral fins, a fine blue.

The upper half of the tail blue; the lower part of its rays yellow.

120. Gib-
bous.

THIS fpecies was taken off *Anglefea*: its length was eight inches; the greateft depth three: it was of a very deep and elevated form, the back being vaftly arched, and very fharp or ridged.

From the beginning of the head to the nofe, was a fteep declivity.

The teeth like thofe of the others.

The eyes of a middling fize; above each a dufky femilunar fpot.

The neareft cover of the gills finely ferrated.

The fixteen firft rays of the back ftrong and fpiny; the other nine foft and branched.

The pectoral fins confifted of thirteen, the ventral of fix rays; the firft ray of the ventral fin was ftrong and fharp.

The anal fin confifted of fourteen rays, of which the three firft were ftrongly aculeated.

The tail was large, rounded at the end, and the

rays

GIBBOUS WRASSE.

TRIMACULATED WRASSE.

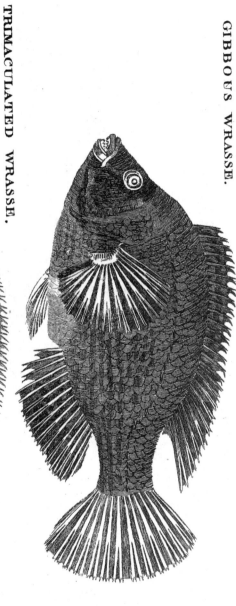

N.º 120.

N.º 118.

Pl. XLVII.   N.º 122.

COMBER WRASSE.

ANTIENT WRASSE ?   N.º 115.

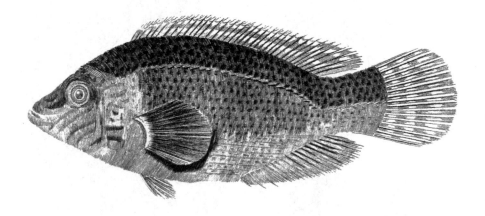

GOLDSINNY.   N.º 121.

rays branched; the ends of the rays extending beyond the webs.

The lateral line was incurvated towards the tail.

The gill covers and body covered with large
fcales.

The firft were moft elegantly fpotted, and ftriped
with blue and orange, and the fides fpotted in the
fame manner; but neareft the back the orange was
difpofed in ftripes: the back fin and anal fin were
of a fea green, fpotted with black.

The ventral fins and tail a fine pea green.

The pectoral fins yellow, marked at their bafe
with tranfverfe ftripes of red.

Goldfinny *Cornubienfium*, Mr. *Jago. Raii fyn. pifc.* 163.
fig. 3.

THIS and the two following fpecies were difcovered by Mr. *Jago* on the coaft of *Cornwal:* we never had an opportunity of examining
them, therefore are obliged to have recourfe to his
defcriptions, retaining their local names.

In the whole form of the body, lips, teeth, and
fins, it refembles the *Wraffe:* it is faid never to
exceed a palm in length: near the tail is a remarkable

able black fpot: the firft rays of the dorfal fin are tinged with black.

The *Melanurus* of *Rondeletius* (adds he) takes its name from the black fpot near the tail; but in many inftances it differs widely from this fpecies, the tail of the firft is forked, that of the *Goldfinny* is even at the end.

I fufpect that this fpecies was fent to me from *Cornwal.* Befides the fpot near the tail, there was another near the vent.

In the dorfal fin were fixteen fpiny, and nine foft rays: in the pectoral fourteen: in the anal three fpiny, eleven foft: in the ventral fix. The tail almoft even at the end.

122. Comber.

Comber *Cornub. Raii fyn. pifc.* 163. *fig.* 5 ?

I RECEIVED this fpecies from *Cornwal,* and fuppofe it to be the *Comber* of Mr. *Jago.*

It was of a flender form. The dorfal fin had twenty fpiny, eleven foft rays: the pectoral fourteen: the ventral five: the anal three fpiny, feven foft. The tail round.

The color of the back, fins, and tail, red: the belly yellow: beneath the lateral line ran parallel
a fmooth,

a fmooth, even ftripe from gills to tail, of a filvery color.

Cook (*i. e.* Coquus) *Cornubienfium. Raii fyn. pifc.* 163.   123. Cook.
fig. 4.

THIS fpecies, Mr. *Jago* fays, is fometimes taken in great plenty on the *Cornifh* coafts. It is a fcaly fifh, and does not grow to any great fize. The back is purple and dark blue; the belly yellow. By the figure it feems of the fame fhape as the Comber, and the tail rounded.

Befides thefe fpecies we recollect feeing taken at the *Giant's Caufeway* in *Ireland*, a moft beautiful kind of a vivid green, fpotted with fcarlet; and others at *Bandooran*, in the county of *Sligo*, of a pale green. We were at that time inattentive to this branch of natural hiftory, and can only fay they were of a fpecies we have never fince feen.

The

XXVI.
PERCH.

The edges of the gill-covers ferrated.

Seven branchioftegous rays.

Body covered with rough fcales.

Firft dorfal fin fpiny ; the fecond foft *.

124. Com-
MON.

Πέρκη *Arift. Hift. an. Lib.* VI. *c.* 14.
Perca *Aufonii Mofella*, 115.
Une Perche de riviere. *Belon*, 291
Perca fluviatilis. *Rondel. fluviat.* 196. *Gefner pifc.* 698.
Ein Barfs. *Schonevelde*, 55.
A Perch. *Wil. Iĉth.* 291. *Raii fyn. pifc.* 97.
Perca lineis utrinque fex tranfverfis nigris, pinnis ventralibus rubris. *Arted. fynon.* 66.
Perca fluviatilis. P. pinnis dorfalibus diftinĉtis, fecunda radiis fedecim *Lin. fyft.* 481. *Gronov. Zooph.* No. 301.
Abboree. *Faun. Suec.* No. 332.
Perfchling, Barfchieger. *Kram.* 384. *Wulff Boruf.* No. 27.

THE perch of *Ariftotle* and *Aufonius* is the fame with that of the moderns. That mentioned by *Oppian*, *Pliny*, and *Athenæus* †, is a fea-fifh probably of the *Labrus* or *Sparus* kind, being enumerated by them among fome congene-

---

* The *Ruffe* is an exception, having only one dorfal fin, but the fourteen firft rays of it are fpiny.

† *Oppian Halieut.* I. 124. *Plinii Lib.* IX. *c.* 16. *Athenæus Lib.* VII. *p.* 319.

rous

Pl. XLVIII.

PERCH.

SEA PERCH.

N.º 126.

rous fpecies.   Our perch was much efteemed by
the *Romans*:

Nec te delicias menfarum Perca, *filebo*
Amnigenos inter pifces dignande marinis.   Ausonius,

It is not lefs admired at prefent as a firm and delicate fifh; and the *Dutch* are particularly fond of
it when made into a difh called *Water Souchy*.

It is a gregarious fifh, and loves deep holes and
gentle ftreams.   It is a moft voracious fifh, and
eager biter: if the angler meets with a fhoal of
them, he is fure of taking every one.

It is a common notion that the pike will not
attack this fifh, being fearful of the fpiny fins
which the perch erects on the approach of the
former.   This may be true in refpect to large fifh;
but it is well known the fmall ones are the moft
tempting bait that can be laid for the pike.

The perch is a fifh very tenacious of life: we
have known them carried near fixty miles in dry
ftraw, and yet furvive the journey.

Thefe fifh feldom grow to a large fize: we once   Size.
heard of one that was taken in the *Serpentine*
river, *Hyde-Park*, that weighed nine pounds, but
that it is very uncommon.

The body is deep: the fcales very rough: the   Descrip.
back much arched: fide-line near the back.

The irides golden: the teeth fmall, difpofed in
the jaws and on the roof of the mouth: the edges
of

of the covers of the gills ferrated: on the lower end of the largeft is a fharp fpine.

The firft dorfal fin confifts of fourteen ftrong fpiny rays: the fecond of fixteen foft ones: the pectoral fins are tranfparent, and confift of fourteen rays; the ventral of fix; the anal of eleven.

The tail is a little forked.

COLOR.

The colors are beautiful: the back and part of the fides being of a deep green, marked with five broad black bars pointing downwards: the belly is white, tinged with red: the ventral fins of a rich fcarlet; the anal fins and tail of the fame color, but rather paler.

CROOKED PERCH.

In a lake called *Llyn Raithlyn*, in *Merionethfhire*, is a very fingular variety of perch: the back is quite hunched, and the lower part of the back bone, next the tail, ftrangely diftorted: the color, and in other refpects, it refembles the common kind, which are as numerous in the lake as thefe deformed fifh. They are not peculiar to this water, for *Linnæus* takes notice of a fimilar variety found at *Fahlun*, in his own country. I have alfo heard that it is to be met with in the *Thames* near *Marlow*.

Λάϐραξ ?

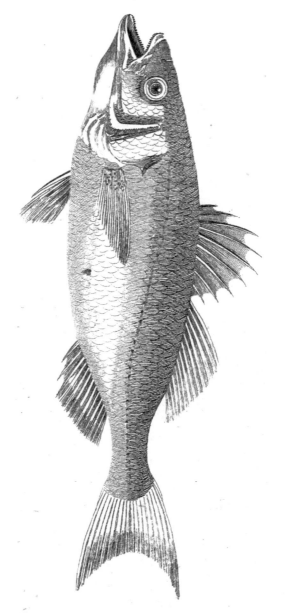

BASSE.

125. BASSE.

Λάϐϱαξ? *Arift. Hift. an. lib.*
　IV. *c.* 10. &c.
Lupus? *Ovid. Halieut.* 112.
Le Bar, le Loup. *Belon,* 113.
Lupus. *Rondel.* 268. *Gefner*
　pifc. 506.
A Baffe. *Wil. Icth.* 271.

Raii fyn. pifc. 83.
Perca radiis pinnæ dorfalis fe-
　cundæ tredecim, ani qua-
　tuordecim. *Arted. fynon.* 69.
Perca Labrax. *Lin. fyft.* 482.
　Gronov. Zooph. No. 300.

THE baffe is a ftrong, active, and voracious
fifh : *Ovid* calls them *rapidi lupi,* a name continued to them by after-writers.

That which we had an opportunity of examining *Size.*
was fmall; but they are faid to grow to the weight
of fifteen pounds.

The irides are filvery : the mouth large : the teeth
are fituated in the jaws, and are very fmall : in the
roof of the mouth is a triangular rough fpace, and
juft at the gullet are two others of a roundifh form.

The fcales are of a middling fize, are very thick
fet, and adhere clofely.

The firft dorfal fin has nine ftrong fpiny rays,
of which the firft is the fhorteft, the middlemoft
the higheft; the fecond dorfal fin confifts of thirteen rays, the firft fpiny, the others foft.

The pectoral fins have fifteen foft rays; the ventral fix rays, the firft fpiny : the anal fourteen rays,
the three firft fpiny, the others foft : the tail is a
little forked.

Vol. III,　　　S　　　　The

The body is formed fomewhat like that of a fal-mon.

The color of the back is dufky, tinged with blue.

The belly white. In young fifh the fpace above the fide line is marked with fmall black fpots.

It is efteemed a very delicate fifh.

126. Sea.　Une Perche de mer. *Belon,* 163.　Perca marina. P. pinnis dor-falibus unitis XV. fpin fis,
Perca marina. *Salvian,* 225.　XIV. muticis, corpore litu-
Rondel. 182. *Wil. Icth.* 327.　ris variegato. *Lin. fyft.* 483.
Raii fyn. pifc. 140.

THIS fpecies is about a foot long: the head large and deformed: eyes great: teeth fmall and numerous. On the head and covers of the gills are ftrong fpines. The dorfal fin is furnifh-ed with fifteen ftrong fpiny rays, and fourteen foft: the pectoral with eighteen: the ventral with one fpiny, and five foft: the anal with three fpiny, and eight foft: the tail, even at the end: the lateral line parallel to the back. The color red, with a black fpot on the covers of the gills, and fome tranfverfe dufky lines on the fides.

It is a fifh held in fome efteem at the table.

Cernua

Cernua. *Belon*, 186.
Percæ fluviatilis genus minus.
  Gefner pifc. 701.
Afpredo. *Caii opufc.* 107.
Ein ftuer, ftuerbarfs. *Schone-*
  velde, 56.
Cernua fluviatilis. *Wil. Icth.*
  334.
Ruffe. *Raii fyn. pifc.* 143.
Perca dorfo monopterygio, ca-
  pite cavernofo. *Arted. fyn.* 68.

Perca cernua. P. pinnis dor- 127. RUFFE,
  falibus unitis radiis 27.
  fpinis 15. cauda bifida.
  Lin. fyft. 487. *Gronov.*
  Zooph. No.
Giers, Snorgers. *Faun. Suec.*
  No. 119.
Schroll, Pfaffenlaus. *Schaeff.*
  pifc. 37. *Tab.* II. *Wulff*
  Borufs. No. 35.

THIS fifh is found in feveral of the *Englifh*
ftreams : it is gregarious, affembling in large
fhoals, and keeping in the deepeft part of the water.

It is of a much more flender form than the perch,
and feldom exceeds fix inches in length.

The teeth are very fmall, and difpofed in rows.

It has only one dorfal fin extending along the
greateft part of the back : the firft rays, like thofe
of the perch, are ftrong, fharp, and fpiny; the
others foft.

The pectoral fins confift of fifteen rays; the
ventral of fix; the anal of eight; the two firft
ftrong and fpiny : the tail a little bifurcated.

The body is covered with rough compact fcales.

The back and fides are of a dirty green, the
laft inclining to yellow, but both fpotted with
black.

The dorfal fin is fpotted with black: the tail
marked with tranfverfe bars.

S 2   The

MR. *Jago* has left ſo brief a deſcription of this fiſh, that we find difficulty in giving it a proper claſs : it agrees with the *Ruffe* in the form of the body, and the ſmallneſs of the teeth, in having a ſingle extenſive fin on the back, a forked tail, and being of that ſection of bony fiſh, termed *Thoracic :* theſe appear by the figure, the teeth excepted. The other characters muſt be borrowed from the deſcription.

    " It is ſmooth, with very ſmall thin ſcales, fif-
" teen inches long, three quarters of an inch
" broad ; head and noſe like a peal or trout ;
" little mouth ; very ſmall teeth, beginning from
" the noſe four inches and three quarters, near
" ſix inches long ; a forked tail ; a large double
" noſtril. Two taken at *Loo, May* 26, 1721, in
" the *Sean,* near the ſhore, in ſandy ground with
" ſmall ore weed."

Three

STICKLEBACKS.

N.º 131.

Nº 130.

Nº 129.

Pl. L.

Three branchioftegous rays.

The belly covered with bony plates.

One dorfal fin, with feveral fharp fpines between it and the head.

La Grande Efpinoche, un Epinard, une Artiere. *Belon*, 328.
Pifciculi aculeati prius genus. *Rondel. fluviat.* 206. *Gefner pifc.* 8.
Stickleback, Banftickle, or Sharpling. *Wil. Icth.* 341. *Raii fyn. pifc.* 145.

Gafterofteus aculeis in dorfo tribus. *Arted. fynon.* 80.
Gafterofteus aculeatus. *Lin. fyft.* 489. *Gronov. Zooph.* No. 406.
Spigg, Horn-fifk. *Faun. Suec.* No. 336.
Stichling, Stachel-fifch. *Wulff Boruff.* No. 37.

THESE are common in many of our rivers, but no where in greater quantities than in the *Fens* of *Lincolnfhire*, and fome of the rivers that creep out of them. At *Spalding* there are, once in feven or eight years, amazing fhoals that appear in the *Welland*, and come up the river in form of a vaft column. They are fuppofed to be the multitudes that have been wafhed out of the fens by the floods of feveral years, and collected in fome deep hole, till overcharged with numbers, they are periodically obliged to attempt a change of place. The quantity is fo great, that they are ufed to manure the land, and trials have been made

S 3

to

to get oil from them. A notion may be had of this vaſt ſhoal, by ſaying that a man employed by the farmer to take them, has got for a conſiderable time four ſhillings a day by ſelling them at a half-penny per buſhel.

DESCRIP.

This ſpecies ſeldom reaches the length of two inches: the eyes are large: the belly prominent: the body near the tail ſquare: the ſides are covered with large bony plates, placed tranſverſely.

On the back are three ſharp ſpines, that can be raiſed or depreſſed at pleaſure: the dorſal fin is placed near the tail: the pectoral fins are broad: the ventral fins conſiſt each of one ſpine, or rather plate, of unequal lengths, one being large, the other ſmall; between both is a flat bony plate, reaching almoſt to the vent: beneath the vent is a ſhort ſpine, and then ſucceeds the anal fin.

The tail conſiſts of twelve rays, and is even at the end.

The color of the back and ſides is an olive green; the belly white; but in ſome the lower jaws and belly are of a bright crimſon.

130. TEN SPINED.

La petite Eſpinoche. *Belon,* 328.
Piſciculi aculeati alterum ge-nus. *Rondel. fluviat.* 206. *Geſner piſc.* 8.
Leſſer Stickleback. *Wil. Icth.* 342. *Raii ſyn. piſc.* 145.

Gaſteroſteus aculeis in dorſo decem. *Arted. ſynon.* 80.
Gaſteroſteus pungitius. *Lin. ſyſt.* 491. *Gronov. Zooph.* No. 405.
Benunge, Gaddſur, Gorquad. *Faun. Suec.* No. 337.

THIS ſpecies is much ſmaller than the former, and of a more ſlender make.

The

The back is armed with ten fhort fharp fpines, which do not incline the fame way, but crofs each other.

The fides are fmooth, not plated like thofe of the preceding: in other particulars it refembles the former.

The color of the back is olive: the belly filvery.

Aculeatus, five Pungitius ma-rinus longus, Stein-bicker, Ersfkruper. *Schonevelde*, 10. *Tab.* IV. *Sib. Scot.* III. 24. *Tab.* 19. Aculeatus marinus major. *Wil. Icth.* 340. *App.* 23. *Raii*	*fyn. pifc.* 145. Gafterofteus aculeis in dorfo quindecim. *Arted. fynon.* 81. Gafterofteus fpinachia. *Lin. fyft.* 492. *Gronov. Zooph.* No. 407. *Faun. Suec.* No. 338.	131. Fifteen Spined.

THIS fpecies inhabits the fea, and is never found in frefh water.

Its length is above fix inches: the nofe is long and flender: the mouth tubular: teeth fmall.

The fore part of the body is covered on each fide with a row of bony plates, forming a ridge; the body afterwards grows very flender, and is quadrangular.

Between the head and the dorfal fin are fifteen fmall fpines: the dorfal fin is placed oppofite the anal fin: the ventral fins are wanting.

The tail is even at the end.

The color of the upper part is a deep brown: the belly white.

S 4   Seven

XXIX.
MACKREL.

Seven branchioftegous rays.

Several fmall fins between the dorfal fin and the tail.

132. Com-
mon.

Σκόμβρος. *Arift. Hift. an. Lib.* VI. *c.* 17. IX. *c.* 2. *Athenæus, Lib.* III. 121. VII. 321. *Oppian Halieut.* I. 142. Scomber. *Ovid Halieut.* 94. *Plinii Lib.* IX. *c.* 15. XXXI. *c.* 8.
Macarello, Scombro. *Salvian.* 241 *.
Le Macreau. *Belon,* 197.
Scomber. *Rondel.* 233. *Gefner pifc.* 841. (pro 861.)

Makerel. *Schonevelde,* 66.
Mackrell, or Macarel. *Wil. Icth.* 181. *Raii fyn. pifc.* 58.
Scomber pinnulis quinque in extremo dorfo, polypterygio, aculeo brevi ad anum. *Arted. fynon.* 48.
Scomber Scomber. *Lin. fyft.* 492. *Gronov. Zooph.* No. 304.
Mackrill. *Faun. Suec.* No. 339.

THE mackrel is a fummer fifh of paffage that vifits our fhores in vaft fhoals. It is lefs ufeful than other fpecies of gregarious fifh, being very tender, and unfit for carriage; not but that it may be preferved by pickling and falting, a method, we believe, practifed only in *Cornwall*†, where it proves a great relief to the poor during winter.

GARUM.

It was a fifh greatly efteemed by the *Romans,*

* This is the firft opportunity we have had of looking into *Salvianus,* whofe *Italian* fynonyms we make ufe of.

† *Borlafe Cornwall,* 269.

becaufe

Pl. LI.

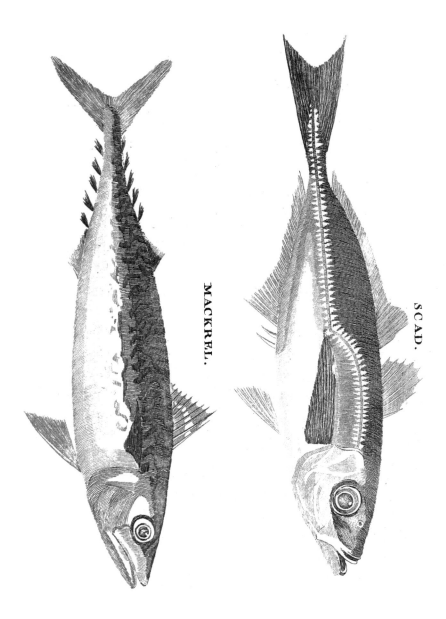

SCAD.

MACKREL.

becaufe it furnifhed the pretious *Garum*, a fort of pickle that gave a high relifh to their fauces, and was befides ufed medicinally. It was drawn from different kinds of fifh, but that made from the mackrel had the preference: the beft was made at *Carthagena*, vaft quantities of mackrel being taken near an adjacent ifle, called from that circumftance, *Scombraria* *; and the *Garum*, prepared by a certain company in that city, bore a high price, and was diftinguifhed by the title of *Garum Sociorum* †.

This fifh is eafily taken by a bait, but the beft time is during a frefh gale of wind, which is thence called a *mackrel* gale.

In the fpring the eyes of *mackrel* are almoft covered with a white film; during which period they are half blind. This film grows in winter, and is caft the beginning of fummer.

It is not often that it exceeds two pounds in weight, yet we heard that there was one fold laft fummer in *London* that weighed five and a quarter.

SIZE.

The nofe is taper and fharp-pointed: the eyes large: the jaws of an equal length: the teeth fmall, but numerous.

DESCRIP.

The form of this fifh is very elegant.

The body is a little compreffed on the fides: towards the tail it grows very flender, and a little angular.

* *Strabo Lib.* III. 109.
† *Plinii Lib.* XXXI. *c.* 8.

The

The firſt dorſal fin is placed a little behind the pectoral fin, is triangular, and conſiſts **of** nine or ten ſtiff rays; the ſecond lies at a diſtance from the other, and has twelve ſoft rays; the pectoral twenty; the ventral ſix: at the baſe of the anal fin is a ſtrong ſpine.

Between the laſt dorſal fin and the tail, are five ſmall fins, and the ſame number between the anal fin and the tail.

COLOR.

The tail is broad and ſemilunar: the color of the back and ſides above the lateral line, is a fine green, varied with blue, marked with black lines, pointing downwards; beneath the line the ſides and belly are of a ſilvery color.

It is a moſt beautiful fiſh when alive; for nothing can equal the brilliancy of its color, which death impairs, but does not wholly obliterate.

133. TUNNY. Θύννος. *Ariſt. Hiſt. an. Lib.* II. *c.* 13. &c. *Athenæus, Lib.* VII. 301. *Oppian Halieut.* III. 620.
Thunnus. *Ovid Halieut.* 95. *Plinii Lib.* IX. *c.* 15.
Tonno. *Salvian.* 123.
Le Thon. *Belon,* 99.
Thunnus. *Rondel.* 241. *Geſner piſc.* 957.
Thunnus vel orcynus. *Schonevelde,* 75.

Tunny fiſh, or Spaniſh Mackrell. *Wil. Icth.* 176. *Raii ſyn. piſc.* 57. *Sibbald Scot.*
Scomber pinnulis octo vel novem in extremo dorſo, ex fulco ad pinnas ventrales. *Arted. ſynon.* 49.
Scomber Thunnus. Sc. pinnulis utrinque octo. *Lin. ſyſt.* 493. *Gronov. Zoopb.* No. 305.

THE tunny was a fiſh well known to the antients, it made a conſiderable branch of commerce;

Pl. LII.

TUNNY.

merce ; the time of its arrival into the *Mediterra-nean* from the ocean was obferved, and ftations for taking them eftablifhed in places it moft frequent-ed ; the eminencies above the fifhery were ftyled Θυννοσκοπεῖα *, and the watchmen that gave notice to thofe below of the motions of the fifh, Θυννοσκόποι †. From one of the former the lover in *Theocritus* threatened to take a defperate leap, on account of his miftrefs's cruelty.

ἐκ επακκεις ?

Τάν ϛαίταν ἀποδὺς εἰς κυματα τηνα ἀλευμαι

Ωπερ τὰς ΘΥΝΝΩΣ σκοπιάζεται Ὀλπις ο γριπεύς.

Do you not hear ? then, rue your Goat-herd's fate,
For, from the rock where *Olpis* doth defcry
The numerous *Thunny*, I will plunge and die.

The very fame ftation, in all probability, is at this time made ufe of, as there are very confider-able thunny fifheries on the coaft of *Sicily*, as well as feveral other parts of the *Mediterranean*‡, where they are cured, and make a great article of pro-vifion in the adjacent kingdoms. They are caught

---

* *Strabo Lib.* V. 156.

† *Oppian Halieut.* III. 638. This perfon anfwers to what the *Cornifh* call a *Huer*, who watches the arrival of the pil-chards.

‡ Many of them are the fame that were ufed by the antients, as we learn from *Oppian* and others.

in

in nets, and amazing quantities are taken, for they come in vaft fhoals, keeping along the fhores.

They frequent our coafts, but not in fhoals like the Tunnies of the *Mediterranean*. They are not uncommon in the *Lochs* on the weftern coaft of *Scotland*; where they come in purfuit of herrings; and, often during night, ftrike into the nets, and do confiderable damage. When the fifhermen draw them up in the morning, the Tunny rifes at the fame time towards the furface, ready to catch the fifh that drop out. On perceiving it, a ftrong hook baited with a herring, and faftened to a rope, is inftantly flung out, which the Tunny feldom fails to take. As foon as hooked, it lofes all fpirit; and after a very little refiftance, fubmits to its fate. It is dragged to the fhore and cut up, either to be fold frefh to people who carry it to the country markets, or is preferved falted in large cafks.

The pieces, when frefh, look exactly like raw beef; but when boiled turn pale, and have fomething of the flavor of falmon.

One, which was taken when I was at *Inveraray* in 1769, and was weighed for my information, weighed 460 pounds.

The fifh, I examined, was feven feet ten inches long: the greateft circumference five feet feven; the leaft near the tail one foot fix. The body was round and thick, and grew fuddenly very flender towards the tail; and near that part was angular. The

The *irides* were of a pale green: the teeth very minute.

The firſt dorſal fin conſiſted of thirteen ſtrong ſpines; which, when depreſſed, were ſo concealed in a deep ſlit in the back, as to be quite inviſible till very cloſely inſpected. Immediately behind this fin was another, tall and falciform: almoſt oppoſite to it, was the anal fin, of the ſame form. The ſpurious fins were of a rich yellow color: of theſe there were eleven above, and ten below.

The tail was in form of a creſcent; and two feet ſeven inches between tip and tip.

The ſkin on the back was ſmooth, very thick, and black. On the belly the ſcales were viſible. The color of the ſides and belly ſilvery, tinged with cærulean and pale purple: near the tail marbled with grey.

They are known on the coaſt of *Scotland* by the name of *Mackrelſture: Mackrel,* from being of that genus; and *ſture,* from the *Daniſh, ſtor,* great.

Sauro. *Salvian.* 79.
Un Sou, Macreau baſtard. *Belom,* 186.
Trachurus. *Rondel.* 233.
Lacertus *Bellonii. Geſner piſc.* 467.
Muſeken, Stocker. *Schonevelde,* 75.
Scad, Horſe-mackrell. *Wil.*

Icth. 290. *Raii ſyn. piſc.* 92. Scomber linea laterali aculeata, pinna ani oſſiculorum 30. *Arted. ſynon.* 50. Scomber Trachurus. Sc. pinnis unitis, ſpina dorſali recumbente, linea laterali loricata. *Lin. ſyſt.* 494. *Gronov. Zooph.* No. 308.

134. SCAD

THAT which we examined was ſixteen inches long: the noſe ſharp; the eyes very large; the

the irides filvery: the lower jaw a little longer than the upper : the edges of the jaws were rough, but without teeth.

On the upper part of the covers of the gills was a large black fpot.

The fcales were large and very thin : the lower half of the body quadrangular, and marked each fide with a row of thick ftrong fcales, prominent in the middle, extending to the tail.

The firft dorfal fin confifted of eight ftrong fpines : the fecond lay juft behind it, and confifted of thirty-four foft rays, and reached almoft to the tail. The pectoral fins narrow and long, and compofed of twenty rays: the ventral of fix branched rays.

The vent was in the middle of the belly; the anal fin extended from it to the tail, which was greatly forked.

The head and upper part of the body varied with green and blue : the belly filvery.

This fifh was taken in the month of *October*; was very firm and well tafted, having the flavor of mackrel.

Head

Head compreſſed, ſteep, and covered with ſcales.
Two branchioſtegous rays.
Body covered with large ſcales, eaſily dropping off.

135. Red.

Τρίγλη? *Ariſt. Hiſt. an. Lib.*
II. *Oppian Halieut.* I. 590.
Τρίγλη Σώφρων. *Athenæus,*
Lib. VII. 325.
Mullus. *Ovid Halieut.* 123.
Plinii Lib. IX. *c.* 17.
Triglia. *Salvian.* 235.
Le Rouget barbé, Surmurlet.
Belon, 170.
Mullus barbatus. *Rondel.* 290.
Geſner piſc. 565.

Petermanneken, Goldeken.
Schonevelde, 47.
Mullus *Bellonii.* *Wil. Icth.*
285. *Raii ſyn. piſc.* 90.
Trigla capite glabro, cirris
geminis in maxilla inferiore.
Arted. ſynon. 71.
Mullus cirris geminis, corpore
rubro. *Lin. ſyſt.* 495. *Gro-*
nov. Zooph. No. 286.

THIS fiſh was highly eſteemed by the *Romans,*
and bore an exceeding high price. The
capricious epicures of *Horace*'s * days, valued it in
proportion to its ſize ; not that the larger were
more delicious, but that they were more difficult
to be got. The price that was given for one in the
time of *Juvenal,* and *Pliny,* is a ſtriking evidence
of the luxury and extravagance of the age :

Mullum *ſex millibus emit*
Æquantem ſane paribus ſeſtertia libris †.

* *Sat. Lib.* II. *ſ.* II. 33.
† *Juvenal Sat.* IV. 48 l. 8 s. 9 d.

The

The lavifh flave
Six thoufand pieces for a Mullet gave,
A fefterce for each pound.   DRYDEN.

But *Afinius Celer* *, a man of confular dignity,
gave a ftill more unconfcionable fum, for he did
not fcruple beftowing eight thoufand nummi, or
fixty-four pounds eleven fhillings and eight-pence,
for a fifh of fo fmall a fize as the mullet ; for according to *Horace*, a *Mullus trilibris*, or one of three
pounds, was a great rarity ; fo that *Juvenal*'s fpark
muft have had a great bargain in comparifon of
what *Celer* had.

But *Seneca* fays that it was not worth a farthing,
except it died in the very hand of your gueft :
that fuch was the luxury of the times, that there
were ftews even in the eating rooms, fo that the
fifh could at once be brought from under the table,
and placed on it : that they put the mullets in
tranfparent vafes, that they might be entertained
with the various changes of its rich color while
it lay expiring †. *Apicius* ‡, a wonderful genius

---

* *Plin. Lib.* IX. *c.* 17.

† *In cubili natant pifces : et fub ipfa menfa capitur, qui
ftatim transferitur in menfam : parum videtur recens mullus
nifi qui in convivæ manu moritur. Vitreis ollis inclufi offeruntur,
et obfervatur morientium color, quem in multas mutationes luctante
fpiritu vertit.* Seneca Nat. Quæft. *Lib.* III. *c.* 16.

‡ *Ad omne luxus ingenium mirus.*

for

for luxurious inventions, firft hit upon the method of fuffocating them in the exquifite *Carthaginian* * pickle, and afterwards procured a rich fauce from their livers. This is the fame gentleman whom *Pliny*, in another place, honors with the title of *Nepotum omnium altiffimus gurges* †, an expreffion too forcible to be rendered in our language.

We have heard of this fpecies being taken on the coaft of *Scotland*, but had no opportunity of examining it; and whether it is found in the weft of *England* with the other fpecies, or variety, we are not at this time informed. *Salvianus* makes it a diftinct fpecies, and fays, that it is of a purple color, ftriped with golden lines, and that it did not commonly exceed a palm in length: no wonder then that fuch a prodigy as one of fix pounds fhould fo captivate the fancy of the *Roman* epicure.

Mr. *Ray* eftablifhes fome other diftinctions, fuch as the firft dorfal fin having nine rays, and the color of that fin, the tail, and the pectoral fins, being of a very pale purple.

On thefe authorities we form different fpecies of thefe fifh, having only examined what *Salvianus* and Mr. *Ray* call the *Mullus major*, which we defcribe under the title of

* *Garum Sociorum*, vide p. 222.

† *Lib.* X. *c.* 48.

136. STRIP-
ED.

Mullus major. *Salvian.* 236.
Mullus major noster et *Sal-*
viani. 95. *Cornubienfibus.*
A Surmullet. *Wil. Icth.* 285.
Raii fyn. pifc. 91.
Trigla capite glabro, lineis

utrinque quatuor luteis,
longitudinalibus, parallelis.
Arted. fynon. 72.
Mullus cirris geminis lineis
luteis longitudinalibus. *Lin.*
fyft. 496.

THIS fpecies was communicated to us by Mr. *Pitfield* of *Exeter:* its weight was two pounds and an half; its length was fourteen inches; the thickeft circumference eleven. It appears on the coaft of *Devonfhire* in *May*, and retires about *November*.

The head fteep: the nofe blunt: the body thick: the mouth fmall: the lower jaw furnifhed with very fmall teeth: in the roof of the mouth is a rough hard fpace: at the entrance of the gullet above is a fingle bone, and beneath are a pair, each with echinated furfaces, that help to comminute the food before it paffes down.

From the chin hung two beards, two inches and a half long.

The eyes large: the irides purple: the head and covers of the gills very fcaly.

The firft dorfal fin was lodged in a deep furrow, and confifted of fix ftrong, but flexible rays; the fecond of eight; the pectoral fins of fixteen; the ventral of fix branched rays; the anal of feven: the tail is much forked.

The

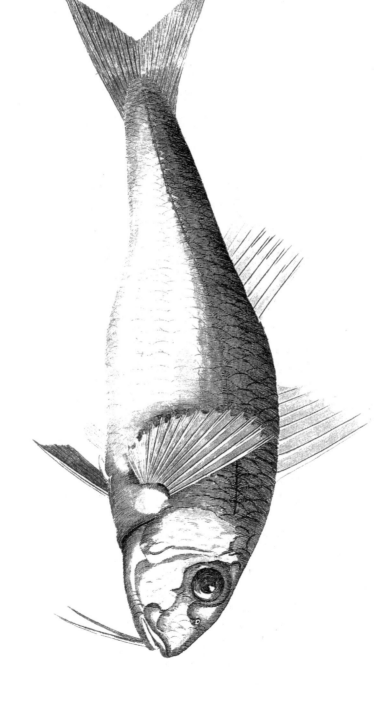

The body very thick, and covered with large fcales; beneath them the color was a moft beautiful rofy red*; the changes of which, under the thin fcales, gave that entertainment to the *Roman* epicures as above mentioned: the fcales on the back and fides were of a dirty orange; thofe on the nofe a bright yellow: the tail a reddifh yellow.

The fides were marked lengthways with two lines of a light yellow color: thefe, with the red color of the dorfal fins, and the number of their rays, Mr. *Ray* makes the character of the *Cornifh Surmullet*: thefe are notes fo liable to vary by accident, that till we receive further information from the inhabitants of our *weftern* coafts, where thefe fifh are found, we fhall remain doubtful whether we have done right in feparating this from the former, efpecially as *Doctor Gronovius* has pronounced them to be only varieties.

* This color is moft vivid during fummer.

T 2 Note

XXXI.
GURNARD.

Nofe floping.

Head covered with ftrong bony plates.

Seven branchioftegous rays.

Three flender appendages at the bafe of the pectoral fins.

137. GREY.

Gurnatus feu Gurnardus grifeus, the Grey Gurnard.
Wil. Icth. 279. *Raii fyn.*
pifc. 88.
Trigla vario roftro diacantho,
aculeis geminis ad utrum-
que oculum. *Arted. fynon,*
74.
Trigla Gurnardus, Tr. digitis
ternis dorfo maculis nigris
rubrifque. *Lin. fyft.* 497.
Gronov. Zooph. No. 283.

THE nofe pretty long, and floping: the end bifurcated, and each fide armed with three fhort fpines.

The eyes very large; above each were two fhort fpines: the forehead and covers of the gills filvery; the laft finely radiated.

The teeth fmall, placed in the lower and upper jaws, in the roof of the mouth, and bafe of the tongue.

Noftrils minute, and placed on the fides of the nofe.

On the extremity of the gill covers was a ftrong, fharp, and long fpine: beneath that, juft above the pectoral fins, another.

The

Pl. LIV.

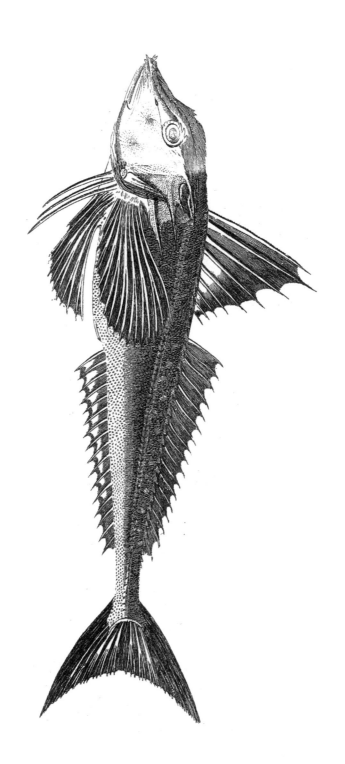

GREY GURNARD.

The firft dorfal fin confifted of eight fpiny rays; the fides of the three firft tuberculated.

The fecond dorfal fin of nineteen foft rays: both fins lodged in a groove, rough on each fide, but not ferrated.

The pectoral fins do not extend as far as the anal fins, are tranfparent, and fupported by ten rays, bifurcated from their middle: the three beards at their bafe as ufual.

The ventral fins had fix rays, the firft fpiny, and the fhorteft of all.

The anal fin nineteen, each foft.

The tail bifurcated.

The lateral line very prominent, ftrongly ferrated, and of a filvery color.

The back, tail, and a fmall fpace beneath the fide line, were of a deep grey, covered with fmall fcales, and in parts fpotted with white and yellow; the belly filvery.

Thefe fifh are ufually taken with the hook, in deep water, bite eagerly even at a red rag; and fometimes are fond of fporting near the furface. They are often found of the length of two feet and a half.

T 3 Κὸκκυξ

138. Red.

Κὸκκυξ ? *Ariſt Hiſt. an. lib.*
IV. *c.* 9. *Oppian Halieut.* I.
97.
Κὸκκυξ ἐρυθρος, *Athenæus lib.*
VII. 309.
Peſce capone, Cocco, Orga-
no. *Salvian.* 191.
Le Rouget. *Belin,* 199.
Cuculus. *Rondel.* 287. *Geſner
piſc.* 305.
Smiedecknecht, Kurre-fiſche.

Schonevelde, 32.
Red Gurnard, or Rotchet.
Wil. Icth. 281. *Raii ſyn.
piſc.* 89.
Trigla tota rubens, roſtro parum bicorni, operculis bran-
chiarum ſtriatis. *Arted ſynon.* 74.
Trigla cuculus. Tr. digitis
ternis, linea laterali mutica.
Lin. ſyſt. 497.

THIS ſpecies agrees in its general appearance
with the tub fiſh; but in theſe particulars
differs.

The covers of the gills are radiated: the ſpines
are longer and ſlenderer in thoſe of the red
gurnard. The noſe armed on each ſide with two
ſharp ſpines.

The fins and body are of a fuller red: the
ſcales are larger: head leſs and narrower: the
pectoral fins are edged with purple, not with
blue: are much ſhorter, for when extended they
do not reach to the anal fin. The ſide line is
ſtrongly ſerrated: the top of the back leſs ſo than
that of the tub fiſh. The tail red and almoſt even
at the end.

Λυρα;

PIPER.

Pl. IV.

Λυρα ?   *Arift. Hift. an. lib.*   Trigla roftro longo diacantho,   139. Piper.
    IV. *c. 9.*   naribus tubulofis. *Arted.*
Lyra. *Rondel.* 298.   *Gefner*   *fyn.* 74.
    pifc. 516.   Trigla Lyra. Tr. digitis ter-
The Piper. *Wil. Icth.* 282.   nis, naribus tubulofis. *Lin.*
    Raii fyn. pifc. 89.   *fyft.* 496.

THIS fpecies is frequently taken on the weftern coafts of this kingdom, and efteemed an excellent fifh. It is alfo found off *Anglefea*.

The weight of one which was communicated to us by Mr. *Pitfield* *, was three pounds and an half; the thickeft circumference thirteen inches, the left, which was next the tail, only three: the length near two feet.

The head was very large, and that part of the body next to it very thick: the nofe divided into two broad plates, each terminated with three fpines: on the inner corner of each eye is a ftrong fpine: the bony plates of the head terminate on each fide with another.

The covers of the gills are armed with one very fharp and ftrong fpine, and are prettily ftriated: immediately over the pectoral fin is another fpine very large and fharp pointed.

---

* We have been informed, that this fifh is found at all times of the year on the weftern coafts, and is taken in nets.

T 4   The

The noftrils very minute : the eyes large.

The lower jaw much fhorter than the upper : the teeth in both very minute.

The firft dorfal fin confifted of nine very ftrong fharp fpines, the fecond of which is the longeft; the fecond fin begins juft behind the firft, and confifts of eighteen foft rays : the pectoral fins were long, and had twelve branched rays; the ventral fins fix, very ftrong and thick : the anal eighteen, the firft fpiny : the tail fmall, in proportion to the fize of the fifh, and forked.

The back on each fide the dorfal fin was armed with a fet of ftrong and very large fpines, pointing towards the tail like the teeth of a faw.

The fcales were fmall, but very hard and rough : the lateral line bent a little at its beginning, that went ftrait to the tail, and was almoft fmooth.

140. Sapphi-rine.	Hirundo *Aldrov*. The Tub-fifh, *Cornub. Wil Icth.* 280. *Raii fyn. pifc.* 88. Trigla capite aculeato, appendicibus utrinque tribus ad pinnas pectorales. *Arted. fynon.* 73.	Trigla hirundo. Tr. digitis ternis, linea laterali aculeata. *Lin. fyft.* 497. Knorrhane, Knoding, Knot, Smed. *Faun. Suec.* No. 340.

THIS fpecies is of a more flender form than the preceding.

The pupil of the eye is green : on the inner cor-

ner

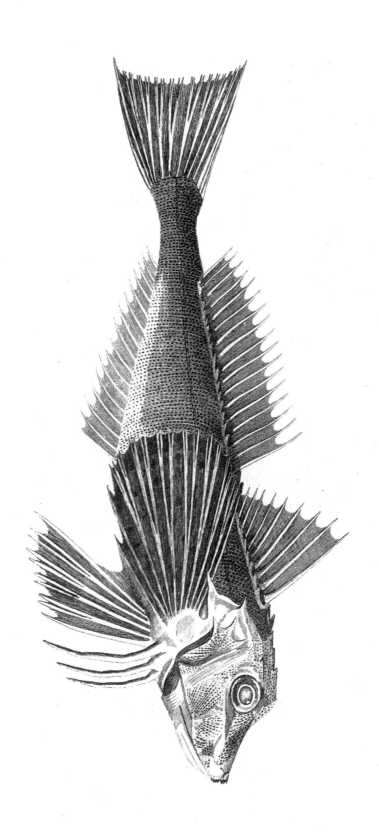

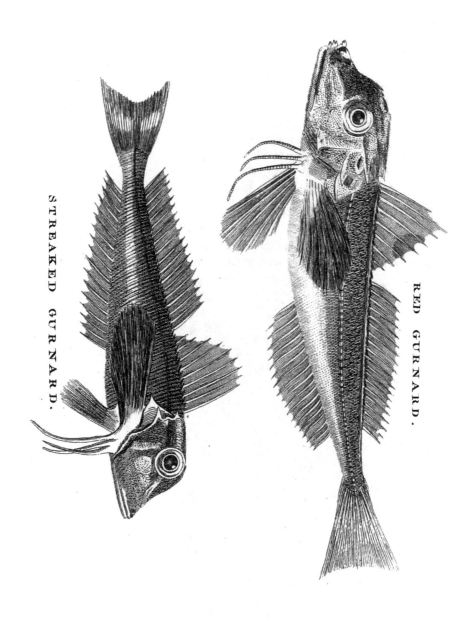

RED GURNARD.

STREAKED GURNARD.

ner of each are two small spines. But what at once distinguishes this from the other species is the breadth and colors of the pectoral fins, which are very broad and long, of a pale green, most beautifully edged, and spotted with rich deep blue.

The dorsal fins are lodged between two rows of spines, of a serrated form: the back is of a greenish cast: the side line is rough: the sides are tinged with red; the belly white.

These fish are found on the coast of *Cornwall*. We have also taken them off *Anglesea*.

Cuculus lineatus, the Streaked Gurnard. *Raii syn. pisc.* 165. 141. STREAK-*fig.* 11. ED.

THIS is one of the *Cornish* fish communicated to Mr. *Petiver* by Mr. *Jago*. He says the head is large, and distinguished with stellated marks; the eyes great; the covering of the gills thorny; the mouth small, and without teeth. By the figure the nose seems not to be bifurcated. The pectoral fins large, and spotted, beneath them three filaments: the color of the body red: the belly white, marked with many streaks, pointing downwards, from the back.

Mr. *Jago* imagines it to be the *Mullis imberhis* of *Rondeletius*. *Wil. Icth.* 278.

SECT.

SECT. IV.   ABDOMINAL.

XXXII.
L O C H E.

Eyes in the upper part of the head.
Aperture to the gills clofed below.
Several beards on the end of the upper jaw.
Body of almoft an equal thicknefs.
One dorfal fin.

142. BEARD-
ED.

La Loche franche. *Belon*, 321.
Cobitis barbatula. *Rondel.
fluviat*. 204.
Cobitis fluviatilis barbatula.
Gefner pifc. 404.
Smerling, Smerle. *Schone-
velde*, 31.
Loche, or Groundling. *Wil.
Icth*. 265. *Raii fyn. pifc.*
124.

Cobitis tota glabra maculofa,
corpore fubtereti. *Arted.
fyncn*. 2.
Cobitis Barbatula. C. cirris
fex capite inermi compreffo.
Lin. fyft. 499. *Gronov.
Zooph*. No. 202.
Gronling. *Faun. Suec*. No. 341.
Grundel. *Kram*. 396. *Wulff.
Boruff*. No. 40.

THE loche is found in feveral of our fmall rivers, keeping at the bottom on the gravel, and is on that account, in fome places, called the *Groundling* : it is frequent on the ftream near *Amefbury*, in *Wiltfhire*, where the fportfmen, through frolick, fwallow it down alive in a glafs of white wine.

The largeft we ever heard of was four inches and three quarters in length, but they feldom arrive to that fize.

The

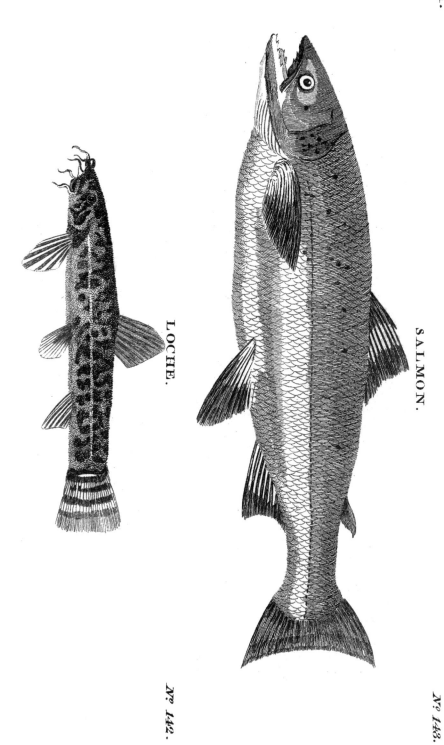

Pl. LVIII.

LOCHE.

SALMON.

N.º 142.

N.º 143.

The mouth is small, placed beneath, and has no teeth : on the upper mandible are six small beards, one at each corner of the mouth, and four at the end of the nose.

The dorsal fin consists of eight rays; the pectoral of eleven; the ventral of seven; the anal of six : the tail is broad, and has sixteen or seventeen rays.

The body is smooth and slippery, and almost of the same thickness : the color of the head, back, and sides, is in some white, in others of a dirty yellow, very elegantly marked with large spots, consisting of numberless minute black specks : the pectoral, dorsal, and caudal fins are also spotted : the belly and ventral fins of a pure white : the tail broad, and a little rounded.

COLOR.

Branchi-

XXXIII.
SALMON.

Branchioftegous rays unequal in number.

Two dorfal fins; the fecond thick, and without rays.

* With teeth.

143. Salmon. Salmo *Plinii Lib.* IX. *c.* 18. *Aufonius Mofel.* 97.
Salmone. *Salvian.* 100.
Le Saulmon. *Belon,* 271.
Salmo. *Rondel. fluviat.* 167. *Gefner pifc.* 824. *Schone-velde,* 64.
Salmon. *Wil. Icth.* 189. *Raii fyn. pifc.* 63.

Salmo roftro ultra inferiorem maxillam fæpe prominente. *Arted. fynon.* 22.
Salmo Salar *Lin. fyft.* 509. *Gronov. Zooph.* No. 369.
Lax. *Faun. Suec.* No. 122.
Lachfs. *Wulff. Borufs.* No. 42.

THE falmon is a northern fifh, being unknown in the *Mediterranean* fea, and other warm climates: it is found in *France* in fome of the rivers that empty themfelves into the ocean *, and north as far as *Greenland*; they are alfo very common in *Newfoundland*, and the northern parts of *North America*. Salmons are taken in the rivers of *Kamtfchatka* †, but whether they are of the

* *Rondel. fluviat.* 167.

† *Hift. Kamtfch.* 143.

fame

fame fpecies with the *European* kind is not very certain.

They are in feveral countries a great article of commerce, being cured different ways, by falting, pickling, and drying : there are ftationary fifheries in *Iceland*, *Norway* *, and the *Baltic*, but we believe no where greater than thofe at *Colraine* in *Ireland*; and in *Great Britain* at *Berwick*, and in fome of the rivers of *Scotland*.

The falmon was known to the *Romans*, but not to the *Greeks : Pliny* fpeaks of it as a fifh found in the rivers of *Aquitaine : Aufonius* enumerates it among thofe of the *Mofel*.

Nec te puniceo rutilantem vifcere Salmo
Tranfierem, latæ cujus vaga verbera caudæ
Gurgite de medio fummas referuntur in undas,
Occultus placido cum proditur æquore pulfus.
Tu loricato fquamofus pectore, frontem
Lubricus, et dubiæ facturus fercula cænæ,
Tempora longarum fers incorrupta morarum,
Præfignis maculis capitis, cui prodiga nutat
Alvus, opimatoque fluens abdomine venter.

Nor I thy fcarlet belly will omit,
O Salmon, whofe broad tail with whifking ftrokes
Bears thee up from the bottom of the ftream
Quick to the furface ; and the fecret lafh
Below, betrays thee in the placid deep.
Arm'd in thy flaky mail, thy gloffy fnout

* There was, about the year 1578, a pretty confiderable falmon fifhery at *Gola*, in *Ruffian Lapland. Hackluyt. voy.* I. 416.

Slippery

Slippery efcapes the fifher's fingers ; elfe
Thou makeft a feaft for niceft judging palates :
And yet long uncorrupted thou remaineft :
With fpotted head remarked, and wavy fpread,
Of paunch immenfe o'erflowing wide with fat.

ANONYMOUS.

ASCENDS
RIVERS.

The falmon is a fifh that lives both in the falt
and frefh waters, quitting the fea at certain feafons
for the fake of depofiting its fpawn in fecurity, in the
gravelly beds of rivers remote from their mouths.
There are fcarce any difficulties but what they
will overcome, in order to arrive at places fit for
their purpofe: they will afcend rivers hundreds of
miles, force themfelves againft the moft rapid
ftreams, and fpring with amazing agility over ca-
taracts of feveral feet in height. Salmon are frequently taken in the *Rhine* as high up as *Bafil*;

SALMON
LEAPS.

they gain the fources of the *Lapland* rivers* in
fpite of their torrent-like currents, and furpafs the
perpendicular falls of *Leixflip* †, *Kennerth* ‡, and
Pont aberglaftyn §; thefe laft feats we have been
witnefs to, and feen the efforts of fcores of fifh, fome
of which fucceeded, others mifcarried during the
time of our ftay.

* *Scheff. Lap.* 139.

† Near *Dublin*.

‡ On the *Tivy* in *South Wales*, which *Michael Drayton* celebrates in his *Polyolbion* on this account.

§ Amidft *Snowdon* hills, a wild fcene in the ftyle of *Salvator
Rofa*.

It

It may here be proper to contradict the vulgar error of their taking their tail in their mouth when they attempt to leap; such as we saw, sprung up quite straight, and with a strong tremulous motion.

Other particulars relating to the natural history of this fish, we shall relate in our accounts of the fisheries, either from our own observations, or from such as have been communicated to us from different places: the fullest we have been favoured with, is from the late Mr. *Potts*, of *Berwick*, to whom the public is indebted for the following very curious history of the salmon fishery on the *Tweed*.

At the latter end of the year, or in the month of *November*, the salmon begin to press up the rivers as far as they can reach, in order to spawn; when that time approaches they search for a place fit for the purpose: the male and female unite in forming a proper receptacle for it in the sand or gravel, about the depth of eighteen inches; in this the female deposits her spawn, the male his milt, which they cover carefully, as it is said, with their tails, for after spawning they are observed to have no skin on that part.

Spawning.

The spawn lies buried till spring, if not disturbed by violent floods; but the salmon hasten to sea as soon as they are able, to purify and cleanse themselves, and to recover their strength; for after spawning they become very poor and lean, and then are called *Kipper*.

When the salmon first enter the fresh water, they

are

are obferved to have abundance of infects adhering to them, efpecially above the gills : thefe are the *Lernææ Salmoneæ* of *Linnæus*, and are figns that the fifh are in high feafon. Thefe animals die and drop off, foon after the falmon have left the fea.

About the latter end of *March* the fpawn begins to exclude the young, which gradually increafe to the length of four or five inches, and are then termed 'Smelts* or *Smouts* : about the beginning of *May* the river is full of them ; it feems to be all alive ; there is no having an idea of the numbers without feeing them ; but a feafonable flood then hurries them all to the fea, fcare any or very few being left in the river.

About the middle of *June* the earlieft of the fry begin to drop, as it were, into the river again from the fea, at that time about twelve, fourteen, or fixteen inches, and by a gradual progrefs, increafe in number and fize till about the end of *July*, which is at *Berwick* termed the height of *Gilfe* time, the name given to the fifh at that age : the end of *July*, or beginning of *Auguft* they leffen in number, but increafe in fize, fome being fix, feven, eight, or nine pounds in weight ; this appears to be a furprifing quick growth, yet we have received from a gentleman at *Warrington*, an inftance ftill more fo : a kipper falmon weighing 7 lb. three quarters, taken on the 7th of *February*, being marked with a fciffars, on the back, fin, and tail, and turned into the river, was again

QUICK
GROWTH.

taken

taken on the 17th of *March* following, and then was found to weigh 17 lb. and a half.

All fifhermen agree, that they never find any food in the ftomach of this fifh. It is likely they may neglect their food entirely during the time of fpawning, as *fea lions* and *fea bears* are known to do for months together during their breeding feafon: and it may be obferved, that like thofe animals, the falmons return to the fea lank and lean, and come from the falt water in good condition. It is evident that at times their food is both fifh and worms, for the angler ufes both with good fuccefs; as well as a large, gaudy, artificial fly, which probably the fifh miftakes for a *gay libellula* or dragon fly.

Food uncertain.

The capture in the *Tweed*, about the month of *July*, is prodigious; in a good fifhery, often a boat load, and fometimes near two, are taken in a tide: fome few years ago there were above feven hundred fifh taken at one hawl, but from fifty to a hundred is very frequent: the coopers in *Berwick* then begin to falt both *Salmon* and *Gilfes* in pipes, and other large veffels, and afterwards barrel* them to fend abroad, having then far more than the *London* markets can take off their hands.

Capture.

Moft of the falmon taken before *April*, or to the fetting in of the warm weather, is fent frefh to *Lon-*

---

* The falmon barrel holds above forty-two gallons, wine meafure.

U   *don*

don in baſkets, unleſs now and then the veſſel is diſappointed by contrary winds, of ſailing immediately; in that caſe the fiſh is brought aſhore again to the coopers offices, and boiled, pickled, and kitted, and ſent to the *London* markets by the ſame ſhip, and freſh ſalmon put in the baſkets in lieu of the ſtale ones. At the beginning of the ſea-

Price. ſon, when a ſhip is on the point of ſailing, a freſh clean ſalmon will ſell from a ſhilling to eighteen pence a pound, and moſt of the time that this part of the trade is carried on, the prices are from five to nine ſhillings per ſtone *, the value riſing and falling according to the plenty of fiſh, or the pro-ſpect of a fair or foul wind. Some fiſh are ſent in this manner to *London* the latter end of *September*, when the weather grows cool, but then the fiſh are full of large roes, grow very thin bellied, and are not eſteemed either palatable or wholeſome.

The price of freſh fiſh in the month of *July*, when they are moſt plentiful, has been known to be as low as 8 *d.* per ſtone, but laſt year never leſs than 16 *d.* and from that to 2 *s.* 6 *d.*

Season. The ſeaſon for fiſhing in the *Tweed* begins *November* 30th, but the fiſhermen work very little till after *Chriſtmas*; it ends on *Michaelmas-Day*; yet the corporation of *Berwick* (who are conſervators

---

* A ſtone of ſalmon weighs 18 lb. 10 oz. and half, or in other terms, four ſtones, or fifty-ſix pounds avoirdupoiſe, is only three ſtones, or forty-two pounds, fiſh weight at *Berwick.*

of

of the river) indulge the fishermen with a fortnight paft that time, on account of the change of the ftyle.

There are on the river forty-one confiderable fifheries extending upwards, about fourteen miles from the mouth (the others above being of no great value) which are rented for near 5400*l*. per annum. The expence attending the fervants wages, nets, boats, &c. amount to 5000*l*. more, which together makes up the fum 10400*l*. Now in confequence the produce muft defray all, and no lefs than twenty times that fum of fifh will effect it, fo that 208000 falmon muft be caught there one year with another.

There is a misfortune attending the river *Tweed*, which is worthy a parlementary remedy ; for there is no law for preferving the fifh in it during the fence months, as there is in the cafe of many other *Britifh* rivers. This being the boundary between the two kingdoms, part of it belongs to the city of *Berwick*, and the whole north fide (beginning about two miles from the town) is entirely *Scotch* property. From fome difagreement between the parties they will not unite for the prefervation of the fifh, fo that in fome fifheries on the north fide they continue killing falmon the whole winter, when the death of one fifh is the deftruction of thoufands *.

* I think that this grievance is now removed.

U 2   The

The legiflature began very early to pay attention to this important article: by the 13th *Edward* I. there is an act which prohibits the capture of the falmon from the Nativity of our Lady to St. *Martin*'s Day, in the waters of the *Humber*, *Owfe*, *Trent*, *Done*, *Arre*, *Derwent*, *Wharfe*, *Nid*, *Yore*, *Swale*, and *Tees*; and other monarchs in after-times, provided in like manner for the fecurity of the fifh in other rivers.

Scotland. *Scotland* poffeffes great numbers of fine fifheries on both fides of that kingdom. The *Scotch* in early times had moft fevere laws againft the killing of this fifh; for the third offence was made capital, by a law of *James* IV. Before that, the offender had power to redeem his life *. They were thought in the time of *Henry* VI. a prefent worthy of a crowned head, for in that reign the Queen of *Scotland* fent to the Dutchefs of *Clarence*, ten cafks of falted falmon; which *Henry* directed to pafs duty-free. The falmon are cured in the fame manner as at *Berwick*, and a great quantity is fent to *London* in the fpring; but after that time the adventurers begin to barrel and export them to foreign countries: but we believe that commerce is far lefs lucrative than it was in former times, partly owing to the great encreafe of the *Newfoundland* fifhery, and partly to the general relaxation of the difcipline of abftinence in the *Romifh* church.

* *Regiam Majeftatem.* Stat. *Rob.* III. *c.* 7. *Skene's* Acts. *James* IV. *Parl.* VI.

Ireland

Ireland (particularly the north) abounds with this fish: the most considerable fishery is at *Cranna*, on the river *Ban*, about a mile and an half from *Coleraine*. When I made the tour of that hospitable kingdom in 1754, it was rented by a neighboring gentleman for 620*l.* a year, who assured me that the tenant, his predecessor, gave 1600*l.* per ann. and was a much greater gainer by the bargain for the reasons before-mentioned, and on account of the number of poachers who destroy the fish in the fence months.

The mouth of this river faces the north, and is finely situated to receive the fish that roam along the coast, in search of an inlet into some fresh water, as they do all along that end of the kingdom which opposes itself the northern ocean. We have seen near *Ballicastle*, nets placed in the sea at the foot of the promontories that jut into it, which the salmon strike into as they are wandering close to shore, and numbers are taken by that method.

In the *Ban* they fish with nets eighteen score yards long, and are continually drawing night and day the whole season, which we think lasts about four months, two sets of sixteen men each alternately relieving one another. The best drawing is when the tide is coming in: we were told that at a single draught there were once eight hundred and forty fish taken.

A few miles higher up the river is a ware, where a considerable number of fish that escape the nets

U 3   are

are taken. We were lately informed, that in the year 1760 about 320 tons were taken in the *Cranna* fifhery.

The falmon are cured in this manner: they are firft fplit, and rubbed with fine falt; and after lying in pickle in great tubs, or refervoirs, for fix weeks, are packed up with layers of coarfe brown *Spanifh* falt in cafks, fix of which make a ton. Thefe are exported to *Leghorn* and *Venice* at the price of twelve or thirteen pounds per ton, but formerly from fixteen to twenty-four pounds each.

DESCRIP. The falmon is a fifh fo generally known, that a very brief defcription will ferve. The largeft we ever heard of weighed feventy four pounds. The color of the back and fides are grey, fometimes fpotted with black, fometimes plain: the covers of the gills are fubject to the fame variety: the belly filvery: the nofe fharp pointed: the end of the under jaw in the males often turns up in form of a hook; fometimes this curvature is very confiderable: it is faid that they lofe this hook when they return to the fea.

The teeth are lodged in the jaws and on the tongue, and are flender, but very fharp.

The tail is a little forked.

The

The Grey, i. e. cinereous feu Grifeus. *Wil. Icth.* 193. *Raii fyn. pifc.* 63. Salmo maculis cinereis, caudæ extremo æquali. *Arted. fynon.* 23.

Salmo eriox. *Lin. fyft.* 509. Gralax. *Faun. Suec.* No. 346. Lachfs-forellen mit Schwartz-grauen flecken oder punkt-chens. *Wulff. Boruſs.* No. 43.

144. Grey.

WE are uncertain whether this is not a meer variety of the falmon; but on the autho-rity of Mr. *Ray*, we defcribe them feparate. He fays it is a very ftrong fifh, that it does not afcend the frefh waters till *Auguft*, when it rufhes up with great violence, that it is rarely taken, and not much known.

The inhabitants of the *North* of *England* and of *South Wales* feem extremely well affured, that it is a diftinct fpecies from the falmon. They ap-pear in the *Eſk* in *Cumberland* from *July* to *September*, and are then in fpawn. The lower jaw grows hooked, when they are out of feafon. I was informed they never exceeded thirteen pounds in weight*.

The head is larger in proportion than that of the falmon. In the jaws are four rows of teeth: and on the tongue are eight teeth. The back and fides, above the lateral line, of a deep grey, fpotted

---

* I met with a fifh (I fufpected to be a *Grey*) taken in the fea near *Conway*. It weighed twenty-two pounds.

U 4   with

with number of purplifh fpots.   The belly filvery.
The tail even at the end.

THIS we believe to be the *Sewin*, or *Shewin*
of *South Wales*. The defcription above, was communicated to us by Doctor *Roberts* of *Hereford-
fhire*.

145. SEA.

Trutta taurina, apud nos in *Northumbria* a Bull-trout. *Charlton ex. pifc.* 36.	*Arted. fynon.* 24.
Trutta Salmonata, the Sal-mon-trout, Bull-trout, or Scurf. *Raii fyn. pifc.* 63. *Wil. Icth.* 193.	Salmo trutta. S. ocellis ni-gris, iridibus brunneis, pin-na pectorali punctis fex. *Lin. fyft.* 509. *Gronov. Zooph.* No. 367.
Salmo latus, maculis rubris nigrifque, cauda æquali.	Orlax, Borting. *Faun. Suec.* No. 347.

THIS fpecies migrates like the falmon up feveral of our rivers; fpawns, and returns to the
fea.   That, which I defcribe, was taken in the
Tweed below *Berwick, June* 1769.

The fhape was more thick than the common
trout.   The weight three pounds two ounces. The
irides filvery : the head thick, fmooth, and dufky,
with a glofs of blue and green : the back of the
fame color, which grows fainter towards the fide
line.   The back is plain, but the fides as far as
the lateral line marked with large, diftinct, irregularly fhaped fpots of black : the lateral line ftrait:

the

INT. INI

SAMLET.

TROUT.

the fides beneath the line, and the belly are white.
Tail broad, and even at the end.

The dorfal fin had twelve rays: the pectoral fourteen : the ventral nine : the anal ten.

The flefh when boiled is of a pale red, but well
flavored.

Mr. *Willughby*'s account of the Salmon, Bull,
or Scurf Trout obfcure.   Whether the fame with
this ?

Salar. *Aufonius Mofel.* 88.
Salar et varius, Trotta. *Sal-*
vian. 96.
La Truitte. *Belon,* 274.
Trutta fluviatilis. *Rondel. flu-*
viat. 169. *Gefner pifc.* 1002.
Foren, Forellen. *Schonevelde,*
77.

A Trout. *Wil. Icth.* 199. *Raii*
fyn. pifc. 65.
S. maculis rubris, maxilla
inferiore longiore. *Arted.*
fynon. 23.
Salmo Fario. *Lin. fyft.* 509.
Laxoring, Forell, Stenbit.
Faun. Suec. No. 348.

146. **TROUT.**

IT is matter of furprize that this common fifh
has efcaped the notice of all the antients, except *Aufonius :* it is alfo fingular, that fo delicate a
fpecies fhould be neglected at a time when the
folly of the table was at its height; and that the
epicures fhould overlook a fifh that is found in
fuch quantities in the lakes of their neighborhood,
when they ranfacked the univerfe for dainties. The
milts of *Muranæ* were brought from one place;
the

the livers of *Scari* from another *; and *Oyfters* even from fo remote a fpot as our *Sandwich* †: but there was, and is a fafhion in the article of good living. The *Romans* feem to have defpifed the trout, the piper, and the doree; and we believe Mr. *Quin* himfelf would have refigned the rich paps of a pregnant fow ‡, the heels of camels §, and the tongues of *Flamingos* ‖, though dreffed by *Heliogabalus*'s cooks, for a good jowl of falmon with lobfter fauce.

When *Aufonius* fpeaks of this fifh, he makes no euloge on its goodnefs, but celebrates it only for its beauty.

Purpureifque SALAR *ftellatus Tergore guttis.*

With purple fpots the *Salar*'s back is ftained.

Thefe marks point out the fpecies he intended: what he meant by his *Fario* is not fo eafy to determine: whether any fpecies of trout, of a fize between the *falar* and the falmon; or whether the falmon itfelf, at a certain age, is not very evident.

---

* *Suetonius, vita* Vitellii.

† *Juvenal Sat.* IV. 141.

‡ *Martial, Lib.* XIII. *Epig.* 44.

§ *Lamprid. vit. Heliogab.*

‖ *Martial, Lib.* XII. *Epig.* 71.

Teque

Teque inter geminos ſpecies, neutrumque et utrumque,
Qui nec dum Salmo, *nec* Salar *ambiguuſque.*
Amborum medio Fario *intercepte ſub ævo.*

Salmon or *ſalar*, I'll pronounce thee neither;
A doubtful kind, that may be none, or either,
Fario, when ſtopt in middle growth.

In faɛt the colors of the trout, and its ſpots, vary greatly in different waters, and in different ſeaſons; yet each may be reduced to one ſpecies. In *Llyndivi*, a lake in *South Wales*, are trouts called *Coch y dail*, marked with red and black ſpots as big as ſix-pences; others unſpotted, and of a reddiſh hue, that ſometimes weigh near ten pounds, but are bad taſted.

In *Lough Neagh* in *Ireland*, are trouts called there *Buddaghs*, which I was told ſometimes weigh-ed thirty pounds, but it was not my fortune to ſee any during my ſtay in the neighborhood of that vaſt water.

Trouts (probably of the ſame ſpecies) are alſo taken in *Hulſe-water*, a lake in *Cumberland*, of a much ſuperior ſize to thoſe of *Lough Neagh*. Theſe are ſuppoſed to be the ſame with the trout of the lake of *Geneva*, a fiſh I have eaten more than once, and think but a very indifferent one.

In the river *Eynion*, not far from *Machyntleth*, in *Merionethſhire*, and in one of the *Snowdon* lakes, are found a variety of trout, which are naturally deformed, having a ſtrange crookedneſs near the

Crooked
Trouts.

tail,

tail, refembling that of the perch before defcribed. We dwell the lefs on thefe monftrous productions, as our friend the Hon. *Daines Barrington*, has already given an account of them in an ingenious differtation on fome of the *Cambrian* fifh, publifhed in the *Philofophical Tranfactions* of the year 1767.

GILLAROO TROUT *. The ftomachs of the common trouts are uncommonly thick, and mufcular. They feed on the fhell-fifh of lakes and rivers, as well as on fmall fifh. They likewife take into their ftomachs gravel, or fmall ftones, to affift in comminuting the teftaceous parts of their food. The trouts of certain lakes in *Ireland*, fuch as thofe of the province of *Galway*, and fome others, are remarkable for the great thicknefs of their ftomachs, which, from fome flight refemblance to the organs of digeftion in

NAME. birds, have been called gizzards : the *Irifh* name the fpecies that has them, *Gillaroo* trouts. Thefe ftomachs are fometimes ferved up to table, under the former appellation. It does not appear to me, that the extraordinary ftrength of ftomach in the *Irifh* fifh, fhould give any fufpicion, that it is a diftinct fpecies : the nature of the waters might increafe the thicknefs; or the fuperior quantity of fhell-fifh, which may more frequently call for the ufe of its comminuting powers than thofe of our trouts, might occafion this difference. I had opportunity of comparing the ftomach of a great

* *Philofoph. Tranfact.* Vol. LXIV. p. 116. 310.

Gillaroo

Gillaroo trout, with a large one from the *Uxbridge* river. The laft, if I recollect, was fmaller, and out of feafon; and its ftomach (notwithftanding it was very thick) was much inferior in ftrength to that of the former: but on the whole, there was not the left fpecific difference between the two fubjects.

Trouts are moft voracious fifh, and afford excellent diverfion to the angler: the paffion for the fport of angling is fo great in the neighborhood of *London*, that the liberty of fifhing in fome of the ftreams in the adjacent counties, is purchafed at the rate of ten pounds per annum.

Thefe fifh fhift their quarters to fpawn, and, like falmon, make up towards the heads of rivers to depofit their roes. The under jaw of the trout is fubject, at certain times, to the fame curvature as that of the falmon.

A trout taken in *Llynallet*, in *Denbighfhire*, DESCRIP. which is famous for an excellent kind, meafured feventeen inches, its depth three and three quarters, its weight one pound ten ounces: the head thick; the nofe rather fharp: the upper jaw a little longer than the lower; both jaws, as well as the head, were of a pale brown, blotched with black: the teeth fharp and ftrong, difpofed in the jaws, roof of the mouth and tongue, as is the cafe with the whole genus, except the *Gwyniad*, which is toothlefs, and the *Grayling*, which has none on its tongue.

The

The back was dufky; the fides tinged with a purplifh bloom, marked with deep purple fpots, mixed with black, above and below the fide line which was ftrait: the belly white.

The firft dorfal fin was fpotted; the fpurious fin brown, tipped with red; the pectoral, ventral, and anal fins, of a pale brown; the edges of the anal fin white: the tail very little forked when extended.

147. WHITE. THIS fpecies migrates out of the fea into the river *Efk* in *Cumberland* from *July* to *September*, and is called from its color the *Whiting*. When dreffed, their flefh is red, and moft delicious eating. They have, on their firft appearance from the falt water, the *lernæa falmonea*, or falmon loufe, adhering to them. They have both melt and fpawn; but no fry has as yet been obferved. This is the fifh called by the *Scots*, *Phinocs*.

They never exceed a foot in length. The upper jaw is a little longer than the lower: in the firft are two rows of teeth; in the laft, one: on the tongue are fix teeth.

The back is ftrait: the whole body of an elegant form: the lateral line is ftrait; color, between that and the top of the back, dufky and filvery intermixed; beneath the line of an exquifite

fite filvery whitenefs : firft dorfal fin fpotted with black : tail black, and much forked.

The firft dorfal fin has eleven rays; pectoral, thirteen ; ventral, nine ; anal, nine.

Le Tacon ? *Belon*, 275.
Salmulus, *Herefordiæ* Samlet dictus. *Wil. Icth.* 192.
Salmulus, the Samlet *Herefordienfibus*, Branlin et Fin-gerin *Eboracenfibus. Raii fyn. pifc.* 63.
Salmoneta, a Branlin. *Ray's Letters*, 199.

148. S A M-LET,

THE famlet is the left of the trout kind, is frequent in the *Wye*, in the upper part of the *Severn*, and the rivers that run into it, in the north of *England*, and in *Wales*. It is by feveral imagined to be the fry of the falmon; but our reafons for diffenting from that opinion are thefe:

Firft, It is well known that the falmon fry never continue in frefh water the whole year ; but as numerous as they appear on their firft efcape from the fpawn, all vanifh on the firft vernal flood that happens, which fweeps them into the fea, and leaves fcarce one behind.

Secondly, The growth of the falmon fry is fo quick and fo confiderable, as fuddenly to exceed the bulk of the largeft famlet : for example, the fry that have quitted the frefh water in the fpring, not larger than gudgeons, return into it again a foot or more in length.

Thirdly,

Thirdly, The falmon attain a confiderable bulk before they begin to breed: the famlets, on the contrary, are found male and female*, (diftinguifhed by the milt and roe) of their common fize.

Fourthly, They are found in the frefh waters in all times of the year, and even at feafons when the falmon fry have gained a confiderable fize. It is well known, that near *Shrewfbury* (where the are called *Samfons*) they are found in fuch quantities in the month of *September*, that a fkilful angler, in a coracle, will take with a fly from twelve to fixteen dozen in a day.

They fpawn in *November* and *December*, at which time thofe of the *Severn* pufh up towards the head of that fair river, quitting the leffer brooks, and return into them again when they have done.

They have a general refemblance to the trout, therefore muft be defcribed comparatively.

Firft, The head is proportionably narrower, and the mouth lefs than that of the trout.

Secondly, Their body is deeper.

Thirdly, They feldom exceed fix or feven inches in length: at moft, eight and a half.

Fourthly, The pectoral fins have generally but one large black fpot, though fometimes a fingle fmall one attends it; whereas the pectoral fins of the trout are more numeroufly marked.

Fifthly, The fpurious or fat fin on the back is

* It has been vulgarly imagined, that there were no other than males of this fpecies.

never

Pl. LX.

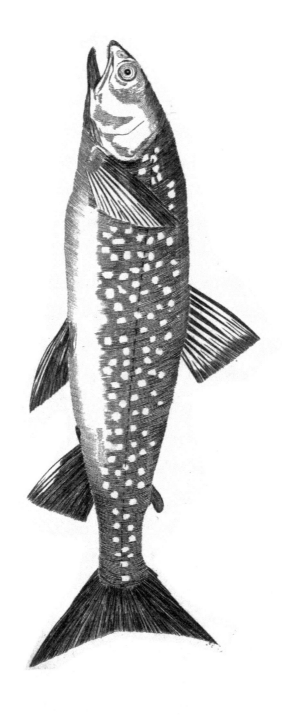

CHARR.

never tipped with red; nor is the edge of the anal fin white.

Sixthly, The fpots on the body are fewer, and not fo bright.

It is alfo marked from the back to the fides with fix or feven large bluifh bars; but this is not a certain character, as the fame is fometimes found in young trouts.

Seventhly, The tail of the famlet is much more forked than that of the trout.

Thefe fifh are very frequent in the rivers of *Scotland*, where they are called *Pars*. They are alfo common in the *Wye*, where they are known by the name of *Skirlings*, or *Lafprings*.

L'Omble, ou Humble. *Belon*, 281.
Umbla feu Humble *Belonii Gefner pifc*. 1005.
Umbla minor. *Gefner pifc*. 1013.
Torgoch *Wallis*. *Weftmorlandis* Red Charre *Lacus Winander mere*. *Wil. Icth*. 196. *Raii fyn. pifc*. 65.

Salmo vix pedalis, pinnis ventralibus rubris, maxilla inferiore longiore. *Arted. fyn*. 25.
Salmo alpinus. *Lin. fyft*. 516. *Gronov. Zooph. No*. 372.
Roding, *Lapponibus* Raud. *Faun. Suec. No*. 124.
Charr-fifh. *Phil. Tranf*. 1755. 210.

149. CHARR.

THE charr is an inhabitant of the lakes of the north, and of thofe of the mountanous parts of *Europe*. It affects clear and pure waters, and is very rarely known to wander into running

VOL. III.　　　　X　　　　　　ftreams,

ftreams, except into fuch whofe bottom is fimilar
to the neighboring lake.

It is found in vaft abundance in the cold lakes
on the fummits of the *Lapland Alps*, and is almoft
the only fifh that is met with in any plenty in
thofe regions; where it would be wonderful how
they fubfifted, had not Providence fupplied them
with innumerable *larvæ* of the *Gnat* kind*: thefe
are food to the fifh, who in their turn are a fupport to the migratory *Laplanders* in their fummer
voyages to the diftant lakes.

In fuch excurfions thofe vacant people find a
luxurious and ready repaft in thefe fifh, which
they drefs and eat without the addition † of fauces;
for exercife and temperance render ufelefs the inventions of epicurifm.

---

* A pupil of *Linnæus* remarks in the fourth volume of the
Amæn. Acad. p. 156, that the fame infects which are fuch a
peft to the rein deer, afford fuftenance to the fifh of the vaft
lakes and rivers of *Lapland*. But at the fame time that we
wonder at *Linnæus's* inattention to the food of the birds and
fifh of that country, which abound even to a noxious degree,
we muft, in juftice to that Gentleman, acknowledge an over-
fight of our own in the fecond volume of the *Britifh Zoology*,
p. 522, edition the fecond, where we give the *Lapland* waters
only one fpecies of water plant; for on a more careful review
of that elaborate performance, the *Flora Lapponica*, we difcover three other fpecies, viz. *Scirpus, No.* 18, *Alopecurus,
No.* 38, *Ranunculus, No.* 234; but thofe fo thinly fcattered over
the *Lapland* lakes, as ftill to vindicate our affertion, as to the
fcarcenefs of plants in the waters of alpine countries.

† *Arted. Sp. pifc.* 52.

                                              There

There are but few lakes in our ifland that produce this fifh, and even thofe not in any abundance. It is found in *Winander Mere* in *Weftmorland*; in *Llyn Quellyn*, near the foot of *Snowdon*; and before the difcovery of the copper-mines, in thofe of *Llynberris*, but the mineral ftreams have entirely deftroyed the fifh in the laft lakes *. Whether the waters in *Ireland* afford the charr, we are uncertain, but imagine not, except it has been overlooked by their writers on the natural hiftory of that kingdom. In *Scotland* it is found in *Loch Inch*, and other neighboring lakes, and is faid to go into the *Spey* to fpawn.

The largeft and moft beautiful we ever received were taken in *Winander Mere*, and were communicated to us by the Rev. Mr. *Farrifh* of *Carlifle*, with an account of their natural hiftory. He favored me with five fpecimens, two under the name of the *Cafe Charr*, male and female ; another he called the *Gelt Charr*, i. e. a charr which had not fpawned the preceding feafon, and on that account is reckoned to be in the greateft perfection. The two others were infcribed, the *Red Charr*, the *Silver* or *Gilt Charr*, the *Carpio Lacus* Benaci, *Raii fyn. pifc.* 66, which laft are in *Weftmorland* diftinguifhed by the epithet *red*, by reafon of the flefh affuming a higher color than the other when dreffed.

---

* They are alfo found in certain lakes in *Merionethfhire*.

X 2   On

VARIETIES.

On the clofeft examination, we could not difcover any fpecific differences in thefe fpecimens, therefore muft defcribe them as the fame fifh, fubject only to a flight variation in their form, hereafter to be noted. But there is in another refpect an effential difference, we mean in their œconomy, which is in all beings invariable; the particulars we fhall deliver in the very words of our obliging informant.

SPAWNING OF THE CASE CHARR.

The *Umbla minor*, or cafe charr, fpawns about *Michaelmas*, and chiefly in the river *Brathy*, which uniting with another called the *Rowthay*, about a quarter of a mile above the lake, they both fall into it together. The *Brathy* has a black rocky bottom; the bottom of the *Rowthay* is a bright fand, and into this the charr are never obferved to enter. Some of them however fpawn in the lake, but always in fuch parts of it which are ftony, and refemble the channel of the *Brathy*. They are fuppofed to be in the higheft perfection about *May*, and continue fo all the fummer, yet are rarely caught after *April*. When they are fpawning in the river they will take a bait, but at no other time, being commonly taken, as well as the other fpecies, in what they call *breaft nets*, which are in length about twenty-four fathoms, and about five, where broadeft.

GILT CHARR.

The feafon which the other fpecies fpawns in is from the beginning of *January* to the end of *March*. They are never known to afcend the rivers,

rivers, but always in thofe parts of the lake which are fpringy, where the bottom is fmooth and fandy, and the water warmeft. The fifhermen judge of this warmth, by obferving that the water feldom freezes in the places where they fpawn, except in intenfe frofts, and then the ice is thinner than in other parts of the lake. They are taken in great-eft plenty from the end of *September* to the end of *November* : at other times they are hardly to be met with. This fpecies is much more efteemed for the table than the other, and is very delicate when potted.

We muft obferve, that this account of the fpawn-ing feafon of the *Weftmorland* charrs, agrees very nearly with that of thofe of *Wales*, the laft appear-ing about a month later, keep moving from fide to fide of the pool, and then retire into the deep water, where they are fometimes but rarely taken.

. This remarkable circumftance of the different feafon of fpawning in fifh, apparently the fame (for the red charr of *Winander*, is certainly not the *Carpio Lacus* Benaci) puzzles us greatly, and makes us wifh that the curious, who border on that lake, would pay farther attention to the na-tural hiftory of thefe fifh, and favor us with fome further lights on the fubject.

We fhall now defcribe the varieties by the names afcribed to them in the north.

The length of the red charr to the divifion in its Red Charr. tail, was twelve inches ; its biggeft circumference

X 3   almoft

almoſt ſeven.　The firſt dorſal fin five inches and three quarters from the tip of its noſe, and conſiſted of twelve branched rays: the firſt of which was ſhort, the fifth the longeſt: the fat fin was very ſmall.

Each of the five fiſh had double noſtrils, and ſmall teeth in the jaws, roof of the mouth, and on the tongue.

The head, back, dorſal fin, and tail of each, was of a duſky blue; the ſides rather paler, marked with numbers of bright red ſpots: the bellies of the *Red Charr* were of a full and rich red; thoſe of the *Caſe Charr* rather paler; from this particular the *Welch* call theſe fiſh *Torgoch*, or red belly.

The firſt rays of the anal and ventral fins of each, were of a pure white; the reſt of each fin on the lower part of the body, tinged with red.

The lateral line ſtrait, dividing the fiſh in two equal parts, or nearly ſo.

The jaws of the *Caſe Charr* are perfectly even; on the contrary, thoſe the *Red Charr* were unequal, the upper jaw being the broadeſt, and the teeth hung over the lower, as might be perceived on paſſing the finger over them.

The branchioſtegous rays were, on different ſides of the ſame fiſh, unequal in number, viz. 12,--11, 11,--10, 10--9, except in one, where they were 11,--11.

GELT
CHARR.

The *Gelt*, or *Barren Charr*, was rather more
　　　　　　　　　　　　　　　　ſlender

Pl. LXI.

GRAYLING.

SMELT.

flender than the others, as being without fpawn. The back of a gloffy dufky blue: the fides filvery, mixed with blue, fpotted with pale red: the fides of the belly were of a pale red, the bottom white.

The tails of each bifurcated.

The charrs we have feen, brought from *Snowdon* lakes, were rather fmaller than thofe of *Weftmorland*, their colors paler. The fuppofed males very much refemble the *Gelt Charr*; but that is not a certain diftinction of fex, for the Rev. Mr. *Farrington* *, has told me that the fifhermen do not make that diftinction.

Θυμαλλος *Ælian. de an. lib.* xiv. c. 22.
Umbra *Aufonii Mofella.* 90.
Thymalus, Thymus. *Salvian.* 81. *Belon,* 276.
Thymus, Umbra fluviatilis. *Rondel. fluv.* 187, 172. *Gefner pifc.* 132.
A Grayling, or Umber. *Wil.*

Icth. 187. *Raii fyn. pifc.* 62. Coregonus maxilla fuperiore longiore, pinna dorfi officulorum viginti trium. *Arted. fynon.* 20.
Salmo Thymallus. *Lin. fyft.* 512. *Gronov. Zooph. No.* 375. *Afch. Kram.* 390.

150. GRAYL-ING.

THE grayling haunts clear and rapid ftreams, and particularly fuch that flow through mountanous countries. It is found in the rivers of *Derbyfhire*; in fome of thofe of the north; in

* Who favored the Royal Society with a paper on the *Welch* charr. *Vide Phil. Tranf.* 1755.

X 4   the

the *Tame* near *Ludlow*; in the *Lug*, and other
ftreams near *Leominfter*; and in the river near
Chriftchurch, Hampfhire. It is alfo very common
in *Lapland*; the inhabitants make ufe of the guts
of this fifh inftead of rennet, to make the cheefe
which they get from the milk of the rein deer*.

It is a voracious fifh, rifes freely to the fly, and
will very eagerly take a bait. It is a very fwift
fwimmer, and difappears like the tranfient paffage
of a fhadow, from whence we believe is derived
the name of *Umbra.*

> *Effugienfque oculos celeri levis* Umbra *natatu* †.
> The *Umbra* fwift efcapes the quickeft eye.

Thymalus and *Thymus*, are names beftowed on it
on account of the imaginary fcent, compared by
fome to that of thyme; but we never could perceive any particular fmell.

Descrip.   It is a fifh of an elegant form; lefs deep than
that of a trout : the largeft we ever heard of was
taken near *Ludlow*, which was above half a yard
long, and weighed four pounds fix ounces, but
this was a very rare inftance.

The irides are filvery, tinged with yellow: the
teeth very minute, feated in the jaws and the roof
of the mouth, but none on the tongue: the head
is dufky; the covers of the gills of a gloffy green:

---

* *Flora Lap.* 109. *Amæn Acad.* IV. 159.
† Aufonii Mofel. 90.

the

the back and fides of a fine filvery grey, but when the fifh is juft taken, varied flightly with blue and gold: the fide-line is ftrait.

The fcales large, and the lower edges dufky, forming ftrait rows from head to tail.

The firft dorfal fin has twenty-one rays; the three or four firft are the fhorteft, the others almoft of equal lengths; this fin is fpotted, all the others are plain.

The tail is much forked.

Epelan de mer. *Belon*, 282.
Eperlanus. *Rondel. fuviat.*
  196. *Gefner pifc.* 362.
Spirincus et Stincus. *Gefner Paralip.* 29.
A Spyrling a Sprote. *Turner epift. ad. Gefn.*
Stindt, et Stinckfifch. *Schonevelde*, 70.
A Smelt. *Wil. Icth.* 202.

Raii fyn. pifc. 66.
Ofmerus radiis pinnæ ani feptendecim. *Arted. fynon.* 21.
Salmo eperlanus. S. capite diaphano, radiis pinnæ ani feptendecim. *Lin. fyft.* 511.
Gronov. Zooph. No.
Nors, Slom. *Faun. fuec. No.* 350.

151. SMELT.

THE fmelt inhabits the feas of the northern parts of *Europe*, and we believe never is found as far fouth as the *Mediterranean*: the *Seine* is one of the *French* rivers which receive it, but whether it is found fouth of that, we have not at prefent authority to fay. If we can depend on the obfervations of navigators, who generally have too much to think of to attend to the *minutiæ* of natural hiftory, thefe fifh are taken in the ftraits of *Magellan,*

Magellan*, and of a moft furprifing fize, fome meafuring twenty inches in length, and eight in circumference.

They inhabit the feas that wafh thefe iflands the whole year, and never go very remote from fhore, except when they afcend the rivers. It is remarked in certain rivers that they appear a long time before they fpawn, being taken in great abundance in *November*, *December*, and *January*, in the *Thames* and *Dee*, but in others not till *February*, and in *March* and *April* they fpawn; after which† they all return to the falt water, and are not feen in the rivers till the next feafon. It has been obferved, that they never come into the *Merfey* as long as there is any fnow water in the river.

Thefe fifh vary greatly in fize, but the largeft we ever heard of was thirteen inches long, and weighed half a pound.

They have a very particular fcent, from whence is derived one of their *Englifh* names *Smelt*, i. e. fmell it. That of *Sparling*, which is ufed in *Wales* and the north of *England*, is taken from the *French* *Eperlan*. There is a wonderful difagreement in the opinion of people in refpect to the fcent of this fifh; fome affert it flavors of the violet; the *Ger-*

---

* *Narborough's Voy.* 123.

† In the river *Conway*, near *Llanrwft*, and in the *Merfey* they never continue above three or four weeks.

mans

mans, for a very different reafon, diftinguifh it by the elegant title of *Stinckfifch* *.

Smelts are often fold in the ftreets of *London* fplit and dried. They are called dried *Sparlings*, and are recommended as a relifh to a glafs of wine in the morning.

It is a fifh of a very beautiful form and colour: the head is tranfparent, and the fkin in general fo thin, that with a good microfcope the blood may be obferved to circulate.

The irides are filvery: the pupil of a full black: the under jaw is the longeft: in the front of the upper jaw are four large teeth; thofe in the fides of both are fmall; in the roof of the mouth are two rows of teeth; on the tongue two others of large teeth.

Descrip.

The firft dorfal fin has eleven rays; the pectoral fins the fame number; the ventral eight; the anal fourteen.

The fcales are fmall, and readily drop off: the tail confifts of nineteen rays, and is forked.

The color of the back is whitifh, with a caft of green, beneath which it is varied with blue, and then fucceeds a beautiful glofs of a filvery hue.

---

* And not without reafon, if we may depend on *Linnæus*, who fays there are in the *Baltic* two varieties, the one, which is called *Nors*, *fœtidiffimus*, *ftercoris inftar*, which in the early fpring, when the peafants come to buy it, fills all the ftreets of *Upfal* with the fmell. He adds, that at this feafon agues reign there. *Faun. fuec. p.* 125.

Without

** Without Teeth.

152. GWI-
NIAD.

Le Lavaret. *Belon*, 278.

Lavaretus; Piscis *Lemani* lacus *Bezola* vulgo nuncupatus. Alius Piscis proprius *Lemani* lacus. *Rondel. fluviat.* 162, 163, 164. *Gesner pisc.* 29, 30, 31.

Albula nobilis, Snepel, Helte? *Schonevelde*, 12.

Vandesius et Gevandesius. *Sib. Scot.* 26.

Guiniad *Wallis* piscis lacus *Balensis*, *Ferræ* (ut puto) idem. *Wil. Icth.* 183. *Raii syn. pisc.* 61.

Lavaretus *Allobrogum*, Schelley *Cumberlandis*. *Wil. Icth.*

183. *Raii syn. pisc.* 61.

Albula cærulea. *Scheuchzer it. Alp.* II. 481.

Coregonus maxilla superiore longiore plana, pinna dorsi osficulorum 14. *Arted. synon.* 19.

Salmo Lavaretus. *Lin. syst.* 512.

Sijk, Stor-sijk. *Faun. Suec. No.* 352.

Gwiniad. *Phil. Transf.* 1767. 211.

Adelfisch, Gangfisch, Weiss-fisch, Weisser Blauling, Schnapel. *Wulff Boruss.* 37.

Reinankl. *Kram.* 389.

THIS fish is an inhabitant of several of the lakes of the *Alpine* parts of *Europe*. It is found in those of *Switzerland*, *Savoy*, and *Italy*; of *Norway*, *Sueden*, *Lapland**, and *Scotland*; in

* *Schæffer*, in his history of *Lapland*, p. 140. says, that these fish are caught there of the weight of ten or twelve pounds. We wish *Linnæus* had executed his intention of favoring the world with his *Lachesis Lapponica*, in which he promised a complete history of that country. I once reminded him of it, and it is with true regret, that I give his answer: *Nunc nimis seró inciperem*,

Me quoque debilitat series immensa laborum,
Ante meum tempus cogor et esse senem :
Firma sit illa licet solvetur in æquore navis,
Quæ nunquam liquidis sicca carebit aquis.

those

Pl. LXII.

GWINIAD.

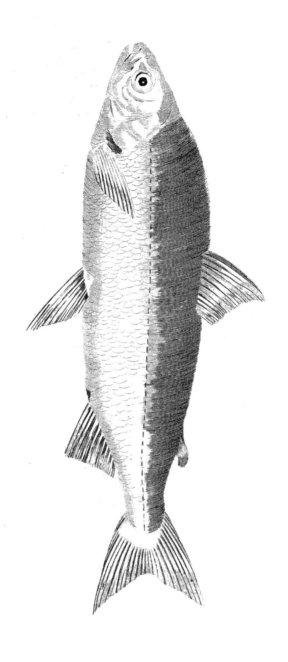

P Mazell Sculp

thofe of *Ireland*, and of *Cumberland*; and in *Wales*, in that of *Llyntegid*, near *Bala*, in *Merionethſhire*.

It is the fame with the *Ferra* of the lake of *Geneva*, the *Schelly* *, of *Hulſe-water*, the *Pollen* of *Lough Neagh*, and the *Vangis* and *Juvangis* of *Loch Mabon*. The *Scotch* have a tradition that it was firſt introduced there by the beauteous queen, their unhappy *Mary Stuart*; and as in her time the *Scotch* court was much frenchified, it feems likely that the name was derived from the *French*, *vendoiſe*, a dace; to which a flight obſerver might be tempted to compare it from the whiteneſs of its ſcales. The *Britiſh* name *Gwiniad*, or whiting, was beftowed on it for the fame reafon.

It is a gregarious fiſh, and approaches the ſhores in vaſt ſhoals in fpring and in fummer, which prove in many places a bleffed relief to the poor of inland countries, in the fame degree as the annual return of the herring is to thofe who inhabit the coaſts. The Rev. Mr. *Farriſh*, of *Carliſle*, wrote me word, that he was affured by a *Hulſe-water* fiſherman, that laft fummer he took between feven and eight thoufand at one draught. I muft not paſs by that gentleman without acknowledging my obligations to him for an account of the *Charrs* and the *Schelly*; he being one of the valuable embelliſhers of this work, for whom I am indebted to the friendſhip of his late worthy prelate.

---

* The inhabitants of *Cumberland* give this name alfo to the chub, from its being a fcaly fiſh.

The

The *Gwiniad* is a fish of an infipid tafte, and muft be eaten foon, for it will not keep long; thofe that choofe to preferve them do it with falt. They die very foon after they are taken. Their fpawning feafon in *Llyntegid* is in *December*.

It has long ago been obferved in *Cambden* *, that thefe fifh never wander into the *Dee*, nor the falmon never ventures into the lake: this muft be allowed to be generally the cafe; but by accident the firft have been known to ftray as far as *Llandrillo*, fix miles down the river, and a falmon has now and then been found trefpafling in the lake †.

The largeft *Gwiniad* we ever heard of weighed between three and four pounds: we have a *Ferra* we brought with us out of *Switzerland*, that is fifteen inches long; but thefe are uncommon fizes: the fifh which we defcribe was eleven inches long, its greateft depth three.

The head fmall, fmooth, and of a dufky hue: the eyes very large: the pupil of a deep blue: the nofe blunt at the end: the jaws of equal length: the mouth fmall and toothlefs: the branchioftegous rays nine: the covers of the gills filvery, powdered with black.

The back is a little arched, and flightly carinated: the color, as far as the lateral line, gloffed with deep blue and purple, but towards the lines affumes

* *Vol.* II. 790.
† *Hon.* D. Barrington's *Letter to Dr.* Watfon. Phil. Tranf. 1767.

a filvery

a filvery caft, tinged with gold, beneath which thofe colors entirely prevale.

The fide line is quite ftrait, and confifts of a feries of diftinct fpots of a dufky hue : the belly is a little prominent, and quite flat on the bottom.

The firft dorfal fin is placed almoft in the middle, and confifts of fourteen branched rays ; the fecond is thin, tranfparent, and not diftant from the tail.

The pectoral fins had eighteen rays, the firft the longeft, the others gradually fhortening ; the ventral fins were compofed of twelve, and the anal of fifteen, all branched at their ends ; the ventral fins in fome are of a fine fky blue, in others as if powdered with blue fpecks ; the ends of the other lower fins are tinged with the fame color.

The tail is very much forked : the fcales large, and adhere clofe to the body.

Upper

XXXIV.
PIKE.

Upper jaw fhorter than the lower.

Body long, flender, compreffed fideways.

One dorfal fin placed near the tail.

153. PIKE.

Lucius. *Aufonii Mofella,* 122.
Luccio. *Salvian.* 94.
Le Brochet. *Belon,* 292. *Itin.* 104.
Lucius. *Rondel. fluviat.* 188. *Gefner pifc.* 500.
Heket, Hecht. *Schonevelde,* 44.

Pike, or Pickerel. *Wil. Icth.* 236. *Raii fyn. pifc.* 112.
Efox roftro plagioplateo. *Art. fynon.* 26.
Efox Lucius. *Lin. fyft.* 516. *Gronov Zooph. No.* 361.
Gjadda. *Faun. Suec. No.* 355.
Hecht. *Kram.* 388.

THE pike is common in moft of the lakes of *Europe,* but the largeft are thofe taken in *Lapland,* which, according to *Schæffer,* are fometimes eight feet long. They are taken there in great abundance, dried, and exported for fale. The largeft fifh of this kind which we ever heard of in *England,* weighed thirty-five pounds.

According to the common faying, thefe fifh were introduced into *England* in the reign of *Henry* VIII. in 1537. They were fo rare, that a pike was fold for double the price of a houfe-lamb in *February,* and a pickerel for more than a fat capon. How far this may be depended on, I cannot fay, for this fifh is mentioned in the *Boke* of *St. Albons,* printed in the year 1496, and is not there fpoke of as a fcarce fifh, as was then the cafe with refpect to the carp. Great numbers of this fifh were dreffed in the year 1466, at the great feaft given by *George Nevil,* Arch-bifhop of *York.*

All writers who treat of this fpecies bring inftances of its vaft voracioufnefs. We have known one that was choaked by attempting to fwallow

one

PIKE.

SEA PIKE.

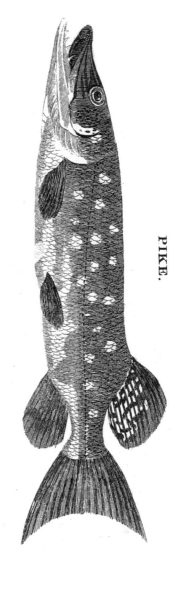

one of its own fpecies that proved too large a mor-
fel.   Yet its jaws are very loofely connected; and
have on each fide an additional bone like the jaw
of a viper, which renders them capable of greater
diftenfion when it fwallows its prey.   It does not
confine itfelf to feed on fifh and frogs; it will
devour the water rat, and draw down the young
ducks as they are fwimming about.   In a manu-
fcript note which we found, p. 244, of our copy
of *Plott's* Hiftory of *Staffordfhire*, is the following
extraordinary fact: " At Lord *Gower's* canal at
" *Trentham*, a pike feized the head of a fwan as
" fhe was feeding under water, and gorged fo
" much of it as killed them both.   The fervants
" perceiving the fwan with its head under water
" for a longer time than ufual, took the boat,
" and found both fwan and pike dead *."

But there are inftances of its fiercenefs ftill more
furprizing, and which indeed border a little on the
marvellous.   *Gefner* † relates, that a famifhed pike
in the *Rhone* feized on the lips of a mule that was
brought to water, and that the beaft drew the fifh
out before it could difengage itfelf.   That people
have been bit by thefe voracious creatures while
they were wafhing their legs, and that they will

---

* This note we afterwards difcovered was wrote by Mr,
Plott, of *Oxford*, who affured me he inferted it on good authority.

† *Gefner pifc.* 503.

even contend with the otter for its prey, and endeavour to force it out of its mouth *.

Small fish shew the same uneasiness and detestation at the presence of this tyrant, as the little birds do at the sight of the hawk or owl. When the pike lies dormant near the surface (as is frequently the case) the lesser fish are often observed to swim around it in vast numbers, and in great anxiety. Pike are often haltered in a noose, and taken while they lie thus asleep, as they are often found in the ditches near the *Thames* in the month of *May*.

In the shallow water of the *Lincolnshire* fens they are frequently taken in a manner peculiar, we believe, to that county, and the isle of *Ceylon* †. The fishermen make use of what is called a crownnet, which is no more than a hemispherical basket, open at top and bottom. He stands at the end of one of the little fenboats, and frequently puts his basket down to the bottom of the water, then poking a stick into it, discovers whether he has any booty by the striking of the fish; and vast numbers of pike are taken in this manner.

LONGEVITY. The longevity of this fish is very remarkable, if we may credit the accounts given of it. *Rzaczynski* ‡ tells us of one that was ninety years old;

* *Walton.* 157.
† *Knox's Hist. Ceylon*, 28.
‡ *Hist. Nat. Poloniæ*, 152.

but

but *Gefner* * relates, that in the year 1497, a pike
was taken near *Hailbrun*, in *Suabia*, with a brazen
ring affixed to it, on which were thefe words in
Greek characters : *I am the fifh which was firft of all*
put into this lake by the hands of the governor of the
univerfe, Frederick *the Second, the 5th of* October,
1230 : fo that the former muft have been an infant
to this *Methufalem* of a fifh.

Pikes fpawn in *March* or *April*, according to
the coldnefs or warmth of the weather. When
they are in high feafon their colors are very fine,
being green, fpotted with bright yellow ; and the
gills are of a moft vivid and full red. When out
of feafon, the green changes to grey, and the yellow fpots turn pale.

The head is very flat ; the upper jaw broad, and   Descrip.
is fhorter than the lower : the under jaw turns up a
little at the end, and is marked with minute punctures.

The teeth are very fharp, difpofed only in the
front of the upper jaw, but in both fides of the
lower, in the roof of the mouth, and often the
tongue. The flit of the mouth, or the gape, is
very wide ; the eyes fmall.

The dorfal fin is placed very low on the back,
and confifts of twenty-one rays ; the pectoral of
fifteen ; the ventral of eleven ; the anal of eighteen.

The tail is bifurcated.

---

* *Icones pifcium,* 316, where a print of the ring is given.

Y 2   Βελόνη.

154. GAR.   Βελόνη, *Ariſt. Hiſt. an.* II. c. 15. &c.

Βελόνη, Ραφις? *Athenæus lib.* VII. 319.

Acus, five Belone *Plinii lib.* IX. c. 51.

Acuchia. *Salvian,* 68.

L'Aguille, ou Orphie. *Belon,* 161.

Acus prima ſpecies. *Rondel.* 227. *Geſner piſc.* 9.

Horn-fiſck. *Schonevelde,* 11.

Horn-fiſh, or Gar-fiſh. *Wil.*

Icth. 231. *Raii ſyn. piſc.* 109.

Eſox roſtro cuſpidato gracili ſubtereti, et ſpithamali. *Arted. ſynon.* 27.

Eſox Belone. E. roſtro utraque maxilla dentata. *Lin. ſyſt.* 517. *Gronov. Zooph.* No. 362.

Nabbgjadda, Horngiall. *Faun. Suec.* No. 156.

See-naadel, Sack-nadel. *Wulff Boruſſ.* No. 70.

THIS fiſh which is found in many places, is known by the name of the *Sea Needle*. It comes in ſhoals on our coaſts in the beginning of ſummer, and precedes the mackrel: it has a reſemblance to it in taſte, but the light green, which ſtains the back bone of this fiſh when boiled, gives many people a diſguſt to it.

DESCRIP.   The common ſea pike, or ſea needle, ſometimes grows to the length of three feet, or more.

The jaws are very long, ſlender, and ſharp pointed; the under extends much farther than the upper, and the edges of both are armed with numbers of ſhort ſlender teeth: the inſide of the mouth is purple: the tongue ſmall: the eyes large: the irides ſilvery: the noſtrils wide and round.

The body is ſlender: the belly quite flat, bounded on both ſides by a rough line.

The

SAURY.

Pl. LXIV.

W. Griffith pinxt.

P. Mazell sculp.

The pectoral fins confift of fourteen rays; the ventral fin fmall, and placed very remote from the head, confifts of feven rays, the firft fpiny.

The dorfal fin lies on the very loweft part of the back, confifts of fixteen rays; the firft are high, the others lower as they approach the tail; the anal fin is of the fame form, and placed oppofite to the other; and has twenty-one rays.   The tail is much forked.

The colors are extremely beautiful when the fifh is in the water : the back of a fine green, beneath that appears a rich changeable blue and purple: the fides and belly are of a fine filvery hue.

Saurus, *Rondel. pifc.* 232.   *fyn. pifc.* 169.   155. Saury.
Skipper, *Cornubienfium. Raii*   The Saury. *Tour Scotland* 1769.

THE length is eleven inches : the nofe flender : the jaws produced like thofe of the fea needle, but of equal length.   The upper mandible a little incurvated.   Their length one inch.

The eyes large; the body anguilliform : but towards the tail grows fuddenly fmaller, and tapers to a very inconfiderable girth.   On the lower part of the back is a fmall fin, and between it and the tail fix fpurious like thofe of the mackrel.   Correfpondent to thefe, below are the anal fin and fix fpurious.   The pectoral and ventral fins very fmall:

Y 3   the

the tail much forked. The back dufky : the belly bright and filvery.

Great numbers of thefe fifh were thrown afhore on the fands of *Leith,* near *Edinburgh,* after a great ftorm in *November* 1768. *Rondeletius* defcribes this fpecies among the fifh of the *Mediterranean*; but fpeaks of it as a rare kind.

Teeth

ARGENTINE

ATHERINE

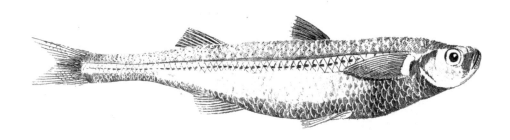

Teeth in the jaws and tongue.

Eight branchiostegous rays.

Vent near the tail.

The ventral fins compofed of many rays.

Sphyræna parva, five fphyrænæ fecunda fpecies. *Rondel.* 227. *Gefner pifc.* 883? Pifciculus *Romæ*, Argentina dictus. *Wil. Icth.* 229. *Raii fyn. pifc.* 108 ?

Argentina. *Arted. fynon.* 17. Argentina *Sphyræna. Lin. fyft.* 518. *Gronov. Zooph. No.* 349 ?

156. SHEPPY.

A LITTLE fifh, which I believe to be of this fpecies, was brought to me in 1769, taken in the fea near *Downing*.

The length was two inches one-fourth: the eyes large; and irides filvery. The lower jaw floped much: the teeth fmall.

The body compreffed, and of an equal depth almoft to the anal fin. The tail forked.

The back was of a dufky green: the fides and covers of the gills as if plated with filver. The lateral line was in the middle and quite ftrait.

On each fide of the belly was a row of circular punctures: above them another, which ceafed near the vent.

Mr. *Willughby* fays, that the outfide of the air bladder of this fifh confifts of a foliaceous filvery fkin, which was made ufe of in the manufacture of artificial pearl.

Y 4

The

The upper jaw a little flat.

Six branchioftegous rays.

A filvery ftripe along the fide.

157. ATHE-
RINE.

Epfetus? *Belon*, 209.
Εψητος, Atherina. *Rondel.*
215, 216. *Boffuet Epig.*
66, 67. *Gefner pifc.* 71,
72.
Pifciculus *Anguella Venetiis*
dictus; forte Hepfetus *Ron-*
deletii, vel *Atherina* ejuf-

dem. *Wil. Icth.* 209.
Raii fyn. pifc. 79.
Atherina. *Arted. fynon. App.*
116.
Atherina Hepfetus. A. pinna
ani radiis fere duodecim.
Lin. fyft. 519. *Gronov.*
Zooph. No. 399.

THIS fpecies is very common in the fea near *Scuthampton*, where it is called a *Smelt.* The higheft feafon is from *March* to the latter end of *May*, or beginning of *June*; in which month it fpawns. It never deferts the place; and is conftantly taken except in hard froft. It is alfo found on other coafts of our ifland.

The length is above four inches one-fourth. The back ftrait: the belly a little protuberant. On the back are two fins. I neglected to count the rays. The tail is much forked.

The fifh is femipellucid, covered with fcales: the color filvery, tinged with yellow: the fide line ftrait: beneath it is a row of fmall black fpots.

Body

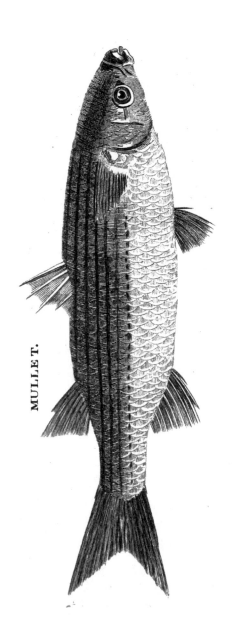

MULLET.

PARR.

Pl. LXVI.

Body and covers of the gills clothed with large
    fcales.

Six incurvated branchioftegous rays.

Teeth on the tongue and in the palate only.

Κεφαλος, Κεϛρεϛ. *Arift. Hift.*   Cephalus. *Rondel.* 260. *Gefner*

Κεϛρεϛ. *Oppian Halieut.* III.   Mullet. *Wil. Icth.* 274. *Raii*
98. *Athenæus lib.* VII. 306.   *fyn. pifc.* 84.

Mugil *Ovid Halieut.* 37. *Pli-*   Mugil. *Arted. fynon.* 52.
nii lib. IX. *c.* 8. 17.   Mugil cephalus. M. pinna

Cephalo. *Salvian,* 75.   dorfali anteriore quinque

Le Mulet. *Belon,* 205.   radiate. *Lin. fyft.* 520.
                         Gronov. Zooph. No. 397.

THE mullet is juftly ranked by *Ariftotle* among the *Pifces Littorales,* or thofe that prefer the fhores to the full fea: they are found in great plenty on feveral of the fandy coafts of our ifland, and haunt in particular thofe fmall bays that have influxes of frefh water. They come in great fhoals, and keep rooting like hogs in the fand or mud, leaving their traces in form of large round holes. They are very cunning, and when furrounded with a net, the whole fhoal frequently efcapes by leaping over it, for when one takes the lead, the others are fure to follow: this circum-ftance is taken notice of by *Oppian*; whether the latter part of his obfervation is true, is what we are uncertain.

                            Κεϛρεϛ

Κεςρευς μεν πλεκτῆσιν ἐν ἀγκοίνῃσι λίνοιο,
ἙλκόμενΘ- δόλον ἔτι περίδρομον ἡγνοίησεν.
Ὑ῎ψι δ᾽ ἀναθρώσκει λελιημένος ὕδατος ἄκρη,
Ὀρθὸς ἄνω σπεύδων ὅσσον σθένος ἅλματι κέφω
Ὀρμῆσαι· βαλῆς δὲ σαόφρονος ἐκ ἑμάτησε.
Πολλάκι γὰρ ριπῆσι καὶ ὕατα πείσματα φελλῶν
Ρηιδίως ὑπερᾶλτο, καὶ ἐξήλυξε μόροιο.
Εἰ δ᾽ ὅγ᾽ ἀνορμηθεὶς πρῶτον ςόλον, αὖτις ὀλισθῆ
Ἑς βρόχον, ἐκ ἔτ᾽ ἔπειτα βιάζεται, ᾿δ᾽ ἀνορύει,
Ἀχνύμενος· πείρη δὲ μαθὼν ἀποπαύεται ὁρμῆς.

The *Mullet**, when encircling feines inclofe,
The fatal threads and treach'rous bofom knows.
Inftant he rallies all his vig'rous powers,
And faithful aid of every nerve implores;
O'er battlements of cork up-darted flies,
And finds from air th' efcape that fea denies.
But fhould the firft attempt his hopes deceive,
And fatal fpace th' imprifon'd fall receive,
Exhaufted ftrength no fecond leap fupplies;
Self-doom'd to death the proftrate victim lies,
Refign'd with painful expectation waits,
'Till thinner elements compleat his fates.　　Jones.

Oppian had good opportunity of examining thefe fifh, for they fwarm during fome feafons on the coafts of the *Mediterranean*. Near *Martegues*, in the fouth of *France*, abundance of mullets are taken in weres made of reeds placed in the fhallows. Of the milts of the males, which are there called

---

* Mr. *Jones*, by miftake, tranflates it the *Barbel*.

Alletants,

Alletants, and of the roes of the females, which are called *Botar*, is made *Botargo*. The materials are taken out entire, covered with falt for four or five hours, then preffed a little between two boards or ftones, wafhed, and at laft dried in the fun for thirteen or fourteen days *.

This fifh was fometimes made the inftrument of a horrible punifhment for unfortunate gallants. It was in ufe both at *Athens* † and at *Rome*; but we doubt much whether it was a legal one: for we rather fufpect it was inflicted inftantaneoufly by the injured and enraged hufband, at a feafon when

> *Furor arma miniftrat.*

Juvenal feems to fpeak of it in that light as well as *Horace :* the former, relating the revenge taken by the exafperated fpoufe, defcribes it as very various ;

> *Necat hic ferro, fecat ille cruentis*
> *Verberibus, quofdam mæchos et* Mugilis *intrat* ‡.

The paffage in *Horace* feems not to have been attended to by the critics ; but when he mentions

---

* Mr. *Willughby*'s notes during his travels. *Vide Harris's Col. Voy.* II. 721.

† *Legibus* Athenienfium *adulteri* εν ἔργω *deprehenfi pæna fuit* ῥαφανόδωσις. *Raphani loco utebantur nonnunquam mugile pifce, interdum fcorpione.* Caufauboni *animadvers. in* Athenæum, *lib.* I.

‡ *Satyr.* X. 316

the

the diftreffes that the invader of another's bed underwent, he moft certainly alludes to this penalty:

Difcinƈtâ tunicâ fugiendum eft, ac pede nudo;
Ne nummi pereant, aut Pyga, *aut denique fama* ***.

The mullet is an excellent fifh for the table,
but at prefent not a fafhionable one.

The head is almoft fquare, and is flat on the
top: the nofe blunt: lips thick. It has no teeth,
only in the upper lip is a fmall roughnefs: between
the eyes and the mouth is a hard callus.

The pupil of the eye is black, encircled with a
fmall filvery line: the upper part of the iris is
hazel; the lower filvery.

The form of the body is pretty thick, but the
back not greatly elevated. The fcales are large
and deciduous.

The firft dorfal fin is placed near the middle of
the back, and confifts of four ftrong fpines; the
fecond of nine foft branching rays; the pectoral
has fixteen, the ventral fix; the firft a ftrong
fpine, the others foft.

The tail is much forked.

The color of the back is dufky, varied with
blue and green: the fides filvery, marked with
broad dufky parallel lines, reaching from head to
tail: the belly is filvery.

* *Satyr.* II. *lib.* I. 132.

Head

Pl. LXVII.

FLYING FISH.

N.º 159.

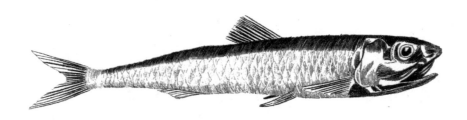

ANCHOVY. *N.º 163.*

Head covered with fcales.

Pectoral fins almoft as long as the body.

Hirundo *Plinii lib.* IX. *c.* 26.
*Εξοκοιτος και Ἀδωνις ? *Athe-
næus lib.* VIII. 332. *Op-
pian Halieut.* I. 157.
χελιδων ? *Oppian* II. 459.
Rondine. *Salvian,* 186.
Hirondelle de mer. *Belon,* 189.
Mugil alatus. *Rondel.* 267.

Gefner pifc. 553. *Wil. Icth.* 159.
233.
Exocætus. *Arted. fynon.* 18.
Exocætus volitans. E. abdo-
mine utrinque carinato.
Lin. fyft. 520. *Amæn. Acad.*
I. 603. *Gronov. Zooph.*
No. 359.

WE can produce but a fingle inftance of this fpecies † being taken on the *Britifh* coafts. In *June* 1765, one was caught at a fmall diftance below *Caermarthen,* in the river *Towy,* being brought up by the tide which flows as far as the town. It is a fifh frequent enough in the *Medi-terranean,* and alfo in the ocean, where it leads a moft miferable life. In its own element it is per-petually haraffed by the *Dorados,* and other fifh of prey. If it endeavors to avoid them by having re-courfe to the air, it either meets its fate from the

---

* *Pliny* mentions it under the fame name, *lib.* IX. *c.* 19.

† This fifh was feen by *John Strange,* Efq; at *Caermarthen,* who was fo obliging as to communicate to me the account of it.

Gulls,

Gulls, or the *Albatrofs,* or is forced down again into the mouth of the inhabitants of water, who below keep pace with its aerial excurfion. Neither is it unfrequent that whole fhoals of them fall on board of fhips that navigate the feas of warm climates : it is therefore apparent, that nature in this creature hath fupplied it with inftruments that frequently bring it into that deftruction it ftrives to avoid, by having recourfe to an element unnatural to it.

The antients were acquainted with this fpecies : *Pliny* mentions it under the name of *Hirundo,* and fpeaks of its flying faculty. It is probable that *Oppian* intended the fame by his Ωκειαι χελιδονες, or the *fwift fwallow* fifh. What *Athenæus* and the laft cited author mean by the Εξοκοιτος and Αδωνις, is not fo evident : they affert it quitted the water and flept on the rocks, from whence it tumbled with precipitation when difturbed by the unfriendly birds : on thefe accounts Icthyologifts feem to have made it fynonymous with the *flying fifh.*

DESCRIP. It refembles the herring in form of the body, but the back is flat : the fcales large and filvery : the dorfal fin is fmall, and placed near the tail : the pectoral fins, the inftruments of flight, are almoft as long as the body : the tail is bifurcated.

Eight

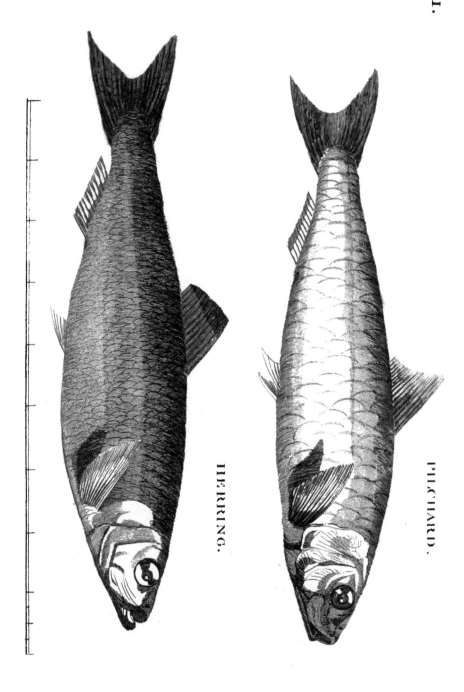

HERRING.

No. 160.

PILCHARD.

No. 161.

Eight branchioftegous rays.
The belly extremely fharp, and often ferrated.

Aringha ex *cimbricis* littori-
bus. *Jovius*, 143.
Hareng, efpece de Chalcis.
Belon, 169.
Harengus. *Rondel.* 222. *Gef-
ner pifc.* 410.
Heringk. *Schonevelde*, 37.
Herring. *Wil. Icth.* 219.
Raii fyn. pifc. 103.
Clupea maxilla inferiore lon-
giore maculis carens. *Arted.*

fynon. 14. α. β.
Clupea Harengus. Cl. im-
maculata, maxilla inferiore
longiore. *Lin. fyft.* 522.
Gronov. Zooph. No. 348.
Sill. *Faun. Suec. No.* 357. α.
Stromming. *Faun. Suec.
No.* 357. β.
Stromling*. *Wulff. Borufs.
No.* 50.

160. Bri-
tish

THE herring was unknown to the antients,
notwithftanding the words χαλκις and μαινις,
are by tranflators rendered *Halec*†, the characters
given of thofe fifh are common to fuch numbers
of different fpecies, as render it impoffible to fay
which they intended.

NAME.

Herrings are found from the higheft northern
latitudes yet known, as low as the northern coafts
of *France*; and excepting one inftance brought by

PLACE.

* The herring of the *Baltic*, in all refpects is like ours, but
fmaller.

† Which word, in fpite of all *lexicographers*, never fignified
any thing but the *garum* or pickle. *Vide p.* 221.

Dod,

Dod *, of a few being once taken in the Bay of *Tangier*, are never found more foutherly.

They are met with in vaft fhoals on the coaft of *America*, as low as *Carolina*. In *Chefapeak* Bay is an annual inundation of thofe fifh, which cover the fhores in fuch quantities as to become a nu-fance †. We find them again in the feas of *Kamtzchatka*, and poffibly they reach *Japan*; for *Kæmpfer* mentions, in his account of the fifh of that country, fome that are congenerous.

The great winter rendezvous of the herring is within the *Arctic* circle: there they continue for many months in order to recruit themfelves after the fatigue of fpawning, the feas within that fpace fwarming with infect food, in a degree far greater than in our warmer latitudes.

MIGRA-TIONS

This mighty army begins to put itfelf in motion in the fpring; we diftinguifh this vaft body by that name, for the word herring is derived from the *German*, *Heer*, an army, to exprefs their num-bers.

They begin to appear off the *Shetland* ifles in *April* and *May*; thefe are only forerunners of the grand fhoal which comes in *June*, and their appear-ance is marked by certain figns by the numbers of birds, fuch as gannets, and others which follow to prey on them: but when the main body ap-proaches, its breadth and its depth is fuch as to

* *Natural Hift. of the Herring, p.* 27.
† *Catefby Carol.* II. XXXIII.

alter

alter the appearance of the very ocean. It is divided into diftinct columns of five or fix miles in length, and three or four in breadth, and they drive the water before them with a kind of rippling: fometimes they fink for the fpace of ten or fifteen minutes, then rife again to the furface, and in bright weather reflect a variety of fplendid colors, like a field of the moft precious gems, in which, or rather in a much more valuable light, fhould this ftupendous gift of Providence be confidered by the inhabitants of the *Britifh* ifles.

The firft check this army meets in its march fouthward, is from the *Shetland* ifles, which divide it into two parts; one wing takes to the eaft, the other to the weftern fhores of *Great Britain*, and fill every bay and creek with their numbers; others pafs on towards *Yarmouth*, the great and antient mart of herrings; they then pafs through the *Britifh* channel, and after that in a manner difappear. Thofe which take to the weft, after offering themfelves to the *Hebrides*, where the great ftationary fifhery is, proceed towards the north of *Ireland*, where they meet with a fecond interruption, and are obliged to make a fecond divifion; the one takes to the weftern fide, and is fcarce perceived, being foon loft in the immenfity of the *Atlantic*; but the other, which paffes into the *Irifh* fea, rejoices and feeds the inhabitants of moft of the coafts that border on it.

Thefe brigades, as we may call them, which are

SEPARA-
TION.

VOL. III. Z thus

thus feparated from the greater columns, are often capricious in their motions, and do not fhew an invariable attachment to their haunts. We have had in our time inftances of their entirely quitting the coafts of *Cardiganfhire*, and vifiting thofe of *Caernarvonfhire* and *Flintfhire*, where they continued for a few years, but in the prefent year have quite deferted our fea, and returned to their old feats. The feafon of their appearance among us was very late, never before the latter end of *November*; their continuance till *February*.

PROVIDEN-
TIAL IN-
STINCT.

Were we inclined to confider this partial migration of the herring in a moral light, we might reflect with veneration and awe on the mighty Power which originally impreffed on this moft ufeful body of his creatures, the inftinct that directs and points out the courfe, that bleffes and enriches thefe iflands, which caufes them at certain and invariable times to quit the vaft polar deeps, and offer themfelves to our expecting fleets. That benevolent Being has never, from the earlieft records, been once known to withdraw this bleffing from the whole, though he often thinks proper to deny it to particulars; yet this partial failure (for which we fee no natural reafon) fhould fill us with the moft exalted and grateful fenfe of his Providence, for impreffing fo invariable and general inftinct on thefe fifh towards a fouthward migration, when the whole is to be benefited, and to withdraw it only when a minute part is to fuffer.

This

This inftinct was given them, that they might remove for the fake of depofiting their fpawn in warmer feas, that would mature and vivify it more affuredly than thofe of the frigid zone.   It is not from defect of food that they fet themfelves in motion, for they come to us full of fat, and on their return are almoft univerfally obferved to be lean and miferable.   What their food is near the pole, we are not yet informed; but in our feas they feed much on the *Onifcus Marinus*, a cruftaceous infect, and fometimes on their own fry.   The herring will rife to a fly.   Mr. *Low* of *Birfa* in the *Orknies* affures me, that he has caught many thoufands with a common trout fly, in a deep hole in a rivulet, into which the tide flows.   He commonly went at the fall of the tide.   They were young fifh, from fix to eight inches in length.

They are in full roe the end of *June*, and continue in perfection till the beginning of winter, when they begin to depofit their fpawn.   The young herrings begin to approach the fhores in *July* and *Auguft*, and are then from half an inch to two inches long : thofe in *Yorkfhire* are called *Herring Sile* *.   Though we have no particular authority for it, yet as very few young herrings are found in our feas during winter, it feems moft certain that they muft return to their parental haunts beneath

---

* The *Suedes* and *Danes* call the old herring *Sill*; but the people of *Slefwick*, from whence the *Anglo-Saxons* came, call the fry *Sylen*,

Z 2   the

the ice, to repair the vaft deſtruction of their race during ſummer, by men, fowl, and fiſh. Some of the old herrings continue on our coaſts the whole year: the *Scarborough* fiſhermen never put down their nets but they catch a few; but the numbers that remain are not worth mention in compariſon to the numbers that return.

Herrings vary greatly in ſize. Mr. *Travis* communicated to me the information of an experienced fiſher, who informed him that there is ſometimes taken near *Yarmouth*, a herring diſtinguiſhed by a black ſpot above the noſe; and that he once ſaw one that was twenty-one inches and an half long. He inſiſted that it was a different ſpecies, and varied as much from the common herring as that does from the pilchard. This we mention in order to incite ſome curious perſon on that coaſt to a farther enquiry.

The eye is very large: the edges of the upper jaw and the tongue are very rough, but the whole mouth is void of teeth: the gill covers are very looſe, and open very wide; which occaſions the almoſt inſtant death of the herring when taken out of the water, which is well known, even to a proverb.

The dorſal fin conſiſts of about ſeventeen rays, and is placed beyond the centre of gravity, ſo that when the fiſh is ſuſpended by it, the head immediately dips down: the two ventral fins have nine

rays;

rays; the pectoral seventeen; the anal fourteen: the tail is much forked.

The lateral line is not apparent, unless the scales are taken off: the sides are compressed: the belly sharply carinated, but the ridge quite smooth, and not in the left serrated.

The scales are large, thin, and fall off with a slight touch.

Color. The color of the back and sides green, varied with blue: the belly silvery.

Fishery. The herring fishery is of great antiquity: the industrious *Dutch* first engaged in it about the year 1164: they were in possession of it for several centuries, but at length its value became so justly to be known, that it gave rise to most obstinate and well-disputed wars between the *English* and them; but still their diligence and skill gives them a superiority over us in that branch of trade.

Our great stations are off the *Shetland* and *Western Isles*, and off the coast of *Norfolk*, in which the *Dutch* also share. *Yarmouth* has long been famous for its herring fair*; that town is obliged, by its charter, to send to the sheriffs of *Norwich* one hundred herrings, to be made into twenty-four pies, by them to be delivered to the lord of the manor of *East Carleton*, who is to convey them to the

---

* This fair was regulated by an act, commonly called the *Statute of Herrings*, in the 31st year of *Edward* III.

king

king *. The facetious Doctor *Fuller* † takes notice of the great repute the county of *Norfolk* was in for this fish, and, with his usual archness, calls a red herring, a *Norfolk Capon.*

In 1195, *Dunwich* in *Suffolk* accounted to the king for their yearly fee farm rent, £120, 1 mark, and 24000 herrings, 12000 for the monks of *Eye*, and 12000 for those of *Ely.*

The *Dutch* are most extravagantly fond of this fish when it is pickled. A premium is given to the first buss that arrives in *Holland* with a lading of this their *ambrosia*, and a vast price given for each keg. We have been in the country at that happy minute, and observed as much joy among the inhabitants on its arrival, as the *Ægyptians* shew on the first overflowing of the *Nile*. *Flanders* had the honor of inventing the art of pickling herrings. One *William Beukelen*, of *Biervlet*, near *Sluys*, hit on this useful expedient: from him was derived the name *pickle*, which we borrow from the *Dutch* and *German*. *Beukelen* died in 1397. The emperor *Charles* V. held his memory in such veneration for the service he did mankind, as to do his tomb the honor of a visit. It is very singular that most nations give the name of their favorite dish to the facetious attendant on every mountebank. Thus the *Dutch* call him Pickle

---

* *Cambden Britan.* I. 458.
† *British Worthies*, 238.

Herring;

Herring; the *Italians*, Macaroni; the *French*, Jean Pottage; the *Germans*, Hans Wurst*; and we dignify him with the title of Jack Pudding.

Pilchard. *Fuller's Brit. Wor-thies*, 194.	223. *Raii ſyn. piſc.* 104.	161. Pil-
Peltzer. *Schonevelde*, 40.	Clupea δ. *Arted. ſynon.* 16.	chard.
The Pilchard. *Wil. Icth.*	Pilchard. *Borlaſe Cornwall*,	
	272.	

THE pilchard appears in vaſt ſhoals off the *Corniſh* coaſts about the middle of *July*, diſappearing the beginning of winter, yet ſometimes a few return again after *Chriſtmas*. Their winter retreat is the ſame with that of the herring, and their motives for migrating the ſame. They affect, during ſummer, a warmer latitude, for they are not found in any quantities on any of our coaſts except thoſe of *Cornwall*, that is to ſay, from *Fowey* harbor to the *Scilly* iſles, between which places the ſhoals keep ſhifting for ſome weeks.

The approach of the pilchard is known by much the ſame ſigns as thoſe that indicate the arrival of the herring. Perſons, called in *Cornwall Huers*, are placed on the cliffs, to point to the boats ſtationed off the land the courſe of the fiſh. By the 1ſt of *James* I. c. 23, fiſhermen are empowered to

* That is, *Jack Sauſage*.

Z 4

go

go on the grounds of others to *bue*, without being liable to actions of trefpafs, which before occafioned frequent lawfuits.

The emoluments that accrue to the inhabitants of that county are great, and are beft expreffed in the words of Doctor *W. Borlafe*, in his account of the *Pilchard* fifhery.

" It employs a great number of men on the fea,
" training them thereby to naval affairs; employs
" men, women, and children, at land, in falting,
" preffing, wafhing, and cleaning, in making
" boats, nets, ropes, cafks, and all the trades de-
" pending on their conftruction and fale. The
" poor is fed with the offals of the captures, the
" land with the refufe of the fifh and falt, the mer-
" chant finds the gains of commiffion and honeft
" commerce, the fifherman the gains of the fifh.
" Ships are often freighted hither with falt, and
" into foreign countries with the fifh, carrying off
" at the fame time part of our tin. The ufual pro-
" duce of the number of hogfheads exported each
" year, for ten years, from 1747 to 1756 inclufive,
" from the four ports of *Fawy, Falmouth, Penzance,*
" and *St. Ives*, it appears that *Fawy* has exported
" yearly 1732 hogfheads; *Falmouth,* 14631 hogf-
" heads and two-thirds; *Penzance* and *Mounts-Bay,*
" 12149 hogfheads and one-third; *St. Ives*, 1282
" hogfheads: in all amounting to 29795 hogfheads.
" Every hogfhead for ten years laft paft, together
" with the bounty allowed for each hogfhead ex-
" ported,

" ported, and the oil made out of each hogfhead,
" has amounted, one year with another at an a-
" verage, to the price of one pound thirteen fhil-
" lings and three-pence; fo that the cafh paid for
" pilchards exported has, at a medium, annually
" amounted to the fum of forty-nine thoufand five
" hundred and thirty-two pounds ten fhillings."

The numbers that are taken at one fhooting out
of the nets, is amazingly great. Dr. *Borlafe* affured
me, that on the 5th of *October*, 1767, there were
at one time inclofed in *St. Ives's Bay* 7000 hogfheads, each hogfhead containing 35000 fifh, in all
245000000.

This fifh has a general likenefs to the herring, but
differs in fome particulars very effentially; we therefore defcribe it comparatively with the other, having one of each fpecies before us, both of them of
the fame length, *viz.* nine inches and an half.

DESCRIP.

The body of the pilchard is lefs compreffed than
that of the herring, being thicker and rounder: the
nofe is fhorter in proportion, and turns up: the
under jaw is fhorter.

The back is more elevated: the belly lefs fharp:
the dorfal fin of the pilchard is placed exactly in
the centre of gravity, fo that when taken up by
it, the body preferves an equilibrium, whereas that
of the herring dips at the head: the dorfal fin of
the pilchard we examined, being placed only three
inches eight tenths from the tip of the nofe; that
of the herring four inches one tenth.

The

The fcales of the pilchard adhere very clofely, whereas thofe of the herring very eafily drop off.

The pilchard is in general lefs than the herring; the fpecimen we defcribe being a very large one.

The pilchard is fatter, or more full of oil.

162. SPRAT. Spratti. *Wil. Icth* 221. *Raii.*
fyn. pifc. 105.
Clupea quadriuncialis, max- illa inferiore, longiore, ven- tre acutiffimo. *Arted. fynon.* 17.

Clupea Sprattus. Cl. pinna dorfali radiis tredecim. *Lin.*
fyft. 523.
Hwufsbuk. *Faun. Suec. No.* 358.

MR. *Willughby* and Mr. *Ray* were of opinion, that thefe fifh were the fry of the herring: we are induced to diffent from them, not only be- caufe on comparing a fprat and young herring of equal fize, we difcovered fome fpecific diffe- rences, but likewife for another reafon: the former vifit our coafts, and continue with us in fhoals in- numerable, when the others in general have retired to the great northern deeps.

They come into the river *Thames*, below bridge, the beginning of *November*, and leave it in *March*, and are, during their feafon, a great relief to the poor of the capital.

At *Gravefend*, and at *Yarmouth*, they are cured like red herrings; they are fometimes pickled, and are little inferior in flavor to the *Anchovy*, but the

bones

bones will not diffolve like thofe of the latter.
Mr. *Forfter* tells me, that in the *Baltic* they preferve them in the fame manner, and call them
Breitling, i. e. the little deep fifh, as being deeper
than the *Stromling*, or *Baltic* herring.

The fprat grows to about the length of five inches : the body is much deeper than that of a young
herring of equal length : the back fin is placed
more remote from the nofe than that of the herring,
and we think had fixteen rays.   But one great diftinction between this fifh, the herring and pilchard,
is the belly : that of the two firft being quite
fmooth, that of the laft moft ftrongly ferrated.
Another is, that the herring has fifty fix vertebrae;
this only forty eight.

DESCRIP.

Εγκρανλος ?   *Arift. Hift. an.*
   Lib. VI. *c.* 15.

Εγκρασίχολος ?   *Athenæus, Lib.*
   VII. *c.* 285.

L'Anchoy ?   *Belon*, 165.

Encraficholus ?   *Rondel.* 211.
   Gefner pifc. 68.

Lycoftomus, fehe mareneken ?
   Schonevelde, 46. *Tab.* 5.

Anchovy.   *Wil. Icth.* 225.
   Raii fyn. pifc. 107.

Clupea maxilla fuperiore longiore. *Arted. fynon.* 17.

Clupea encraficolus. *Lin. fyft.*
   523.

163. ANCHO-
VY.

THE true anchovies are taken in vaft quantities in the *Mediterranean*, and are brought
over here pickled. The great fifhery is at *Gorgona*,
a fmall ifle weft of *Leghorn*.

Mr. *Ray* difcovered this fpecies in the eftuary of
the

the *Dee* above a century ago*. Since that time no notice has been taken of it, till a few were taken near my houfe in 1769.

The length of the largeft was fix inches and an half: the body flender, but thicker in proportion than the herring.

The eyes were large: the irides white, with a caft of yellow: the under jaw much fhorter than the upper: the teeth fmall; a row in each jaw, and another on the middle of the tongue. The tongue doubly ciliated on both fides. The dorfal fin confifted of twelve rays, was tranfparent, and placed nearer the nofe than the tail.

The fcales large and deciduous: back green and femipellucid: fides and belly filvery and opake: edge of the belly fmooth: the tail forked.

164. SHAD. Θϱισσα? *Arift. Hift. an. lib.* IX. *c.* 37. *Strabo lib.* XV. 486. XVII. 566. *Athenæus, lib.* IV. 131. VII. 328. *Oppian Halieut.* I. 244.
Alaufa? *Aufonii Mofella,* 128.
Laccia, chiepa. *Salvian,* 104.
L'Alofe. *Belon,* 307.
Thriffa. *Rondel.* 220. *Gefner pifc.* 20.
Bayeke, Meyfifch. *Schonevelde,* 13.

Shad, or Mother of Herrings. *Wil. Icth.* 227. *Raii fyx. pifc.* 105.
Clupea apice maxilla fuperiore bifido, maculis nigris utrinque. *Arted. fynox.* 15.
Clupea alofa. Cl. lateribus nigro maculatis, roftro bifido. *Lin. fyft.* 523. *Gronov. Zooph. No.* 347.

NEITHER *Ariftotle, Athenæus,* nor *Oppian,* have defcribed their Θϱισσα with fuch pre-

* *Ray's* Letters, 47.

cifion,

SHAD.

N.º 164.

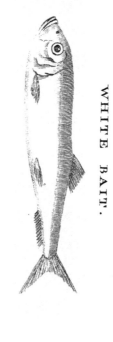

WHITE BAIT.

N.º 176.

cifion, as to induce us to tranflate it the *Shad*, without affixing to it our fceptic mark. *Aufonius* has been equally negligent in refpeft to his *Alaufa :* all he tells us is, that it was a very bad fifh :

Stridentefque focis obfonia plebis ALAUSAS.

Alaufæ crackling on the embers are
Of wretched poverty, th' infipid fare.

But commentators have agreed to render the Θρισσα of the firft, and the *Alaufa* of the laft, by the word *Shad*. Perhaps they were directed by the authority of *Strabo*, who mentions the Θρισσα the fuppofed *Shad*, and the Κιτρευς, or *Mullet*, as fifh that afcend the *Nile* at certain feafons, which, with the *Dolphin** of that river, he fays, are the only kinds that venture up from the fea for fear of the crocodile. That the two firft are fifh of paffage in the *Nile*, is confirmed to us by *Belonius* †, and by *Haffelquift* ‡. The laft fays it is found in the *Mediterranean* near *Smyrna*, and on the coaft of *Ægypt*, near *Rofetto*; and that in the months *December* and *January* it afcends the *Nile*, as high as *Cairo :* that it is ftuffed with pot marjoram, and

---

* This is the *Dolphin* of the *Nile*, a fifh now unknown to us. *Pliny lib.* VIII. *c.* 25. fays, it had a fharp fin on its back, with which it deftroyed the crocodile, by thrufting it into the belly of that animal, the only penetrable place.

† *Belon. Itin.* 98.

‡ *P.* 385. 388. *Suedifh* edition.

when

when dreffed in that manner will very nearly intox-
icate the eater.

In *Great Britain* the *Severn* affords this fifh in
higher perfection than any other river. It makes
its firft appearance there in *May*, but in very
warm feafons in *April*; for its arrival, fooner or
later, depends much on the temper of the air. It
continues in the river about two months, and then
is fucceeded by a variety which we fhall have occafion to mention hereafter.

The *Severn* fhad is efteemed a very delicate fifh
about the time of its firft appearance, efpecially in
that part of the river that flows by *Gloucefter*, where
they are taken in nets, and ufually fell dearer than
falmon: fome are fent to *London*, where the fifh-
mongers diftinguifh them from thofe of the *Thames*,
by the *French* name of *Alofe*.

Whether they fpawn in this river and the *Wye* is
not determined, for their fry has not yet been afcer-
tained. The old fifh come from the fea into the
river in full roe. In the months of *July* and *Auguft*,
multitudes of bleak frequent the river near *Gloucefter*; fome of them are as big as a fmall herring,
and thefe the fifhermen erroneoufly fufpect to be
the fry of the fhad. Numbers of thefe are taken
near *Gloucefter* in thofe months only, but none of
the emaciated fhad are ever caught in their return *.

* *Belon* alfo obferves, that none are taken in their return, *on
les prend en montant contre les rivieres, et jamais en defcendant.*

The

The *Thames* fhad does not frequent that river till
the latter end of *May* or beginning of *June*, and
is efteemed a very infipid coarfe fifh. The *Severn*
fhad is fometimes caught in the *Thames*, though
rarely, and called *Allis* (no doubt *Alofe*, the *French*
name) by the fifhermen, in that river. About the
fame time, and rather earlier, the variety called
near *Gloucefter* the *Twaite*, makes its appearance,
and is taken in great numbers in the *Severn*, and is
held in as great difrepute as the fhad of the *Thames*.
The differences between each variety are as follow:

The true *Shad* weighs fometimes eight pounds,
but their general fize is from four to five.

The *Twaite*, on the contrary, weighs from half
a pound to two pounds, which it never exceeds.

The twaite differs from a fmall fhad only in having one or more round black fpots on the fides; if
only one, it is always near the gill, but commonly
there are three or four, placed one under the
other *.

The other particulars agree in each fo exactly,
that the fame defcription will ferve for both.

The head flopes down confiderably from the
back, which at the beginning is very convex, and
rather fharp: the body from thence grows gradually lefs to the tail.

* I muft here acknowledge my obligations to Doctor *Lyfons*,
of *Gloucefter*, for his communications relating to this fifh, as
well as to feveral other articles relating to thofe of the *Severn*.

The

The under jaw is rather longer than the upper : the teeth very minute.

The dorſal fin is placed very near the centre, is ſmall, and the middle rays are the longeſt : the pectoral and ventral fins are ſmall : the tail vaſtly forked : the belly extremely ſharp, and moſt ſtrongly ſerrated.

The back is of a duſky blue : above the gills begins a line of dark ſpots, which mark the upper part of the back on each ſide ; the number of theſe ſpots is uncertain in different fiſh, from four to ten.

The

CARP.

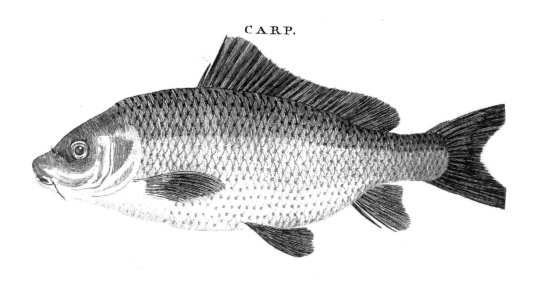

BREAM.

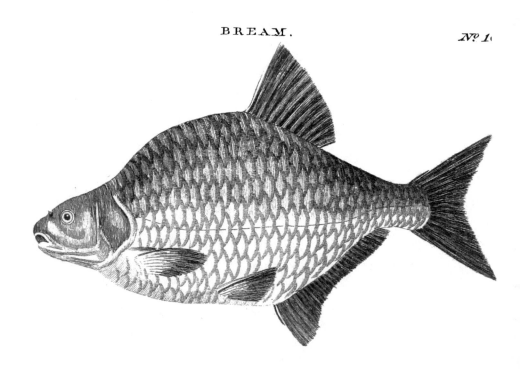

The mouth without teeth.   **XL. CARP.**
Three branchioftegous rays.
One dorfal fin.

* With bearded mouths.

Κυπρινος ? *Arift. Hift. an. lib.* Cyprinus cirris quatuor, offi- 165. **Carp.**
  IV. 8. VI. 40. VIII. 20. II. 30.   culo tertio pinnarum dorfi,
  Oppian Halieut. I. 101. 592.   ac ani uncinulis armato.
Raina Burbara. *Salvian.* 92.   *Arted. fynon.* 3.
La Carpe. *Belon,* 267.   Cyprinus carpio. C. pinna ani
Cyprinus. *Rondel. fluviat.* 150. radiis 9. cirris 4. pinna dor-
  Gefner pifc. 309.   falis radio fecundo poftice
Cyprinus nobilis, edle Karpe, ferrato. *Lin. fyft.* 525. *Gro-*
  Karpffe. *Schonevelde,* 32.   *nov. Zooph. No.* 330.
Carp. *Wil. Icth.* 245. *Raii fyn.* Karp. *Faun. Suec. No.* 359.
  pifc. 115.

THIS is one of the naturalized fifh of our country, having been introduced by *Leonard
Mafchal,* about the year 1514 *, to whom we were
alfo indebted for that excellent apple the *pepin.* The
many good things that our ifland wanted before
that period, are enumerated in this old diftich :

> Turkies, carps, hops, pickerel, and beer,
> Came into *England* all in one year. †

As

* *Fuller's Britifh Worthies, Suffex.* 113.
† I infert this note to fhew that it was known here before.
The extract was made from *the Boke of St. Albon's* printed at
Weftminfter, by *Wynkyn de Worde,* in the year 1496. I think
myfelf much obliged to Mr. *Haworth,* in *Chancery-lane,* not
only for this but feveral other curious remarks.
  ' The carpe is a dayntous fifshe, but there ben but fewe in
' *Englonde,* and therfore I wryte the caffe of him. For he is
' too ftronge enarmyd in the mouthe that there may noo weke
' harnays hold hym. And as touchyne his baytes, I have but
' lytyll knoolege of it, and we were loth to wryte more than

Ruffia wants thefe fifh at this day; *Sweden* has them only in the ponds of the people of fafhion; *Polifh Pruffia* is the chief feat of the carp; they abound in the rivers and lakes of that country, particularly in the *Frifch* and *Curifch-haff*, where they are taken of a vaft fize. They are there a great article of commerce, and fent in well-boats to *Sweden* and *Ruffia*. The merchants purchafe them out of the waters of the *nobleffe* of the country, who draw a good revenue from this article. Neither are there wanting among our gentry, inftances of fome who make good profit of their ponds.

The antients do not feparate the carp from the fea fifh. We are credibly informed that they are fometimes found in the harbor of *Dantzick*, between the town and a fmall place called *Hela*.

Carp are very long-lived. *Gefner* * brings an inftance of one that was an hundred years old. They alfo grow to a very great fize. On our own knowledge we can fpeak of none that exceeded twenty pounds in weight: but *Jovius* † fays, that they were fometimes taken in the *Lacus Larius* (the *Lago di Como*) of two hundred pounds weight: and *Rzaczynfki* ‡ mentions others taken in the *Dniefter* that were five feet in length.

‘ I know and have provyd. But well I wote that the redde
‘ worm and the menow ben good baytyn for him at all
‘ tymes, as I have herd faye of perfones credyble, and alfo
‘ founde wryten in bokes of credence.

    * *Gefner pifc*. 312.   † *De pifcibus Romanis*, 131.
    ‡ *Hift. Nat. Poloniæ*, 142.

They

They are alfo extremely tenacious of life, and will live for a moft remarkable time out of water. An experiment has been made by placing a carp in a net, well wrapped up in wet mofs, the mouth only remaining out, and then hung up in a cellar, or fome cool place: the fifh is frequently fed with white bread and milk, and is befides often plunged into water. Carp thus managed have been known, not only to have lived above a fortnight, but to grow exceedingly fat, and far fuperior in tafte to thofe that are immediately killed from the pond *.

The carp is a prodigious breeder: its quantity of roe has been fometimes found fo great, that when taken out and weighed againft the fifh itfelf, the former has been found to preponderate. From the fpawn of this fifh *Caviare* is made for the *Jews*, who hold this fturgeon in abhorrence. We have forbore in this work to enter into minute calculations of the numbers each fifh may produce. It has already been moft fkilfully performed by Mr. *Harmer*, and printed in the *Philofophical Tranfactions* of the year 1767. We fhall, in our Appendix, take the liberty of borrowing fuch part of his tables of the fœcundity of fifh, as will demonftrate the kind attention of Providence, towards the

Foecundity.

---

* This was told me by a gentleman of the utmoft veracity, who had twice made the experiment. The fame fact is related by that pious Philofopher Doctor *Derham*, in his *Phyfico-Theology*, edit. 9th. 1737. *ch. 1. p. 7. n. e.*

A a 2   preferving

preferving fo ufeful a clafs of animals for the fervice of its other creatures.

Thefe fifh are extremely cunning, and on that
account are by fome ftyled the *river fox.* They
will fometimes leap over the nets, and efcape that
way; at others, will immerfe themfelves fo deep
in the mud, as to let the net pafs over them.
They are alfo very fhy of taking a bait; yet at the
fpawning time they are fo fimple, as to fuffer themfelves to be tickled, handled, and caught by any
body that will attempt it.

This fifh is apt to mix its milt with the roe of
other fifh, from which is produced a fpurious breed:
we have feen the offspring of the carp and tench,
which bore the greateft refemblance to the firft:
have alfo heard of the fame mixture between the
carp and bream.

DESCRIP.　　The carp is of a thick fhape: the fcales very
large, and when in beft feafon of a fine gilded hue.

The jaws are of equal length; there are two
teeth in the jaws, or on the tongue; but at the entrance of the gullet, above and below, are certain
bones that act on each other, and comminute the
food before it paffes down.

On each fide of the mouth is a fingle beard;
above thofe on each fide another, but fhorter: the
dorfal fin extends far towards the tail, which is a
little bifurcated; the third ray of the dorfal fin
is very ftrong, and armed with fharp teeth, pointing

Pl. LXXI.

BARBEL.

ing downwards; the third ray of the anal fin is
conftructed in the fame manner.

Barbus. *Aufonius Mofella,* 94.	quatuor, pinna ani officu- 166.BARBEL.
Barbeau. *Belon,* 299.	lorum feptem. *Arted. fy-*
Barbus, Barbo. *Salvian,* 86.	*non.* 8.
Barbus. *Rondel. fluviat.* 194.	Cyprinus Barbus. C. pinna
Gefner pifc. 123.	ani radiis 7. cirris 4. pinnæ
Barbe, Barble. *Schonevelde,*	dorfi radio fecundo utrin-
29.	que ferrato. *Lin. fyft.* 525.
Barbel. *Wil. Icth.* 259.	*Gronov. Zooph. No.* 331.
Raii fyn. pifc. 121.	Barbe, Barble. *Wulff Borufs,*
Cyprinus oblongus, maxilla	*No.* 52.
fuperiore longiore, cirris	

THIS fifh was fo extremely coarfe, as to be
overlooked by the antients till the time of
Aufonius, and what he fays is no panegyric on it;
for he lets us know it loves deep waters, and that
when it grows old it was not abfolutely bad.

> *Laxos exerces* BARBE *natatus,*
> *Tu melior pejore ævo, tibi contigit uni*
> *Spirantum ex numero non inlaudata feneêlus.*

It frequents the ftill and deep parts of rivers,
and lives in fociety, rooting like fwine with their
nofes in the foft banks.   It is fo tame as to fuffer
itfelf to be taken with the hand; and people have
been known to take numbers by diving for them.
In fummer they move about during night in fearch

A a 3   of

of food, but towards autumn, and during winter, confine themselves to the deepest holes.

They are the worst and coarsest of fresh water fish, and seldom eat but by the poorer sort of people, who sometime boil them with a bit of bacon to give them a relish. The roe is very noxious, affecting those who unwarily eat of it with a nausea, vomiting, purging, and a slight swelling.

Descrip.   It is sometimes found of the length of three feet, and eighteen pounds in weight: it is of a long and rounded form: the scales not large.

Its head is smooth: the nostrils placed near the eyes: the mouth is placed below: on each corner is a single beard, and another on each side the nose.

The dorsal fin is armed with a remarkable strong spine, sharply serrated, with which it can inflict a very severe wound on the incautious handler, and even do much damage to the nets.

The pectoral fins are of a pale brown color; the ventral and anal tipped with yellow: the tail a little bifurcated, and of a deep purple: the side line is strait.

The scales are of a pale gold color, edged with black : the belly is white.

Tinca,

Tinca. *Ausonius Mosella*, 123.
Tinca. *Jovius*, 124.
Tinca, Tenca. *Salvian*, 90.
La Tanche. *Belon*, 325.
Tinca. *Rondel. fluviat.* 157.
  Gesner pisc. 984.
Schley, Slye. *Schonevelde*, 76.
Tench. *Wil. Icth.* 251. *Raii*
  syn. pisc. 117.
Cyprinus mucosus totus ni-
  grescens, extremitate caudæ

æquali. *Arted. synon.* 5.
Cyprinus pinna ani radiis 25,
  cauda integra, corpore mu-
  coso, cirris 2. *Lin. syst.*
  526. *Gronov. Zooph. No.*
  328.
Suture, Linnare, Skomakare.
  Faun. Suec. No. 363.
Schleihe, Schlegen. *Wulff Bo-*
  russ. No. 55.

167. TENCH.

THE tench underwent the fame fate with the barbel, in refpect to the notice taken of it by the early writers; and even *Aufonius*, who firft mentions it, treats it with fuch difrefpect, as evinces the great capricioufnefs of tafte; for that fifh, which at prefent is held in fuch good repute, was in his days the repaft only of the *Canaille*.

> *Quis non et virides vulgi folatia* Tincas
> *Norit ?*

It has been by fome called the *Phyfician* of the fifh, and that the flime is fo healing, that the wounded apply it as a flyptic. The ingenious Mr. *Diaper*, in his *pifcatory* ecloges, fays, that even the voracious pike will fpare the tench on account of its healing powers:

A a 4

The

The *Tench* he fpares a medicinal kind :
For when by wounds diftreft, or fore difeafe,
He courts the falutary fifh for eafe ;
Clofe to his fcales the kind phyfician glides,
And fweats a healing balfam from his fides *.

Whatever virtue its flime may have to the inhabitants of the water, we will not vouch for, but
flefh is a wholefome and delicious food to thofe
  the earth.　The *Germans* are of a different opinion.　By way of contempt, they call it *Shoemaker.*
Gefner even fays, that it is infipid and unwhole-
fome.

It does not commonly exceed four or five pounds
in weight, but we have heard of one that weighed
ten pounds ; *Salvianus* fpeaks of fome that arrived
at twenty pounds.

They love ftill waters, and are rarely found in
rivers : they are very foolifh, and eafily caught.

Descrip.　The tench is thick and fhort in proportion to
its length : the fcales are very fmall, and covered
with flime.

The irides are red : there is fometimes, but not
always, a fmall beard at each corner of the mouth.

The color of the back is dufky ; the dorfal and
ventral fins of the fame color : the head, fides, and
belly, of a greenifh caft, moft beautifully mixed
with gold, which is in its greateft fplendor when
the fifh is in the higheft feafon.

* *Ecl.* II.

The

The tail is quite even at the end, and very broad.

Gobio. *Aufonius Mofella*, 132.
Gobio fluviatilis. *Salvian*, 214.
Goujon de riviere. *Belon*, 322.
Gobio fluviatilis. *Rondel. fluviat.* 206. *Gefner pifc.* 399.
Gudgeon. *Wil. Icth.* 264.

Raii fyn. pifc. 123.
Cyprinus quincuncialis maculofus, maxilla fuperiore longiore cirris duobus ad ꝯ *Arted. fynon.* 2.
Cyprinus pinna ani radiis 2ꝯ *Lin. Syft. Nat.* 526. *Gronov. Zooph. No.* 329.

168. Gud-geon.

*A*RISTOTLE mentions the gudgeon in two places; once as a river fiſh, and again as a ſpecies that was gregarious: in a third place he defcribes it as a fea fiſh; we muſt therefore confider the Κωϐιος he mentions, *lib.* IX. *c.* 2. and *lib.* VIII. *c.* 19. as the fame with our ſpecies*.

This fiſh is generally found in gentle ftreams, and is of a fmall fize: thofe few, however, that are caught in the *Kennet*, and *Cole*, are three times the weight of thofe taken elfewhere. The largeft we ever heard of was taken near *Uxbridge*, and weighed half a pound.

They bite eagerly, and are affembled by raking the bed of the river; to this fpot they immediately crowd in fhoals, expecting food from this difturbance.

* The gudgeon is enumerated among the *Syrian* fiſh, by Dr. *Ruffel*, p. 75.

The

DESCRIP.   The ſhape of the body is thick and round: the
irides tinged with red: the gill covers with green
and ſilver: the lower jaw is ſhorter than the upper: at each corner of the mouth is a ſingle beard:
the back olive, ſpotted with black: the ſide line
ſtrait; the ſides beneath that ſilvery: the belly
white.

The tail is forked; that, as well as the dorſal
fin, is ſpotted with black.

** Without Beards.

169. BREAM. La Bremme. *Belon*, 318.
Cyprinus latus five Brama.
Rondel. fluviat. 154. *Geſner
piſc.* 316, 317.
Braſſem, Brachſem. *Schone-
velde, 33.*
Bream. *Wil. Icth.* 248. *Raii
ſyn. piſc.* 116.
Cyprinus pinnis omnibus ni-
greſcentibus,   pinna ani
officulorum viginti ſeptem.
Arted. ſynon. 4.
Cyprinus Brama. *Lin. ſyſt.*
531. *Gronov. Zooph. No.*
345.
Braxen. *Faun. Suec. No.* 360.
Gareikl. *Kram.* 391. Brek-
meɴ. *Wulff
Boruſs. No.* 66.

THE bream is an inhabitant of lakes, or the
deep parts of ſtill rivers. It is a fiſh that is
very little eſteemed, being extremely inſipid.

It is extremely deep, and thin in proportion to
its length. The back riſes very much, and is very
ſharp at the top. The head and mouth are ſmall:
on

Pl. LXXII.

N.º 171

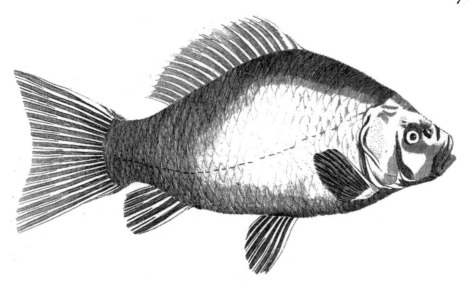

CRUSIAN.

on some we examined in the spring, were abundance of minute whitish tubercles; an accident which *Pliny* seems to have observed befals the fish of the *Lago Maggiore*, and *Lago di Como**. The scales are very large : the sides flat and thin.

The dorsal fin has eleven rays, the second of which is the longest: that fin, as well as all the rest, are of a dusky color; the back of the same hue : the sides yellowish.

The tail is very large, and of the form of a crescent.

Χαϱαξ ? *Athenæus*, lib. VIII. 355. *Oppian Halieut.* I. 174.
La Plestia? *Belon*, 309. La Rosse, 319.
Finscale. *Plot's Oxf.* 184.
Rutilus latior, seu Rubellio fluviatilis a Rud, Roud, or Finscale. *Wil. Icth.* 252. *Raii syn. pisc.* 118.

Cyprinus. *Arted. synon.* 6. No. 8.
Cyprinus erythropthalmus. Cyprinus pinna ani radiis 15. pinnis rubris. *Lin. syst. Nat.* 530.
Sarf. Isarf. *Faun. Suec. No.* 366.

170. RUD.

THIS fish is found in the *Charwell*, near *Oxford*, in the *Witham* in *Lincolnshire*, and in the fens in *Holderness*.

* *Duo Lacus* ITALIÆ *in radicibus* Alpium, LARIUS *et* VERBANUS *appellantur, in quibus pisces omnibus annis* VERGILIARUM *ortu existunt, squamis conspicui crebris atque præacutis, clavorum caligarium effigie : nec amplius quam circa eum mensem, visuntur.* lib. IX. *c.* 18.

Its

Its body is extremely deep, like that of the bream, but much thicker.

DESCRIP.   The head is fmall: the irides yellow, varying in fome almoft to rednefs: the noftrils large: the back vaftly arched, and floping off fuddenly to the head and tail: the fcales very large: the fide line very flightly incurvated.

The dorfal fin confifts of eleven rays; the firft very fhort, the fecond very ftrong, and ferrated on each fide. The pectoral fins confift of feventeen; the ventral of nine; the anal of thirteen rays.

The back is of an olive color: the fides and belly of a gold color, with certain marks of red; the ventral and anal fins, and the tail, generally of a deep red: the tail forked.

We believe this to be the fame with the *Shallow* of the *Cam*; which grows to the length of thirteen inches. It fpawns in *April*.

171. CRUCI-   Cyprinus Caraffius. *Lin. fyft.*   Karaufchen. *Meyer an.* XI. 58.
AN.   Ruda, et Caruffa. *Faun. Suec.*   Karafs. *Gefner pifc. Paralip.*
    N. 364.   16.

THIS fpecies is common in many of the fifh ponds about *London*, and other parts of the fouth of *England*; but I believe is not a native fifh.

It

It is very deep and thick: the back is much arched: the dorfal fin confifts of nineteen rays; the two firft ftrong and ferrated. The pectoral fins have (each) thirteen rays; the ventral nine; the anal feven or eight: the lateral line parallel with the belly: the tail almoft even at the end.

The color of the fifh in general is a deep yellow: the meat is coarfe, and little efteemed.

La Gardon, Rofchie 2. en Angleterre. *Belon*, 316.
Leucifcus. *Rondel. fluviat.* 191.
Rutilus five Rubellus fluviatilis. *Gefner pifc.* 820.
Rottauge. *Schonevelde*, 63.
Roche. *Wil. Icth.* 262. Leucifcus prior. *Rondel.* 260. *Raii fyn. pifc.* 122, 121.

Cyprinus *fargus* dictus. Cyp. 172. ROACH. iride pinnis ventralibus ac ani plerumque rubentibus. *Arted. fynon.* 9, 10.
Cyprinus Rutilus. Cyp. pinna ani radiis 12. rubicunda. *Lin. fyft.* 529.
Mort. *Faun. Suec. No.* 372. Zert. *Wulff Boruff. No.* 59. Altl. *Kram.* 395.

*S*OUND *as a Roach*, is a proverb that appears to be but indifferently founded, that fifh being not more diftinguifhed for its vivacity than many others; yet it is ufed by the *French* as well as us, who compare people of ftrong health to their *Gardon*, our roach.

It is a common fifh, found in many of our deep ftill rivers, affecting, like the others of this genus, quiet waters. It is gregarious, keeping in large fhoals. We have never feen them very large.

Old

Old *Walton* ſpeaks of ſome that weighed two
pounds. In a liſt of fiſh ſold in the *London* markets, with the greateſt weight of each, communicated to us by an intelligent fiſhmonger, is mention of one whoſe weight was five pounds.

The roach is deep, but thin, and the back is
much elevated, and ſharply ridged: the ſcales
large, and fall off very eaſily. Side line bends
much in the middle towards the belly.

173. DACE.	Une vandoiſe, ou Dard. *Be-lon,* 313. Leuciſci ſecunda ſpecies. *Rondel.* 192. *Geſner piſc.* 26. Dace, or Dare. *Wil. Icth.* 260. *Raii ſyn. piſc.* 121. Cyprinus decem digitorum, rutilo longior, et anguſti-	or, pinna ani radiorum de-cem. *Arted. ſynon.* 9. Cyprinus leuciſcus. Cyp. pin-na ani radiis 10. dorſali 9. *Lin. ſyſt.* 528. Laugele. *Meyer's An.* II. *tab.* 97.

THIS, like the roach, is gregarious, haunts
the ſame places, is a great breeder, very
lively, and during ſummer is very fond of frolicking near the ſurface of the water. This fiſh and the
roach are coarſe and inſipid meat.

Its head is ſmall: the irides of a pale yellow:
the body long and ſlender: its length ſeldom above
ten inches, though in the abovementioned liſt is an
account of one that weighed a pound and an half:
the ſcales ſmaller than thoſe of the roach.

The

The back is varied with dufky, with a caft of yellowifh green: the fides and belly filvery: the dorfal fin dufky: the ventral, anal, and caudal fins red, but lefs fo than thofe of the former: the tail is very much forked.

The Graining. *Voy. to the Hebrides,* 11.   174. GRAIN-
ING.

THE *Graining* is found in the *Merfey* near *Warrington*: has much the refemblance of a dace, but is more flender, and the back ftraiter. The ufual length about feven inches and a half. The depth to the length of this is as one to five, of the dace as one to four. The color of the back is filvery, with a bluifh caft. The eyes, ventral, and anal fins are red, but paler than thofe of the dace. The pectoral fin redder.

Capito.

175. Chub. Capito. *Aufon. Mofella*, 85.   Cyprinus oblongus macrolepi-
Squalus, Squaglio. *Salvian*,   dotus, pinna ani officulorum
84.   undecim. *Arted. fynon*. 7.
Le chevefne, Teftard, Vi-   Cyprinus cephalus. **Cyp.**
lain. *Belon*, 315.   pinna ani radiis undecim,
Cephalus fluviatilis. *Ron-*   cauda integra, corpore fub-
del. fluviat. 190.   cylindrico. *Lin. fyft*. 527.
Capito five Cephalus fluvia-   *Gronov. Zooph. No*. 339.
tilis. *Gefner pifc*. 182.   Alte. *Meyer's An*. II. *tab*. 92.
Chub, or Chevin. *Wil. Icth*.   Rapen. *Wulff Borufs. No*. 56.
255. *Raii fyn. pifc*. 119.

SALVIANUS imagines this fifh to have been the *Squalus* * of the antients, and grounds his opinion on a fuppofed error in a certain paffage in *Columella* and *Varro*, where he would fubftitute the word *Squalus* inftead of *Scarus* :. *Columella* fays no more than that the old *Romans* payed much attention to their ftews, and kept even the fea fifh in frefh water, paying as much refpect to the *Mullet* and *Scarus* as thofe of his days did to the *Muræna* and *Bafs*.

That the *Scarus* was not our *Chub*, is very evident; not only becaufe the *Chub* is entirely an inhabitant of frefh waters, but likewife it feems improbable that the *Romans* would give themfelves

---

* A cartilaginous fifh, a fhark. *Vide Plin. lib*. IX. *c*. 24. *Ovid* alfo ranks his *Squalus* with the fea fifh.

Et SQUALUS, *et tenui fuffufus fanguine* MULLUS. *Halieut*. 147.

any

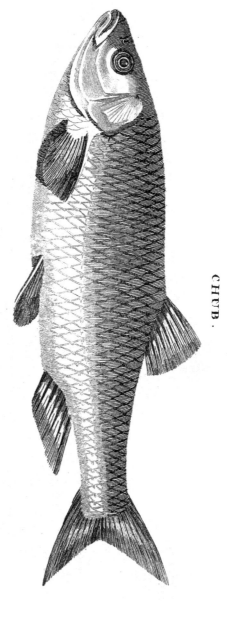

CHUB.

BLEAK.

any trouble about the worſt of river fiſh, when they neglected the moſt delicious kinds; all their attention was directed towards thoſe of the ſea: the difficulty of procuring them ſeems to have been the criterion of their value, as is ever the caſe with effete luxury.

The chub is a very coarſe fiſh and full of bones: it frequents the deep holes of rivers, and during ſummer commonly lies on the ſurface, beneath the ſhade of ſome tree or buſh.   It is a very timid fiſh, ſinking to the bottom on the left alarm, even at the paſſing of a ſhadow, but they will ſoon reſume their ſituation.   It feeds on worms, caterpillars, graſshoppers, beetles, and other coleopterous in-ſects that happen to fall into the water; and it will even feed on cray-fiſh.   This fiſh will riſe to a fly.

This fiſh takes its name from its head, not only in our own, but in other languages: we call it *Chub*, according to *Skinner*, from the old *Engliſh*, *Cop*, a head; the *French*, *Teſtard*; the *Italians*, *Capitone.* -

It does not grow to a large ſize; we have known ſome that weighed above five pounds, but *Salvianus* ſpeaks of others that were eight or nine pounds in weight.

The body is oblong, rather round, and of a pretty equal thickneſs the greateſt part of the way: the ſcales are large.

The irides ſilvery; the cheeks of the ſame color: the head and back of a deep duſky green: the

VOL. III.   B b   ſides

fides filvery, but in the fummer yellow: the belly white: the pectoral fins of a pale yellow: the ventral and anal fins red: the tail a little forked, of a brownifh hue, but tinged with blue at the end.

176. Bleak. Alburnus. *Aufon. Mofella,* 126.
Able ou Ablette. *Belon,* 319.
Alburnus. *Rondel. fluviat.* 208. *Gefner pifc.* 23.
Albula minor. Witinck, Witek and Blike. *Schonevelde,* II. *Tab.* 1.
Bleak. *Wil. Icth.* 263. *Raii fyn. pifc.* 123.
Cyprinus quincuncialis, pinna ani offiaulorum viginti. *Arted. fynon.* 10.
Cyprinus alburnus. *Lin. fyft.* 531. *Gronov. Zooph. No.* 336.
Loja. *Faun. Suec. No.* 373.
Spitflauben, fchneiderfifchl. *Kram.* 395.
Ukeleyen. *Wulff Borufs. No.* 64.

THE taking of thefe, *Aufonius* lets us know, was the fport of children,

     Alburnos *prædam puerilibus hamis.*

· They are very common in many of our rivers, and keep together in large fhoals. Thefe fifh feem at certain feafons to be in great agonies; they tumble about near the furface of the water, and are incapable of fwimming far from the place, but in about two hours recover, and difappear. Fifh thus affected the *Thames* fifhermen call *mad bleaks.* They feem to be troubled with a fpecies of *Gordius* or hair-worm, of the fame kind with

                         thofe

thofe which *Ariftotle* * fays that the *Ballerus* and *Tillo* are infefted with, which torments them fo that they rife to the furface of the water and then die.

Artificial pearls are made with the fcales of this fifh, and we think of the dace.   They are beat in-to a fine powder, then diluted with water, and in-troduced into a thin glafs bubble, which is af-terwards filled with wax.   The *French* were the inventors of this art.   Doctor *Lifter* † tells us, that when he was at *Paris*, a certain artift ufed in one winter thirty hampers full of fifh in this manu-facture.

ARTIFICIAL PEARLS.

The bleak feldom exceeds five or fix inches in length : their body is flender, greatly compreff-ed fideways, not unlike that of the fprat.

DESCRIP.

The eyes are large : the irides of a pale yellow : the under jaw the longeft : the lateral line crooked : the gills filvery : the back green : the fides and belly filvery : the fins pellucid : the fcales fall off very eafily : the tail much forked.

During the month of *July* there appear in the *Thames*, near *Blackwall* and *Greenwich*, innumer-able multitudes of fmall fifh, which are known to the *Londoners* by the name of *White Bait*.   They are efteemed very delicious when fried with fine flour, and occafion, during the feafon, a vaft refort of the lower order of epicures to the taverns con-tiguous to the places they are taken at.

WHITE BAIT.

---

* *Hift. an. lib.* VIII. *c.* 20.
† Journey to *Paris*, 142.

B b 2   There

There are various conjectures about this species, but all terminate in a suppofition that they are the fry of fome fifh, but few agree to which kind they owe their origin. Some attribute it to the fhad, others to the fprat, the fmelt, and the bleak. That they neither belong to the fhad, nor the fprat, is evident from the number of branchioftegous rays, which in thofe are eight, in this only three. That they are not the young of fmelts is as clear, becaufe they want the *pinna adipofa,* or raylefs fin; and that they are not the offspring of the bleak is extremely probable, fince we never heard of the white bait being found in any other river, notwithftanding the bleak is very common in feveral of the *Britifh* ftreams: but as the white bait bears a greater fimilarity to this fifh than to any other we have mentioned, we give it a place here as an appendage to the bleak, rather than form a diftinct article of a fifh which it is impoffible to clafs with certainty.

It is evident that it is of the carp or *Cyprinus* genus: it has only three branchioftegous rays, and only one dorfal fin; and in refpect to the form of the body is compreffed like that of the bleak.

Its ufual length is two inches: the under jaw is the longeft: the irides filvery, the pupil black: the dorfal fin is placed nearer to the head than to the tail, and confifts of about fourteen rays: the fide line is ftrait: the tail forked, the tips black.

The

The head, fides, and belly are filvery; the back tinged with green.

Φοξιν⊙·? *Arift. Hift. an.* VI. c. 13.
Le Veron. *Belon*, 324.
Pifciculus varius. *Rondel. fluviat.* 205.
Phofcium qui vulgo *veronus* (quafi varius) dicitur, *Bellonius. Gefner pifc.* 715.
Elritze, Elderitze. *Schœnevelde*, 57.
Pink, Minim, or Minow.

Wil. Icth. 268. *Raii fyn.* 177. Minow, *pifc.* 125.
Cyprinus tridactylus varius oblongus teretiufculus, pinna ani officulorum octo. *Arted. fynon.* 12.
Cyprinus Phoxinus. Cyp. pinna ani radiis 8. macula fufca ad caudam, corpore pellucido. *Lin. fyft.* 528.

THIS beautiful fifh is frequent in many of our fmall gravelly ftreams, where they keep in fhoals.

The body is flender and fmooth, the fcales being extremely fmall. It feldom exceeds three inches in length.

The lateral line is of a golden color : the back flat, and of a deep olive : the fides and belly vary greatly in different fifh ; in a few are of a rich crimfon, in others bluifh, in others white. The tail is forked, and marked near the bafe with a dufky fpot.

178. Gold-   Kingo, the Gold Fiſh. *Kæm-*   tranſverſa bifurca. *Lin. ſyſt.*
en.   *pfer Hiſt. Japan*, I. 137.   527.   *Faun. Suec.* tab 2.
        Kin-yu.   *Du Halde Hiſt.*   *Gronov. Zooph.* No. 342.
        China. I. 19. 315.   Gold Fiſh.   *Edw.* 209.
        Cyprinus auratus. Cyp. pin-   Kin-yu, five carpio auratus.
        na ani gemina, cauda   *Baſter ſubſec.* II. 78.

THESE fiſh are now quite naturalized in this country, and breed as freely in the open waters as the common carp.

They were firſt introduced into *England* about the year 1691, but were not generally known till 1728, when a great number were brought over, and preſented firſt to Sir *Mathew Dekker*, and by him circulated round the neighborhood of *London*, from whence they have been diſtributed to moſt parts of the country.

In *China* the moſt beautiful kinds are taken in a ſmall lake in the province of *Che-Kyang.* Every perſon of faſhion keeps them for amuſement, either in porcellane veſſels, or in the ſmall baſons that decorate the courts of the *Chineſe* houſes. The beauty of their colors, and their lively motions, give great entertainment, eſpecially to the ladies, whoſe pleaſures, by reaſon of the cruel policy of that country, are extremely limited.

In form of the body they bear a great reſemblance to a carp. They have been known in this iſland

iſland to arrive at the length of eight inches; in their native place they are ſaid * to grow to the ſize of our largeſt herring.

The noſtrils are tubular, and form ſort of appendages above the noſe: the dorſal fin and the tail vary greatly in ſhape: the tail is naturally bifid, but in many is trifid, and in ſome even quadrifid: the anal fins are the ſtrongeſt characters of this ſpecies, being placed not behind one another like thoſe of other fiſh, but oppoſite each other like the ventral fins.

The colors vary greatly; ſome are marked with a fine blue, with brown, with bright ſilver; but the general predominant color is gold of a moſt amazing ſplendor; but their colors and form need not be dwelt on, ſince thoſe who want opportunity of ſeeing the living fiſh, may ſurvey them expreſſed in the moſt animated manner, in the works of our ingenious and honeſt friend Mr. *George Edwards.*

---

* *Du Halde,* 316.

B b 4   APPEN-

APPENDIX.

A P P E N D I X.

THE late Bifhop of *Carlifle* informed me that a tortoife was taken off the coaft of *Scarborough* in 1748 or 1749. It was purchafed by a family at that time there, and a good deal of company invited to partake of it. A gentleman, who was one of the guefts, told them it was a *Mediterranean* turtle, and not wholefome: only one of the company eat of it, and it almoft killed him, being feized with a dreadful vomiting and purging.

TORTOISE, PAGE 7.

Since the printing of that article I have been favored with fome very curious accounts of this reptile, which will give greater light into its natural hiftory than I am capable of, from a moft unphilofophical but invincible averfion to the whole genus. The facts that will appear in the following lines ferve to confirm my opinion of its being an innoxious animal, and, I hope, will ferve to free numbers

TOAD, 13.

numbers from a panic that is carried to a degree of infelicity, and alfo to redeem it from a perfecution which the unmerited ill-opinion the world has conceived, perpetually expofes it to.

The gentlemen I am principally indebted to for my informations are *J. Arfcott*, Efq; of *Tebott*, in *Devonfhire*, and Mr. *Pitfield*, of *Exeter*. Some of thefe accounts were addreffed to Doctor *Milles*, Dean of *Exeter*; others to the worthy Prelate above-mentioned, to whom I owe thefe and many other agreeable correfpondencies; others again to myfelf.

Mr. *Arfcott*'s letters give a very ample hiftory of the nature of the toad: they were both addreffed to Doctor *Milles*, and both were the refult of certain queries I propofed, which the former was fo obliging as to give himfelf the trouble of anfwering in a moft fatisfactory manner.

I fhall firft take the liberty of citing Mr. *Arfcott*'s letter of *September* the 23d, 1768, which mentions fome very curious particulars of this innocent reptile, which, for fuch a number of years, found an afylum from the good fenfe of a family which foared above all vulgar prejudices.

" It would give me the greateft pleafure to be
" able to inform you of any particulars worthy Mr.
" *Pennant*'s notice, concerning the toad who lived
" fo many years with us, and was fo great a favo-
" rite. The greateft curiofity in it was its becom-
" ing fo remarkably tame. It had frequented fome
" fteps

" steps before the hall-door some years before my
" acquaintance commenced with it, and had been
" admired by my father for its size (which was of
" the largest I ever met with) who constantly payed
" it a visit every evening. I knew it myself above
" thirty years, and by constantly feeding it, brought
" it to be so tame that it always came to the can-
" dle, and looked up as if expecting to be taken
" up and brought upon the table, where I always
" fed it with insects of all sorts: it was fondest of
" flesh maggots, which I kept in bran; it would
" follow them, and when within a proper distance,
" would fix its eye, and remain motionless for near
" a quarter of a minute, as if preparing for the
" stroke, which was an instantaneous throwing its
" tongue at a great distance upon the insect, which
" stuck to the tip by a glutinous matter: the mo-
" tion is quicker than the eye can follow*.

" I always imagined that the root of its tongue
" was placed in the fore part of its under jaw, and
" the tip towards its throat, by which the motion
" must be a half circle; by which, when its tongue
" recovered its situation, the insect at the tip would
" be brought to the place of deglutition. I was
" confirmed in this by never observing any internal
" motion in its mouth, excepting one swallow the
" instant its tongue returned. Possibly I might be

* This rapid capture of its prey might give occasion to the
report of its fascinating powers. *Linnæus* says, *Insecta in fauces
fascino revocat,*

" mistaken

" miftaken, for I never diffected one, but content-
" ed myfelf with opening its mouth, and flightly
" infpecting it.

" You may imagine that a toad generally detefted
" (altho' one of the moft inoffenfive of all animals)
" fo much taken notice of and befriended, excited
" the curiofity of all comers to the houfe, who all
" defired to fee it fed, fo that even ladies fo far
" conquered the horrors inftilled into them by
" nurfes, as to defire to fee it. This produced in-
" numerable and improbable reports, making it
" as large as the crown of a hat, &c. &c. This I
" hope will account for my not giving you parti-
" culars more worth your notice. When I firft
" read the account in the papers of toads fucking
" cancerous breafts, I did not believe a word of it,
" not thinking it poffible for them to fuck, having
" no lips to embrace the part, and a tongue fo
" oddly formed ; but as the fact, is thoroughly ve-
" rified, I moft impatiently long to be fully in-
" formed of all particulars relating to it."

Notwithftanding thefe accounts will ferve to point
out fome errors I had adopted, in refpect to this
reptile in my firft fheet, yet it is with much pleafure I lay before the public a more authentic hiftory, collected from Mr. *Arfcott*'s fecond favor; the
anfwer points out my queries, which it is needlefs
to repeat.

Tebott

Tebott, Nov. 1, 1763.

" In refpect to the queries, I fhall here give the
" moft fatisfactory anfwers I am capable of.

" Firft, I cannot fay how long my father had
" been acquainted with the toad before I knew it;
" but when I firft was acquainted with it, he ufed
" to mention it as the old toad I've known fo many
" years; I can anfwer for thirty-fix years.

" Secondly, No toads that I ever-faw appear-
" ed in the winter feafon. The old toad made
" its appearance as foon as the warm weather came,
" and I always concluded it retired to fome dry
" bank to repofe till the fpring. When we new-
" lay'd the fteps I had two holes made in the
" third ftep on each, with a hollow of more than a
" yard long for it, in which I imagine it flept, as
" it came from thence at its firft appearance.

" Thirdly, It was feldom provoked: neither
" that toad (nor the multitudes I have feen tor-
" mented with great cruelty) ever fhewed the left
" defire of revenge, by fpitting or emitting any
" juice from their pimples. Sometimes upon tak-
" ing it up it would let out a great quantity of clear
" water, which, as I have often feen it do the fame
" upon the fteps when quite quiet, was certainly its
" urine, and no more than a natural evacuation.

" Fourthly, A toad has no particular enmity
" for the fpider; he ufed to eat five or fix with his
" millepedes (which I take to be its chief food) that
" I generally provided for it, before I found out
                                            " that

" that flesh maggots, by their continual motion,
" was the most tempting bait; but when offered it
" eat blowing flies and humble bees that come from
" the rat-tailed maggot in gutters, or in short any
" insect that moved. I imagine if a bee was to be
" put before a toad, it would certainly eat it to its
" cost; but as bees are seldom stirring at the same
" time that toads are, they can seldom come in
" their way, as they seldom appear after sun-rising,
" or before sun-set. In the heat of the day they
" will come to the mouth of their hole, I believe,
" for air. I once from my parlour window observed
" a large toad I had in the bank of a bowling-
" green, about twelve at noon, a very hot day, ve-
" ry busy and active upon the grass; so uncommon
" an appearance made me go out to see what it
" was, when I found an innumerable swarm of
" winged ants had dropped round his hole, which
" temptation was as irresistible as a turtle would
" be to a luxurious alderman.

" Fifthly, Whether our toad ever propagated its
" species I know not, rather think not, as it al-
" ways appeared well, and not lessened in bulk,
" which it must have done, I should think, if it
" had discharged so large a quantity of spawn as
" toads generally do. The females that are to
" propagate in the spring, I imagine, instead of
" retiring to dry holes, go into the bottom of
" ponds, and lay torpid among the weeds; for to
" my great surprize in the middle of the winter,
" having

" having for amufement put a long pole into my
" pond, and twifted it till it had gathered a large
" volume of weed, on taking it off I found many
" toads, and having cut fome afunder with my
" knife, by accident, to get off the weed, found
" them full of fpawn not thoroughly formed. I
" am not pofitive, but think there were a few
" males in *March*: I know there are thirty males *
" to one female, twelve or fourteen of whom I have
" feen clinging round a female: I have often dif-
" engaged her, and put her to a folitary male, to
" fee with what eagernefs he would feize her.
" They impregnate the fpawn as it is drawn † out in
" long

* Mr. *John Hunter* has affured me, that during his refi-
dence at *Belleifle*, he diffeded fome hundreds of toads, yet
never met with a fingle female among them.

† I was incredulous as to the *obftetrical* offices of the male
toad, but fince the end is fo well accounted for, and the fad
eftablifhed by fuch good authority, belief muft take place.

Mr. *Demours*, in the Memoirs of the *French* Academy, as
tranflated by Dr. *Templeman*, *vol.* I. 371. has been very particular in refped to the male toad, as ading the part of an
Accoucheur; his account is curious, and clames a place here:

" In the evening of one of the long days in fummer, Mr.
" *Demours* being in the King's garden perceived two toads
" *coupled together* at the edge of an hole, which was formed
" in part by a great ftone at the top.
" Curiofity drew him to fee what was the occafion of the
" motions he obferved, when two fads equally new furprized
" him; the *firft* was the extreme difficulty the female had in
" laying her eggs, infomuch that fhe did not feem capable

" long ftrings, like a necklace, many yards long,
" not in a large quantity of jelly, like frogs fpawn.
" *N. B.* After having held a female fome time in
" my hand, I have, to try if there was any fmell,
" put my finger a foot under water to a male,
" who has immediately feized it, and ftuck to it as
" firmly as if it was a female. *Quere,* Would they
" feize a finger or rag that had touched a can-
" cerous ulcer ?

" Sixthly,

" of being delivered of them without fome affiftance. The
" *fecond* was, that the male was mounted on the back of
" the female, and exerted all his ftrength with his hinder
" feet in pulling out the eggs, whilft his fore-feet embraced
" her breaft.

" In order to apprehend the manner of his working in the
" delivery of the female, the reader muft obferve, that the
" paws of thefe animals, as well thofe of the fore-feet as of
" the hinder, are divided into feveral toes, which can per-
" form the office of fingers.

" It muft be remarked likewife, that the eggs of this fpe-
" cies of toads are included each in a membranous coat that
" is very firm, in which is contained the embryo ; and that
" thefe eggs, which are oblong and about two lines in
" length, being faftened one to another by a fhort but very
" ftrong cord, form a kind of chaplet, the beads of which
" are diftant from each other about the half of their length.
" It is by drawing this cord with his paw that the male
" performs the function of a midwife, and acquits himfelf
" in it with a dexterity that one would not expect from fo
" lumpifh an animal.

" The prefence of the obferver did not a little difcompofe
" the male ; for fome time he ftopped fhort, and threw on
" the

" Sixthly, Infects being their food, I never faw
" any toad fhew any liking or diflike to any plant *.

" Seventhly, I hardly remember any perfons ta-
" king it up except my father and myfelf: I do
" not know whether it had any particular attach-
" ment to us.

" Eighthly, In refpect to its end, I anfwer this
" laft quere. Had it not been for a tame raven, I
" make no doubt but it would have been now liv-
" ing; who one day feeing it at the mouth of its
" hole, pulled it out, and although I refcued it,
" pulled out one eye, and hurt it fo, that notwith-
" ftanding its living a twelvemonth it never enjoyed
" itfelf,

" the *curious impertinent* a fixed look that marked his dif-
" quietnefs and fear; but he foon returned to his work with
" more precipitation than before, and a moment *after* he
" appeared undetermined whether he fhould continue it or
" not. The female likewife difcovered her uneafinefs at the
" fight of the ftranger, by motions that interrupted fome-
" times the male in his operation. At length, whether the
" filence and fteady pofture of the fpectator had diffipated
" their fear, or that the *cafe* was urgent, the male refumed
" his work with the fame vigour, and fuccefsfully performed
" his function."

* This queftion arofe from an affertion of *Linnæus*, that
the toad delighted in filthy herbs. *Delectatur Cotula, Actæa,
Stachyde fœtidi.* The unhappy deformity of the animal
feems to be the only ground of this as well as another mifre-
prefentation, of its conveying a poifon with its pimples, its
touch, and even its breath. *Verrucæ lactefcentes venenatæ
infufæ tactu, anhelitu.*

C c 2

" itſelf, and had a difficulty of taking its food,
" miſſing the mark for want of its eye: before
" that accident had all the appearance of perfect
" health."

What Mr. *Pitfield* communicated to me ſerves
farther to evince the patient and pacific diſpoſition
of this poor animal. If I am thought to dwell too
long on the ſubject, let it be conſidered, that thoſe
who have moſt unprovoked enemies, and feweſt
friends, clame the greateſt pity, and warmeſt vin-
dication. This reptile has undergone all ſorts of
ſcandal; one author makes it the companion of an
atheiſt *; and *Milton* † makes the devil itſelf its
inmate; in a word, all kind of evil paſſions have
been beſtowed on it: It is but juſtice therefore to
ſay ſomething in behalf of an animal that has of
late had ſo many trials of its temper, from experiments occaſioned by the new diſcovery of its
cancer-ſucking qualities. It has born all the han-
dling, teizing, bagging, &c. &c. without the leſt
ſign of a vindictive diſpoſition; but has even made
itſelf a ſacrifice to the diſcharge of its office: this
I know from the reſult of much enquiry; would I
could contradict what is aſſerted, of the inefficacy
of the tryals made of them in the moſt horrible of
diſeaſes; for at this time I myſelf cannot bring one
proof of the ſucceſs. But I would not have any
                                              one

* A great toad was ſaid to have been found in the
lodgings of *Vanini*, at *Toulouſe*. *Vide Johnſon's* Shakeſpear.
† *Paradiſe Loſt*.

one difcouraged from the purfuit of the remedy. Heaven opens to us gradually its favors: the *loadftone* was for ages a meer matter of ignorant amaze at its attractive qualities: *mercury* was a fuppofed poifon, and the terror of phyficians: we now wonder at the powers of electricity, and are ftill but partially acquainted with its ufes: the toad, the object of horror even in the moft enlightened times, is found to be perfectly innocent; it has certainly contributed to the eafe (and as has been faid to the cure) of the unhappy cancered; let the following facts fpeak for themfelves; they come from perfons of undoubted veracity, and will fufficiently eftablifh the truth of the beneficent qualities of this animal.

The firft paper relating to it is very ingenioufly drawn up by Mr. *Pitfield*, for the information of Doctor *Littleton*, Bifhop of *Carlifle* (now happy) who immediately honored me with the copy.

Exon, Auguft 29, 1768.

" Your lordfhip muft have taken notice of a
" paragraph in the papers, with regard to the ap-
" plication of toads to a cancered breaft. A pa-
" tient of mine has fent to the neighborhood of
" *Hungerford*, and brought down the very woman
" on whom the cure was done. I have, with all
" the attention I am capable of, attended the

C c 3   " operation

" operation for eighteen or twenty days, and am
" furprized at the phænomenon. I am in no ex-
" pectation of any great fervice from the applica-
" tion: the age, conftitution, and thoroughly can-
" cerous condition of the perfon, being uncon-
" querable barriers to it. How an ail of that
" kind, abfolutely local, in an otherwife found
" habit, and of a likely age, might be relieved, I
" cannot fay. But as to the operation, thus
" much I can affert, that there is neither pain nor
" naufeoufnefs in it. The animal is put into a
" linen bag, all but its head, and that is held to
" the part. It has generally inftantly laid hold of
" the fouleft part of the fore, and fucked with
" greedinefs until it dropped off dead. It has
" frequently happened that the creature has fwolen
" immenfely, and from its agonies appeared to be
" in great pain. I have weighed them for feveral
" days together, before and after the application,
" and found their increafe of weight, in the dif-
" ferent degrees, from a drachm to near an ounce.
" They frequently fweat exceedingly, and turn
" quite pale: fometimes they difgorge, recover, and
" become lively again. I think the whole fcene
" is furprifing, and a very remarkable piece of na-
" tural hiftory. From the conftant inoffenfivenefs
" which I have obferved in them, I almoft queftion
" the truth of their poifonous fpitting. Many peo-
" ple here expect no great good from the applica-
" tion of toads to cancers; and where the diforder

is

" is not abfolutely local, none is to be expected;
" where it is, and feated in any part, not to be
" well come at for extirpation, I think it is hardly
" to be imagined, but that the having it fucked
" clean as often as you pleafe, muft give great
" relief. Every body knows, that dogs licking of
" fores cures them, which is, I fuppofe, chiefly by
" keeping them clean. If there is any credit to be
" given to hiftory, poifons have been fucked out,

——*Pallentia Vulnera lambit*
Ore Venena trahens.

" are the words of *Lucan* on the occafion: if the
" people to whom thefe words are applied, did
" their cure by immediately following the injection
" of the poifon, the local confinement of another
" poifon brings the cafe to a great degree of fimi-
" larity.

" I hope I have not tired your lordfhip with my
" long tale, as it is a true one, and in my appre-
" henfion a curious piece of natural hiftory, I could
" not forbear communicating it to you. I own I
" thought the ftory in the papers to be an inven-
" tion, and when I confidered the inftinctive prin-
" ciple in all animals of felf prefervation, I was
" confirmed in my difbelief; but what I have re-
" lated I faw, and all theory muft yield to fact.
" It is only the *Rubeta*, the land toad, which has
" the property of fucking; I cannot find any the

C c 4   " left

" left mention of the property in any one of the
" old naturalifts. My patient can bear to have
" but one applied in twenty-four hours: the wo-
" man who was cured had them on day and
" night, without intermiffion, for five weeks. Their
" time of hanging at the breaft has been from one
" to fix hours."

The other is of a woman who made the experiment, which I give, as delivered to me from
" undoubted authority.

About fix years * ago a poor woman received a
crufh on her breaft by the fall of a pail; a complaint in that part was the refult.

Laft year her diforder increafed to an alarming
degree; fhe had five wounds on her breafts, one
exceeding large, from which fragments of bone
worked out, giving her vaft pain; and at the fame
time there was a great difcharge of thin yellow
matter: fhe was likewife reduced to a meer fkele-
ton.

All her left fide and ftomach was much fwelled; her fingers doughy and difcolored.

On the 25th of *September*, 1768, the firft toad
was applied; between that and the 29th fhe ufed
feven, and had that night better reft. She fwal-
lowed with greater eafe, for before that time there
was fome appearance of tumor in her neck, and
a difficulty of getting any thing down.

* *i. e.* from 1769.

October

October 16th, the patient better. It was thought proper as winter was coming on, and of courfe it would be very difficult to procure a number of toads, to apply more at a time, fo three were put on at once. The fwelling in the arm abated, and the woman's reft was good.

During thefe tryals fhe took an infufion of *Wa-ter Parfnep* with *Pulvis Cornacchini.*

December 18th, continued to look ill, but finds herfelf better: two of the wounds were now healed.

She was always moft eafy when the toads were fucking, of which fhe killed vaft numbers in the operation.

January 1769. The laft account that was received, informing that the patient was better.

The remarks made on the animals are thefe:

Some toads died very foon after they had fucked; others lived about a quarter of an hour, but fome lived much longer: for example, one that was applied about feven o'clock fucked till ten, and died as foon as it was taken from the breaft; another that immediately fucceeded continued till three o'clock, but dropped dead from the wound, each fwelled exceedingly, and turned of a pale color.

Thefe toads did not feem to fuck greedily, and would often turn their heads away; but during the time of fucking were heard to fmack their lips like a young child.

As thofe reptiles are apt by their ftruggles to get

out

out of the bag, the open end ought to be made with an open hem, that the ſtring may run the more readily, and faſten tightly about the neck.

It would be improper to quit the ſubject without mentioning the origin of this ſtrange diſcovery, which was owing to a woman near *Hungerford*, who labored under a cancerous complaint in her breaſt, which had long baffled all applications.

The account ſhe gives of the manner in which ſhe came by her knowledge is ſingular, and I may ſay apocryphal. She ſays of herſelf, that in the height of her diſorder ſhe went to ſome church where there was a vaſt crowd : on going into a pew, ſhe was accoſted by a ſtrange clergyman, who, after expreſſing compaſſion for her ſituation, told her that if ſhe would make ſuch an application of living toads * as abovementioned, ſhe would be well.

This dark ſtory is all we can collect relating to the affair. It is our opinion that ſhe ſtumbled upon the diſcovery by accident, and that having ſet up for a cancer doctreſs, ſhe thought it neceſſary to

---

* I have been told that ſhe not only made uſe of living toads, but permitted the dead ones to remain at her breaſt, by way of cataplaſms, for ſome weeks.

I have been informed that the relation of this ſtrange method of cure was brought over a few years ago by one of our foreign miniſters ; and that there is alſo notice taken of it in *Wheeler's* Travels.

amuſe

amufe the world with this myfterious relation*. For it feems very unaccountable, that this unknown gentleman fhould exprefs fo much tendernefs for this fingle fufferer, and not feel any for the many thoufands that daily languifh under this terrible diforder: would he not have made ufe of this invaluable noftrum for his own emolument, or at left, by fome other means have found a method of making it public for the good of mankind ?

Here I take leave of the fubject, which I could not do without exprefling my doubts, as to the method of the woman's obtaining her information ; but in refpect to the authenticity of this new-difcovered property of the toad, facts eftablifh it beyond difpute. Let the humane wifh for fpeedy proofs of the efficacy; and for the fatisfaction of the world, let thofe who are capable of giving indifputable proofs of the fuccefs, take the earlieft opportunity of making the public acquainted with fo interefting an affair.

' I have now given without alteration the whole
' of the facts as ftated in my former edition. They
' are too curious to be loft; as they may ferve to

---

* Mr. *Valentine Greatraks*, who about the year 1664, perfuaded himfelf that he could cure difeafes, by ftroking them out of the parts affected with his hand ; and the famous *Bridget Boftock*, of *Chefhire*, who worked cures by virtue of her fafting fpittle, both came by their art in a manner fupernatural, but by *faith many were made whole.*

' give

' give to after-times a proof of the belief of the
' age, and the fair tryal made of a moſt diſtaſt-
' ful remedy in the moſt dreadful of complaints.'

GLAIN
NAIDR, 30.

This reminds me of another *Welch* word that is
explanatory of the cuſtoms of the antients, ſhewing
their intent in the uſe of the plant *Vervaine* in
their luſtrations; and why it was called by *Dioſco-*
rides Hierobotane, or the ſacred plant, and e-
ſteemed proper to be hung up in their rooms.

The *Britiſh* name *Cas gan Cythrawl,* or the
Devil's averſion, may be a modern appellation,
but is likewiſe called *Y Dderwen fendigaid,* the
holy oak, which evidently refers to the *Druids*
groves.

Pliny informs us, that the *Gauls* uſed it in their
incantations, as the *Romans* and *Greeks* did in
their luſtrations. *Terence,* in his *Andria,* ſhews us
the *Verbena* was placed on altars before the doors
of private houſes in *Athens*; and from the ſame
paſſage in *Pliny* *, we find the *Magi* were guilty
of the moſt extravagant ſuperſtition about this
herb. Strange it is that ſuch a veneration ſhould
ariſe for a plant endued with no perceptible qualities; and ſtranger ſtill it ſhould ſpread from the
fartheſt north to the boundaries of *India.* So general a conſent, however, proves the cuſtom aroſe
before the different nations had loſt all communi-
cation with each other.

* *Lib.* XXV. *cap.* 9.

Her

Her Grace the Dutchefs Dowager of PORT-
LAND did me the honor of communicating the
following fpecies.

This is a new kind of SUCKER found near
Weymouth, which ought to be placed after No. 59.
and may be called the

THE HEAD is flat and tumid on each fide.
The BODY taper.

BIMACU-
LATED.

The PECTORAL fins placed unufually high. It
has only one DORSAL fin ; placed low, or near the
tail.

The TAIL is even at the end.

The color of the head and body is of a fine
pink : of the fins whitifh. On each fide of the
engine of adherence on the belly, is a round black
fpot.

It is figured in Plate **XXII.** of the natural fize.

Another will add a new genus to the *Britifh*
fifh, being of that which *Linnæus* calls *Ophidium*.
It muft find a place after the LAUNCE, *Sand Eel*
or *Ammodytes*, under the trivial name of

BEARD-

BEARDLESS. Ophidium imberbe. *Lin. Syſt.* 431. *Faun. Suec.* No 319.
Ophidium flavum et imberbe. *Schonevelde,* 53 ? *Wil. Icth.*
113. *Raii ſyn. piſc.* 39.

THIS was taken at the ſame place with the former. I have not at this time had opportunity of deſcribing it, therefore am obliged to refer the reader to the writers above cited for the deſcription.

No.

No. II.

OF THE PROLIFICNESS OF FISH.

Fifh.	Weight.		Weight of Spawn.	Fæcundity.	Time.
	oz.	dr.	grains.		
Carp	25.	5.	2571.	203109.	April 4.
Codfifh			12540.	3686760.	Dec. 23.
Flounder	24.	4.	2200.	1357400.	March 14.
Herring	5.	10.	480.	36960.	Oct. 25.
Mackrel	18.	0.	1223½.	546681.	June 18.
Perch	8.	9.	765½.	28323.	April 5.
Pike	56.	4.	5100½.	49304.	April 25.
Roach	10.	6½.	361.	81586.	May 2.
Smelt	2.	0.	149½.	38278.	March 21.
Sole	14.	8.	542½.	100362.	June 13.
Tench	40.	0.		383252*.	May 28.

* Some part of the fpawn of this fifh was by accident loft, fo that the account here is below the reality. *Vide Phil. Tranf.* 1767.

No.

No. III.

Of the method of making ISINGLASS in ICELAND, from the SOUNDS of COD and LING.

THE founds of cod and ling bear general likeness to thofe of the *Sturgeon* kind of *Linnæus* and *Artedi*, and are in general fo well known, as to require no particular defcription. The *Newfound land* and *Iceland* fifhermen fplit open the fifh as foon as taken, and throw the back-bones, with the founds annexed, in a heap; but previous to putrefaction, the founds are cut out, wafhed from their flimes, and falted for ufe. In cutting out the founds, the parts between the ribs are left behind, which are much the beft; the *Iceland* fifhermen are fo fenfible of this, that they beat the bones upon a block with a thick ftick, till the *Pockets*, as they term them, come out eafily, and thus preferve the found entire. If the founds have been cured with falt, that muft be diffolved by fteeping them in water, before they are prepared for *Ifinglafs*. The frefh found muft then be laid upon a block of wood, whofe furface is a little elliptical, to the end of which a fmall hair brufh is

nailed,

nailed, and with a faw-knife, the membranes on each fide of the found muft be fcraped off. The knife is rubbed upon the brufh occafionally, to clear its teeth, the pockets are cut open with fciffars, and perfectly cleanfed of the mucous matter with a coarfe cloth: the founds are afterwards wafhed a few minutes in lime-water, in order to abforb their oily principle; and laftly, in clear water. They are then laid upon nets, to dry in the air; but, if intended to refemble foreign *Ifinglafs*, the founds of cod will only admit of that called book, but thofe of ling both fhapes. The thicker the founds are, the better the *Ifinglafs*, color excepted; but that is immaterial to the brewer, who is its chief confumer.

VOL. III.   D d   No.

No. IV.

CATALOGUE of the ANIMALS DESCRIBED IN THIS VOLUME, WITH THEIR BRITISH NAMES.

R E P T I L E S.

1.	CORIACEOUS Tortoife,	Melwioges.
2.	Common Frog,	Llyffant melyn.
3.	Edible Frog,	Llyffant melyn cefn grwm.
4.	Toad,	Llyffant du, Llyffant daf-adenog.
5.	Natter Jack.	
6.	Great Frog.	
7.	Scaly Lizard.	
8.	Warty Lizard,	Genau goeg ddafadenog.
9.	Brown Lizard,	frech.
10.	Little Lizard,	leiaf.
11.	Anguine Lizard,	naredig.
12.	Viper,	Neidr, Neidr du, Gwiber.
13.	Snake,	Neidr fraith, Neidr y to-menydd.

It is to *Richard Morris*, Efq. that the public is indebted for the *Britiſh* names.

14. Aber-

14. *Aberdeen* Snake.

15. Blind-worm, or Slow-worm,   Pwl dall. Neidr y defaid.

F I S H.

16. COmmon Whale, Morfil Cyffredin.

17. C Pike-headed Whale,   Penhwyad.

18. Fin fifh,   Barfog.

19. Round-lipped Whale, Trwngrwn.

20. Beaked Whale.

21. Blunt-headed Cachalot.

22. Round-headed,   Pengrwn.

23. High-finned,   Uchel aden.

24. Dolphin,   Dolffyn.

25. Porpeffe,   Llamhydydd.

26. Grampus,   Morfochyn.

27. Lamprey, Sea,   Llyfowen bendol, Llamprai.

28. Leffer Lamprey,   Lleprog.

29. Pride.

30. Skate,   Cath fôr, morcath, Rhaien.

31. Sharp-nofed Ray,   Morcath drwynfain.

32. Rough Ray.

33. Fuller Ray.

34. Shagreen Ray.

35. Whip Ray.

D d 2   36. Electric

36. Electric Ray,   Swithbyfg.
37. Thornback,   Moreath bigog.
38. Sting Ray,   Morcath cefn.
39. Angel fifh,   Maelgi.
40. Picked Dog fifh,   Ci Pegod, Picewd.
41. Bafking Shark.
42. White Shark,   Morgi gwin.
43. Blue Shark,   Morgi glas, y Sietc.
44. Long-tailed Shark,   Llwynog mor.
45. Tope,   Ci glas.
46. Spotted Dog fifh,   Ci yfgarmes, morgi mawr.
47. Leffer Dog fifh.
48. Smooth Hound,   Ci Llyfn.
49. Porbeagle.
50. *Beaumaris* Shark.
51. Angler, common,   Morlyffant.
52. Long Angler,   Morlyffant hir.
53. Sturgeon,   Iftwrfion.
54. Oblong Diodon,   Heulbyfg.
55. Short Diodon.
56. Globe Diodon.
57. Lump Sucker,   Jar-fôr.
58. Unctuous Sucker,   Môr falwen.
59. *Jura* Sucker.
60. Longer Pipe fifh.
61. Shorter.
62. Little,   Mor Neidr.
63. Eel,   Llyfowen.
64. Conger,   Mor Llyfowen, Cyngyren.

65. Wolf

65. Wolf fiſh,	Morflaidd.
66. Launce,	Llamrhiaid, Pyſgod by- chain.
67. Morris,	Morys.
68. Sword fiſh,	Cleddyfbyſg.
69. Dragonet, gemmeous.	
70. Dragonet, ſordid.	
71. Weever,	Mor wiber, Pigyn aſtrus.
72. Great Weever.	
73. Common Cod fiſh,	Codſyn.
74. Hadock,	Hadoc.
75. Whiting Pout,	Cod lwyd.
76. Bib,	Deillion.
77. Poor,	Cwdyn ebrill.
78. Coal fiſh,	Chwetlyn glâs.
79. Pollack,	Morlas.
80. Whiting,	Chwitlyn gwyn.
81. Hake,	Cegddu.
82. Forked Hake.	
83. Left Hake.	
84. Trifurcated Hake.	
85. Ling,	Honos.
86. Burbot,	Llefen, Llefenan.
87. Three bearded Cod.	
88. Five bearded Cod.	
89. Torſk.	
90. Creſted Blenny.	
91. Gattorugine.	
92. Smooth Blenny.	

D d 3   93. Spotted

93. Spotted Blenny.
94. Viviparous Blenny.
95. Black Goby.
96. Spotted Goby.
97. Bull Head, River,   Pentarw, Bawd y melinydd.
98. Armed Bull Head,   Penbwl.
99. Father Lasher.
100. Doree,   Sion dori.
101. Opah.
102. Holibut,   Lleden ffreinig.
103. Plaise,   Lleden frech.
104. Flounder,   Lleden 'ddu.
105. Dab,   Lleden gennog, Lleden dwfr croyw.
106. Smear Dab.
107. Sole,   Tafod yr hydd, Tafod yr ych.
108. Smooth Sole.
109. Turbot,   Lleden chwith, Torbwt.
110. Pearl,   Perl.
111. Whiff.
112. Gilt Head,   Peneuryn, Eurben.
113. Red Gilt Head,   Brôm y môr.
114. Toothed Gilt Head.
115. Wrasse, antient,   Gwrach.
116. Ballan.
117. Bimaculated.
118. Trimaculated.
119. Striped.

120. Gibbous.

120. Gibbous.
121. Goldfinny.
122. Comber.
123. Cook.
124. Perch, common,   Perc.
125. Baffe,   Draenog, Gannog.
126. Sea Perch.
127. Ruffe.
128. Black Ruffe.
129. Three fpined Stickle-
    back,   Sil y dom, Pyfgod y gath.
130. Ten fpined,   Pigowgbyfg.
131. Fifteen fpined,   Silod y môr.
132. Mackrel, common, Macrell.
133. Tunny,   Macrell Sopaen.
134. Scad.
135. Red Surmullet,   Hyrddyn coch.
136. Striped Surmullet.
137. Grey Gurnard,   Penhaiarn llwyd, Penhai-
      ernyn.
138. Red Gurnard,   Penhaiarn coch.
139. Piper,   Pibyd.
140. Sapphirine Gurnard, Yfgyfarnog y môr.
141. Streaked Gurnard.
142. Loche, bearded,   Crothell yr afon.
143. Salmon,   Gleifiedyn, Eog, Maran
      Taliefin.
144. Grey,   Penllwyd, Adfwlch.
145. Sea Trout.

D d 4   146. Trout,

146. Trout.   Brithyll.
147. White Trout.
148. Samlet,   Brith y gro.
149. Charr,   Torgoch.
150. Grayling,   Brithyll rheftrog, Glafgan-
                         gen.

151. Smelt,   Brwyniaid.
152. Gwiniad,   Gwiniedyn.
153. Pike,   Penhwyad.
154. Gar Pike,   Môr nodwydd, Corn big.
155. Saury Pike.
156. Argentine.
157. Atherine.
158. Mullet,   Hyrddyn, Mingrwn.
159. Flying Fifh.
160. Herring,   Pennog, yfgaden.
161. Pilchard,   Pennog mair.
162. Sprat,   Coeg Bennog.
163. Anchovy.
164. Shad,   Herlyn, Herling.
165. Carp,   Carp, Cerpyn.
166. Barbel,   Barfbyfg, y Barfog.
167. Tench,   Gwrachen, Ifgretten.
168. Gudgeon,   Crothel.
169. Bream,   Brêm.
170. Rud,   Rhuddgoch.
171. Crucian.
172. Roach,   Rhyfell.
173. Dace,   Darfen, Golenbyfg.

                               174. Graining.

174. Graining.
175. Chub,   Penci, Cochgangen.
176. Bleak,   Gorwynbyfg.
177. Minow,   Crothel y dom, Bychan
                              byfg.

178. Gold Fifh.

A P P E N D I X.

179. Bimaculated Sucker.
180. Beardlefs Ophidium.

INDEX.

I N D E X.

A.

B.

B.

C.

VOL. III. E e Ling,

N.

N.

O.

P.

E e 2   PIPE-

Q.

R.

E e 3 SHAR.

T,

T.

U.

T H E   E N D.

ERRATA.

Page 5, line 25, *for* Cærulæa *read* Cœruleæ. P. 6, l. 8, *for* naturalifts *read* naturalift. P. 8, l. 5, *for* twelve *read* eleven. P. 15, l. 9, *for* horor *read* horror. *Ibid.* l. 14, *for* intrails *read* entrails. P. 57, l. 10, *for* penniformi *read* pinniformi. P. 78, note, *for* tripatinam *read* tripatinum. *Ibid.* *for* appellabatur, fumma &c. *read* appellabatur fumma &c. P. 79, l. 10, *for* Lampetra *read* Lampetræ. P. 85, l. 17, *for* fhire of   *read* fhire of Rofs. P. 86, l. 16, *for* fpiney *read* fpiny. P. 87, l. 9, *(and paffim) for* encreafes *read* increafes. P. 89, l. 18, *for* κι *read* και. P. 91, l. 1, *for* acknowledgements *read* acknowlegements. P. 98, l. 21, *for* in *read* is. P. 105, l. 29, *for* fedement *read* fediment. P. 114, note, *for* 130 *read* 176. P. 129, l. 16, *for* ορθραγοροκ☉ *read* ορθραγορικ☉. P. 131, *for* DIADON *read* DIODON. P. 141, l. 8, *for* ferpentinum *read* ferpentinus. P. 185, l. 18, *for* nufance *read* nuifance. P. 204, l. 1, *for* favoured *read* favored. P. 212, l. 12, *for* reft *read* rays. P. 215, l. 11, and 230, l. 9, *for* fappharine *read* fapphirine. P. 216, l. 3, *for* alepedotus *read* alepidotus. P. 217, l. 16, *for* verrucofo *read* verrucis. *Ibid.* l. 17. *for* bifidis *read* bifido. P. 239, l. 3, *for* on the   fide *read* on the left fide. P. 254, margin, *for* XXVI *read* XXVII. P. 273, note *, *for* p. 222 *read* 265. P. 276, l. 10, *for* vario *read* varia. P. 281, l. 24, *for* Mullis *read* Mullus. P. 286, l. 21, *for* Aberglaftyn *read* Aberglaflyn. P. 288, l. 29, *for* back, fin *read* back-fin. P. 295, l. 1, *for* cinereous *read* cinereus. P. 329, l. 12, *for* radiate *read* radiata. P. 353, l. 9, *for* Cyyrinus *read* Cyprinus. *Ibid.* l. 10, *for* pinna *read* pinnæ. *Ibid.* l. 15, *for* 162 *read* 245. P. 355, l. 18, *for* this *read* the. P. 587, l. 28, *for* fætidi *read* fætidi.

CORRECTIONS of the SPELLING in the WELSH NAMES of Reptiles and Fishes, in Vol. III. BRITISH ZOOLOGY, with fome Additions thereto, by RICHARD MORRIS, Esq.

No.

5. Lyffant gwyllt.
6. Llyffant mawr.
7. Genau goeg gennog.
11. *For* naredig, *read* nadreddig.
12. *For* du, *read* ddu.
14. Neidr Aberdeen.
19. *For* Trwngrwn, *read* Trwyngrwn.
20. ——— trwynfain.
21. ——— penbwl.
23. *For* aden, *read* adain.
25. *For* Llamhydydd, *read* Llamhidydd.
27. *For* bendol, *read* bendoll.
29. Llamprai'r llaid.
32. Moreath arw.
33. Ceffyl Gwyn.
34. Morcath ffreinig.
35. ——— gynffon gwialen.
36. *For* Swithbyfg, *read* Swrthbyfg.
38. *For* Cefn, *read* lefn.
40. *For* Pegod, Picewd, *read* Pigog, Piccwd.

No.

41. Heulgi.
42. *For* gwin, *read* gwyn.
46. *dele* Morgi mawr.
47. Morgi lleiaf.
49. Corgi môr.
50. Morgi mawr.
53. *For* Iftwrfion, *read* Yftwrfion.
55. ——— byrr.
56. ——— crothog.
59. ——— leiaf.
60. Pibellbyfg hir.
61. ——— byrr.
64. *For* Llyfowen, *read* Lyfowen.
69. Morddraig emmog.
70. ——— falw.
72. ——— fawr.
78. *For* chwetlyn, *r.* chwitlyn.
82. ——— fforchogfarf
83. ——— lleiaf.
86. *For* Llefen, Llefenan, *read* Llofen, Llofenen.
87. Codfyn farf tcirfforch.
88. ——— pumfforch.

90. Lly-

CORRECTIONS OF THE SPELLING.

No.

90. Llyfnafeddbyfg cribog.
92. Cleirach gwymmon.
93. Gwrachen fair.
95. Craigbyfg du.
96. ——— brych.
99. Sarph y môr.
101. Brenhinbyfg.
104. *For* 'ddu, *read* ddu.
106. Lleden iraidd.
108. Llefn Dafod yr Hydd.
111. Lleden arw fafnrwth.
113. *For* Brôm, *read* Brêm.
114. Eurben danheddog.
118. Gwrach rengog.
119. ——— gefngrwm.
123. Côgwrach.
126. Perc y môr.
127. Y Garwberc.
128. ——————— du.
133. *For* Sopaen, *read* Yf-
      paen.
134. Macrell y meirch.
136. ——— rhengog.
139. *For* Pibyd, *read* Pibydd.
141. Penhaiarn rheftrog.
143. *Dele* Taliefin.
145. Brithyll y môr.
147. ——— gwyn.
152. *For* Gwiniad, *read* Gwy-
      niad.
156. Arianbyfg.
159. Ehedbyfg.
167. *For* Ifgretten, *read* Yf-
      gretten.
168. *For* Crothel, *read* Cro-
      thell.

No.

173. *For* Golenbyfg, *read*
      Goleubyfg.
177. *For* Crothel, *read* Cro-
      thell.
178. Eurbyfg.
179. Môrfalwen ddeufann.

———

Page.

246. *For* Fernch, *read* French.
256. *For* Raithlyn, *r.* Rhaith-
      lyn.
299. *For* Llyndivi, *read* Llyn-
      teifi.
——— *For* Eynion, *r.* Einion.
——— *For* Machyntleth, *read*
      Machynllaeth.
——— *For* Merionethfhire, *read*
      Montgomeryfhire.
301 *For* Llynallet, *read* Llyn-
      aled.
307. *For* Llynberis, *read*
      Llynperis.
316. *For* Gwiniad, *read* Gwy-
      niad.
317.
318. } Ditto, and in running
      title of four pages,
      and index.
396. *For* Cythrawl, *read*
      Gythraul.
414. Dragonet, 264, *read*
      164.
      For Welch, *read* Welfh,
      throughout.
      White Bait, not in Cata-
      logue. Yr Abwyd
      Gwyn.

ADDITIONAL ERRATA.

Vol. I. P. [xxvi] l. 10, *for* two *read* three. *Ibid.* l. 12, *for* three *read* two. P. 210, l. 9, *for* freſh meat *read* freſh mice. P. 212, l. 2, *for* paper color *read* paler color. P. 217, l. 20, *after* middle feathers, *add* of the tail. P. 231, l. 7, *after* prodigious height, *add* of the ſingle ſtones of. P. 311, l. 14, *after* diſſection, *add* in April. P. 350, l. 23, *for* riſe, *read* riſing. P. 354, l. 12, *after* Wood-lark, *add* and Tit-lark. P. 408, l. 18, *for* ſtanding, *read* ſtunted. P. 411, l. 19, *for* nook *read* noon. **Vol. III.** P. 359, l. 24, *for* Mr. Diaper, *read* Moſes Browne.

C A N C E L S.

Vol. I. E 3	-	- Pages	53,	54.
M 3	-	-	157,	158.
P 3	-	-	205,	206.
Q 4	- -	-	223,	224.
C c 3	-	-	381,	382.
Vol. III. M 6	-	-	171,	172.
X 8	-	-	319,	320.
Z 8	-	-	351,	352.
A a	- -	-	353,	354.

In MAY *next will be publifhed,*

BRITISH ZOOLOGY,

CLASS V.

By THOMAS PENNANT, *Efq*;

CONTAINING ABOUT

NINETY ELEGANT PLATES

OF THE

Shell and Cruftaceous Animals of *Great Britain,*

WITH DESCRIPTIONS.

N. B. This Work will be publifhed both in QUARTO and OCTAVO.

BOOKS of NATURAL HISTORY,

Printed for Benjamin White.

	FOLIO.	l.	s.	d.
1	CATESBY's Carolina, 2 vol. *coloured*, bound —— ——	16	16	0
2	Linnæan Index to Catefby ——	0	2	6
3	Britifh Zoology, by Pennant, *coloured*, half-bound —— ——	11	11	0
4	Borlafe's Natural Hift. and Antiquities of Cornwall, 2 vol. *bound* ——	3	13	6
5	Flora Danica, Oederi, Fafciculi XI. *fut.*	7	4	0
6	Idem Liber, figuris depictis			
7	Chr. Friis Rotboll, Defcript. & Icones Plant. *fut.* ——	0	16	0
8	Jof. Miller's Botanical Prints, 14 Numbers, *coloured* —— ——	14	14	0
9	Curtis's Flora Londinenfis, N° 1, 2, 3, 4, 5, 6, 7, 8, 9, 10, 11, *coloured, at* 5s. *each, to be continued*			
10	The fame, *plain,* 2s. 6d. *each.*			
	2 *UARTO.*			
11	Pennant's Tour to Scotland in 1769, and Voyage to the Hebrides 1772, *with beautiful cuts,* 3 vol. *in boards*	3	13	6
12	Wilkes's Englifh Moths and Butterflies, *coloured* —— ——	9	0	0
13	Drury's Exotic Infects, 2 vol. *coloured,* half-bound ——	5	5	0
14	Brown's Illuftrat. of Zoology, *with* 50 *coloured plates* —— ——	3	3	0
15	Ph. Miller's Gardener's Dict. abridged, *bound* —— ——	1	5	0
16	The Naturalift's Journal, *ftitched*	0	1	6
17	Vandelli Fafciculus Plantarum, *fewed*	0	2	6
18	Milne's Inftitutes of Botany, 2 Parts, *fewed*	0	12	0

B O O K S.

		l.	s.	d.
19	Forſter, (Joan. Rein. & Georg.) nova Genera Plantarum, *cum* 78 *Tabulis, ſemic.*	1	7	0

O C T A V O.

		l.	s.	d.
20	Da Coſta's Elements of Conchology, *cuts, boards*	0	7	6
21	The ſame, *cuts coloured, boards*	0	15	0
22	Pennant's Synopſis of Quadrupeds, *boards*	0	9	0
23	Forſteri Novæ Species Inſectorum, *ſewed*	0	2	6
24	Forſter's Catal. of N. American Animals	0	1	0
25	Oſbeck's and Toreen's Voyage to China, 2 vol. *bound*	0	12	0
26	Catcott's Treatiſe on the Deluge, *bound*	0	6	0
27	Curtis's Fundamenta Entomologiæ, *ſewed*	0	2	6
28	———— Inſtruct. for collecting Inſects, *ſtitched*	0	1	0
29	Randi Index Horti Chelſeiani, *bound*	0	4	0
30	Martyn Catal. Horti Botanici Cantab. & Mantiſſa, *ſut.*	0	4	6
31	Vandelli de Arbore Draconis, *ſut.*	0	1	6
32	Oederi Nomenclator Botanicus, *ſut.*	0	4	0
33	Oederi Enumeratio Plant. Floræ Danicæ, *ſut.*	0	2	0
34	Rob. Sibbaldi [Eq. Aur.] Phalainologia nova, *ſut.*	0	3	0
35	Luidii Lithophylacium Britannicum, *bound*	0	7	0
36	Martyn's firſt Lecture on Botany, *ſtitched*	0	1	6
37	Martyn's Elements of Natural Hiſtory, *ſtitched*	0	1	6
38	Edwards's Elements of Foſſilogy, *ſewed*	0	2	6
39	Lee's Introduction to Botany, 3*d edit. bound*	0	7	6

D U O D E C I M O.

		l.	s.	d.
40	Ph. Miller's Gardener's Kalendar, 16*th edit. bound*	0	3	6

BRITISH ZOOLOGY.

VOL. IV.

CRUSTACEA. MOLLUSCA.

TESTACEA.

O MARE, O LITTUS, verum fecretumque

Μυσεον.' quam multa invenitis, quam multa

dictatis.'

LONDON,

Printed for Benj. White,

MDCCLXXVII.

TO THE

DUTCHESS DOWAGER

OF

PORTLAND,

THIS WORK IS DEDICATED,

AS A GRATEFUL ACKNOWLEGEMENT

OF THE MANY FAVORS

CONFERRED BY HER GRACE

ON HER MOST OBLIGED,

AND MOST OBEDIENT

HUMBLE SERVANT,

Downing,
March 1, 1777.

THOMAS PENNANT.

a

ADVERTISEMENT.

I WISH it had been in my power to have given a perfect conclufion to the ZOOLO-GY of our country: but my fmall acquaintance with INSECTS, and the fourth divifion of the VIth clafs, *Lithophyta* and *Zoophyta*, forbad me to meddle with them. The Public has little reafon to regret this omiffion, fince the univerfal genius JOHN REINHOLD FORSTER, has hinted * a defign of undertaking the firft; and my late worthy friend Mr. ELLIS, (whom LINNÆUS fo juftly ftiles *Lynceus)* has in a great meafure executed the laft.

IN my arrangement of the prefent work, I have taken the liberty of making a diftinct clafs of the CRUSTACEOUS ANIMALS ; and feparated them from INSECTS, among which they are ufually placed.

* Catalogue of Britifh Infects. 2.

I HAVE

I HAVE paid implicit refpect to the *Swedifh* NATURALIST, in my claffing of the VERMES and SHELLS. I have on another occafion *, given my fentiments of that wonderful man, (after RAY) the greateft illuminator of the ftudy of Nature. I have borrowed from him the *Latin* trivial names; fometimes given tranflations of them; fometimes given other *Englifh* names, when I thought them more apt.

GRATITUDE prompts me to mention a moft irreparable lofs in my amiable friend BENJAMIN STILLINGFLEET, Efquire, in whom were joined the beft heart and the ableft head. Benevolence and innocence were his infeparable companions. Retirement his choice, from the moft affectionate of motives ‡. How great, yet how unneceffary was his diffidence in public! How ample, his inftruction in private! How clear his infor-mation! How delicate the conveyance! The pupil received advantage, edified by the hu-mility of the mafter. Thoroughly imbued in Divine Philofophy, he had an uncommon

* *Synopfis* of Quadrupeds, Preface vii.
‡ Mr. GRAY's Letters, 288.

a 3

infight

infight into the ufes of every object of Natural Hiftory; and gave fanction to thofe ftudies, which by trivial obfervers were held moft contemptible. The end of his labors was the GOOD OF MANKIND. He attempted to deftroy the falfe fhame that attended the devotee to Ornithology, the chace of the Infect, the fearch after the Cockle, or the poring over the Grafs. He proved every fubject to be of the greateft fervice to the world, by the proper remarks that might be made on them. The traveller, the failor, the hufbandman might, if they pleafed, draw the moft ufeful conclufions from them. The reader may receive the proof from his tranflations of various effays, the productions of the LINNÆAN fchool; his own CALENDAR of FLORA, and Obfervations on GRASSES. How much to be lamented is this fhort catalogue of the works of fo great, fo good a man! I fpeak not of his Effay on Mufic, as foreign to the fubject. Some of his remarks appear in my *Britifh Zoology*. He thought me fo far deferving of his encouragement, as to dedicate part of his time to farther acts of friendfhip. I received the unfinifhed tokens of his regard by virtue of his promife;

a 3   the

the only papers that were refcued from the flames, to which his modefty had devoted all the reft.

DEFENDED by fo great an example, (howfoever unequally I may follow it) there is hardly any need for an apology for the fubject of the following fheets. But if any fhould require one, I take the liberty of delivering it in the words of my ever regretted friend :

' FROM a partial confideration of things,
' we are very apt to criticife what we ought
' to admire; to look upon as ufelefs what
' perhaps we fhould own to be of infinite
' advantage to us, did we fee a little farther;
' to be peevifh where we ought to give thanks;
' and at the fame time to ridicule thofe, who
' employ their time and thoughts in exa-
' mining what we were, i. e. fome of us moft
' affuredly were, created and appointed to
' ftudy. In fhort, we are too apt to treat
' the Almighty worfe than a rational man
' would treat a good mechanic; whofe
' works he would either thoroughly exa-
' mine, or be afhamed to find any fault with
' them. This is the effect of a partial confi-
' deration of Nature; but he who has can-

5   ' dour

‘ dour of mind, and leifure to look farther,
‘ will be inclined to cry out :

‘ How wond’rous is this fcene ! where all is form’d
‘ With number, weight, and meafure ! all defign’d
‘ For fome great end ! where not alone the plant
‘ Of ftately growth ; the herb of glorious hue,
‘ Or food-full fubftance ; not the laboring fteed,
‘ The herd, and flock that feed us ; not the mine
‘ That yields us ftores for elegance, and ufe ;
‘ The fea that loads our table, and conveys
‘ The wanderer man from clime to clime, with all
‘ Thofe rolling fpheres, that from on high fhed down
‘ Their kindly influence ; not thefe alone,
‘ Which ftrike ev’n eyes incurious, but each mofs,
‘ Each fhell, each crawling infect holds a rank
‘ Important in the plan of Him, who fram’d
‘ This fcale of beings ; holds a rank, which loft
‘ Wou’d break the chain, and leave behind a gap
‘ Which Nature’s felf would rue.   Almighty Being,
‘ Caufe and fupport of all things, can I view
‘ Thefe objects of my wonder ; can I feel
‘ Thefe fine fenfations, and not think of thee ?
‘ Thou who doft thro’ th’ eternal round of time ;
‘ Doft thro’ th’ immenfity of fpace exift
‘ Alone, fhalt thou alone excluded be
‘ From this thy univerfe ? Shall feeble man
‘ Think it beneath his proud philofophy
‘ To call for thy affiftance, and pretend
‘ To frame a world, who cannot frame a clod ?——
‘ Not to know thee, is not to know ourfelves——
‘ Is to know nothing—nothing worth the care
‘ Of man’s exalted fpirit—all becomes
‘ Without thy ray divine, one dreary gloom ;
‘ Where lurk the monfters of phantaftic brains,

a 4   ‘ Order

' Order bereft of thought, uncaus'd effects,
' Fate freely acting, and unerring Chance.
' WHERE meanless matter to a chaos finks
' Or something lower still, for without thee
' It crumbles into atoms void of force,
' Void of resistance—it eludes our thought.
' WHERE laws eternal to the varying code
' Of self-love dwindle. Interest, passion, whim
' Take place of right, and wrong, the golden chain
' Of beings melts away, and the mind's eye
' Sees nothing but the present. All beyond
' Is visionary guess—is dream—is death.'

SYSTEMATIC

SYSTEMATIC INDEX

O F

PLATES.

Class V. CRUSTACEOUS.

Genus I.   CRABS.

SYSTEMATIC INDEX OF PLATES.

O N I S C I,

SYSTEMATIC INDEX OF PLATES.

ONISCI, &c.

Plate XVIII. N° 1. O. Pſora.
    2. Linearis.
    3. Marinus.
    4. Oceanicus.
    5. Entomon.
    6. Oeſtrum.
    7. PHALANGIUM Balænæ.
     SCOLOPENDRA Marina. *No.*
     Tab. **xxv.**

CLASS VI.   W O R M S.

Div. I.   I N T E S T I N E.

Plate XIX. N° 6. Greater DEW-WORM.
    6.A. Leſſer DEW-WORM.
    7. LUG-WORM.
  XX. 3. Marine HAIR-WORM.
    10. Naked TUBE-WORM.
    13. Geometrical LEECH, from
     Roeſel's Inſects.
    14. Tuberculated LEECH.
    15. Glutinous HAG.

         Div.

SYSTEMATIC INDEX OF PLATES.

Div. II.   S O F T.

3

CLASS

SYSTEMATIC INDEX OF PLATES.

Class VI. Div. III. S H E L L S.

* M U L T I V A L V E.

** B I V A L V E.

Plate

SYSTEMATIC INDEX OF PLATES.

PLATE

SYSTEMATIC INDEX OF PLATES.

*** TURBINATED SHELLS.

PLATE

SYSTEMATIC INDEX OF PLATES.

PLATE

SYSTEMATIC INDEX OF PLATES.

b   **** UNI-

SYSTEMATIC INDEX OF PLATES

****   UNIVALVE SHELLS not turbinated.

In Plate **LXXIX** is engraven the BUCCINUM DECUSSATUM from *Weymouth.* It is a young ſhell. When old, the lip is revolute and granulated.

In Plate **LXIV.** at the bottom, are three etchings of a *Mytilus,* from *Weymouth :* a new ſpecies.

10

BRITISH ZOOLOGY,

CLASS V.

CRUSTACEA.

CRUSTACEOUS ANIMALS.

BRITISH ZOOLOGY.

CLASS V.

CRUSTACEOUS ANIMALS.

With eight feet, or ten; rarely fix.   *CANCER.*
Two of the feet clawed.   CRAB.

Two eyes, remote; for the moft part fixed on a
     ftalk, moveable.

Tail foliated, and fhort, lodged in a groove in the
     body.

C. *Lin. Syft.* 1039.   *Pifum.*
     1. PEA.

Cᴀ. WITH rounded and fmooth thorax,
entire and blunt. With a tail of
the fize of the body, which commonly is of the bulk of a pea.

Inhabits the muffel, and unjuftly has acquired
the repute of being poifonous. The fwelling after
eating of muffels is wholly conftitutional; for one
that is affected by it, hundreds remain uninjured.

CRABS, either of this kind, or allied to them, the antients believed to have been the confentaneous inmates of the *pinnæ*, and other bivalves ; which being too ftupid to perceive the approach of their prey, were warned of it by their vigilant friend. *Oppian* tells the fable prettily *.

Οσρακον αυ βυζιας, &c.

In clouded deeps below the *Pinna* hides,
And thro' the filent paths obfcurely glides ;
A ftupid wretch, and void of thoughtful care,
He forms no bait, nor lays the tempting fnare.
But the dull fluggard boafts a *Crab* his friend,
Whofe bufy eyes the coming prey attend.
One room contains them, and the partners dwell
Beneath the convex of one floping fhell ;
Deep in the wat'ry vaft the comrades rove,
And mutual int'reft binds their conftant love ;
That wifer friend the lucky juncture tells,
When in the circuit of his gaping fhells
Fifh wand'ring enter ; then the bearded guide
Warns the dull mate, and pricks his tender fide ;
He knows the hint, nor at the treatment grieves,
But hugs th' advantage, and the pain forgives :
His clofing fhells the *Pinna* fudden joins,
And 'twixt the preffing fides his prey confines ;
Thus fed by mutual aid, the friendly pair
Divide their gains, and all the plunder fhare.

* *Halieut. lib.* ii. He calls the crab Πιννοφυλαξ, *cuftos Pinnæ.*

C. *Lin.*

C. *Lin. Syſt.* 1040. *Gronov. Zooph. No.* 962.   *Minutus.*
Baſter, ii. *p.* 26.   *tab.* iv. *f.* 1. 2.   2. Minute.

Cr. with a ſmooth and ſomewhat ſquare thorax;
the edges ſharp; horns ſhort; leſs than the laſt.
Inhabits our ſhores among *Algæ.*

C. *Lin. Syſt.* 1040. *Gronov. Zooph. No.* 968.   *Longicornis.*
Baſter, ii. *p.* 26. *tab.* iv. *f.* 3.   3. Long-
HORNED.

Cr. with a round ſmooth thorax; with large
claws; very long horns; ſize of the laſt.
Inhabits our ſhores.

Cancer latipes. *Rondel,* 565. *Gronov. Zooph. No.* 954.   *Latipes.*
Cancer latipes parvus oblongus variegatus. *Plancus,* 34.   4. Broad-
tab. iii. *fig.* 7.   FOOT.

Cr. with a ſub-cordated body; ſhort feelers;
angular claws; five ſmall teeth on each ſide; the
hind legs ovated.

C. *Lin. Syſt.* 1043. *Baſter,* ii. *tab.* ii. *f.* 1.   *Mænas.*
Faun. Suec. No. 2026. *Gronov. Zooph.* 955.   5. Common.

Cr. with three notches on the front; five ſerrated
teeth on each ſide; claws ovated; next joint,
                    B 2   toothed;

toothed ; hind feet fubulated ; dirty green color ; red when boiled.

Inhabits all our fhores ; and lurks under the *Algæ*, or burrows under the fand. Is fold ; and eaten by the poor of our capital.

Depurator.
6. CLEAN-
SER.

C. *Lin. Syft.* 1043. *No.* 23.
Seb. Muf. iii. *tab.* xviii. *fig.* 9.

CR. with a fub-cordated body ; thorax on each fide quinque-dentated ; front indented ; claws an-gulated ; fecond joint fpined ; hind legs have the two laft joints ovated and ciliated.

A. vi. Variety with a tuberculated furface. *Vide* tab. iv.

Inhabits generally the deeps ; feeds on dead fifh : hence called *the purifier* or *cleanfer,* as caufing the removal of putrid bodies.

Pagurus.
7. BLACK-
CLAWED.

C. *Lin. Syft.* 1044.   *Gronov. Zooph. No.* 967.
Belon. aquat. 368.   *Rondel. pifc.* 560.   *Faun. Suec. No.* 2028.
Merret's Pinax.

CR. with a crenated thorax ; fmooth body ; quin-que-dentated front ; fmooth claws with black tips ; hind feet fubulated.

6   Inhabits

Inhabits the rocky coasts; the most delicious meat of any; casts its shell between *Christmas* and *Easter*.

The tips of the claws of this species are used in medicine; intended to absorb acidities in the stomach and bowels.

CR. with the thorax quinque-dentated; body covered with short brown velvet-like pile; claws covered with minute tubercles; small spines round the top of the second joint; hind legs broadly ovated. This is among the species taken notice of by *Aristotle* * on account of the broad feet, which, he says, assist them in swimming: as web-feet do the water-fowl.

Velutinus.
8. VELVET.

Inhabits the western coasts of *Anglesea*.

CR. with the thorax quinque-dentated; serrated; body wrinkled transversely; claws furnished with a single spine on the first and second joint; fangs serrated; last pair of legs ovated.

Corrugatus.
9. WRINK-LED.

Found on the shores of *Skie*, opposite to *Loch Jurn*.

CR. with a rectangular body; the thorax armed near the corner with two spines; the claws very

Angulatus.
10. ANGU-LAR.

* *De Part. Anim. lib.* iv. *c.* 8.

B 3   long;

long; the upper fangs black; legs flender and fubulated.

Weymouth. From the Portland cabinet.

Hirtellus.　　C, *Lin. Syft.* 1045. *Faun. Suec. No.* 2029.
11. Bristly. Cancer hirfutus. *Rondel.* 568.

Cr. with a hairy thorax; on both fides flightly quinque-dentated; claws ovated, flightly echinated, and hairy; feet, briftly and fubulated. A fmall fpecies; of a reddifh color.

Found beneath ftones.

Platy-cheles.　Cr. with a tridentated front; thorax entire; claws
12. Great-
clawed.　　of a large fize; depreffed, and greatly ciliated on the outfide; only three fubulated legs on each fide; body little bigger than a horfe-bean, and almoft round: *Antennæ* very long and turning back, when not in ufe.

Inhabits the *Algæ* on the coaft of *Anglefea* and the *Hebrides.*

Caffivelaunus.　Cr. with bifurcated front; a fpine at the corner
13. Long-
clawed.　　of each eye; another on each fide of the thorax towards the tail; body ovated and fmooth; *Antennæ* of the length of the body; the claws above; as long again as the body; feet fubulated. The fuppofed female; of the fame form; only the claws not half fo long.

Inhabits

Inhabits the deep near *Holyhead* and *Red-Wharf Anglefea*. Dredged up.

Cancer. *Lin. Syft.* 1047.   *Horridus.*
C. fpinofus. *Seb. Muf.* iii. *tab.* xxii. *f.* 1. *Gronov. Zooph.* 14. HORRID.
  No. 976.
Fans, Trold Crabber. *Pontop. Norway* ii. 176. *tab. p.* 177.

CR. with a projecting bifurcated fnout, the end diverging ; body heart-fhaped ; and with the claws and legs covered with long and very fharp fpines. A large fpecies.

Inhabits the rocks on the eaftern coaft of *Scotland.* Common to *Norway* and *Scotland,* as many of the marine animals and birds are.

CR. with a quadri-furcated fnout ; the two middle   *Tetra-odon.*
fpines the longeft ; thorax fpiny ; body heart-   15. FOUR-
fhaped and uneven ; claws long ; legs flender.   FORKED.

Inhabits the *Ifle of Wight.*

Cancer. *Lin. Syft.* 1044.
Faun. Suec. No. 2030. *Jonfton Exang. tab.* v. *fig.* 13.   *Araneus.*
                                                              16. SPIDER.

CR. with a bifid fnout ; briftly thorax ; body, heart-fhaped, and tuberculated ; claws long and oblongly ovated ; legs flender, long and fubulated.

Inhabits our fhores. Often covered with a *byffus,* as in fpecimen xvi. A.

B 4   CR. with

Phalangium.
17. Slen-
der-leg'd.

Cr. with a bifid fnout; heart-fhaped, fmall tuberculated body; long claws; legs of a vaft length,
very flender, and hairy.

Inhabits the depths on the coafts of *Anglefea*.

Dorfettenfis.
18. Wey-
mouth.

Cr. with a cordated body, rugged and bent, with
a few fpines; very thick, and long claws; and
very flender legs, the firft pair much longer than
the reft.

Weymouth.   From the Portland cabinet.

Tuberofus.
19. Uneven.

Cr. with a tuberous, fmooth back; fmall claws,
and fhort legs; fnout flightly bifid.

From the fame cabinet.

Afper.
20. Rough.

Cr. with a cordated body; bifid fnout; legs and
claws fhort; thofe and the body rough and fpiny.

From the fame cabinet.

Cylindric

Cylindric body.
Long antennæ.
Long tail.

Cancer.   *Lin. Syſt.* 1050. *No.*
Aſtacus.   *Rondel.* 538.

Gammarus.
21.Vulgar.

L. WITH a ſmooth thorax; ſhort ſerrated ſnout; very long *antennæ*; and between them two ſhorter, bifid; claws and fangs, large, the greater tuberculated, the leſſer ſerrated on the inner edge; four pair of legs; ſix joints in the tail; caudal fins rounded.

Inhabits all the rocky ſhores of our iſland; but chiefly where there is a depth of water. In *Llyn*, in *Caernarvonſhire*, a certain ſmall lobſter, nothing different except in ſize, burrows in the ſand.

Brought in vaſt quantities from the *Orkney* iſles, and many parts of the eaſtern coaſt of *Scotland*, to the *London* markets. Sixty or ſeventy thouſand are annually brought, in well-boats, from the neighborhood of *Montroſe* alone *.

Lobſters fear thunder; and are apt to caſt their claws on a great clap. I am told they will do the ſame on firing a great gun; and that when men of

* Tour in *Scotland,* 1772. *part.* ii. *p.* 146.

war

war meet a lobfter-boat, a jocular threat is ufed, That, if the mafter does not fell them good lobfters, they will *falute him*.

The habitation of this fpecies is in the cleareft water; at the foot of rocks that impend over the fea. This has given opportunity of examining more clofely into the natural hiftory of the animal, than many others who live in an element that prohibits moft of the human refearches, and limits the inquiries of the moft inquifitive. Lobfters are found on moft of the rocky coafts of *Great Britain*. Some are taken by the hand; but the greater quantity in pots, a fort of trap formed of twigs, and baited with garbage; they are formed like a wire moufe-trap, fo that when the lobfter gets in, there is no return. Thefe are faftened to a cord funk into the fea, and their place marked by a buoy.

They begin to breed in the fpring, and continue breeding moft part of the fummer. They propagate *more humano*; and are extremely prolific. Doctor *Bafter* fays he counted 12,444 eggs under the tail, befides thofe that remained in the body, unprotruded. They depofit thefe eggs in the fand, where they are foon hatched.

Lobfters change their cruft annually. Previous to their putting off their old one, they appear fick, languid, and reftlefs. They totally acquire a new coat in a few days; but during the time that they remain defencelefs they feek fome very lonely place,

place, for fear of being attacked and devoured by
fuch of their brethren that are not in the fame weak
fituation.

It is alfo remarkable, that Lobfters and Crabs
will renew their claws, if by accident they are torn
off; and it is certain they will grow again in a few
weeks.

They are very voracious animals, and feed on
fea-weeds, on garbage, and on all forts of dead
bodies.

Additional to this, I beg leave to give an accurate account of the natural hiftory of this animal,
communicated to me by the ingenious Mr. *Travis*,
furgeon, at *Scarborough*.

' *Scarborough*, 25th *Oct.* 1768.

' S I R,

' W E have vaft numbers of fine Lobfters
' on the rocks, near our coaft. The large ones
' are in general in their beft feafon from the middle
' of *October* till the beginning of *May*. Many
' of the fmall ones, and fome few of the larger
' fort are good all the fummer. If they be four
' inches and a half long or upwards, from the tip
' of the head to the end of the back fhell, they
' are called *fizeable Lobfters*. If only four inches,
' they are efteemed half fize; and when fold, two
' of them are reckon'd for one of fize. If they
' be under four inches, they are called *pawks*, and
' are

' are not faleable to the carriers, though, in reality,
' they are in the fummer months fuperior to the
' large ones in goodnefs. The pincers of one of
' the lobfters large claws are furnifhed with
' knobs, and thofe of the other claw are always
' ferrated. With the former it keeps firm hold of
' the ftalks of fubmarine plants, and with the
' latter it cuts and minces its food very dextroufly.
' The knobbed or numb claw, as the Fifhermen
' call it, is fometimes on the right and fometimes
' on the left, indifferently. It is more dangerous
' to be feized by them with the cutting claw than
' the other; but in either cafe, the quickeft way
' to get difengaged from the creature is to pluck
' off its claw. It feems peculiar to the Lobfter
' and Crab, when their claws are pulled off, that
' they will grow again, but never fo large as at
' firft.

   ' The Female or Hen Lobfter does not caft
' her fhell the fame year that fhe depofits her *ova*,
' or, in the common phrafe, is in *berry*. When
' the *ova* firft appear under her tail, they are very
' fmall and extremely black; but they become in
' fucceffion almoft as large as ripe elder-berries
' before they be depofited, and turn of a dark
' brown color, efpecially towards the end of the
' time of her depofiting them. They continue full
' and depofiting the *ova* in conftant fucceffion, as
' long as any of that black fubftance can be found
                    ' in

' in their body, which, when boiled, turns of a
' beautiful red color, and is called their *coral*.
' Hen Lobſters are found in *berry* at all times of
' the year, but chiefly in winter. It is a common
' miſtake, that a berried Hen is always in perfection
' for the table. When her berries appear large
' and browniſh, ſhe will always be found exhauſted,
' watery, and poor. Though the *ova* be caſt at
' all times of the year, they ſeem only to come to
' life during the warm ſummer months of July and
' Auguſt. Great numbers of them may then be
' found, under the appearance of tad-poles, ſwim-
' ming about the little pools left by the tides among
' the rocks, and many alſo under their proper
' form, from half an inch to four inches in length.

' In caſting their ſhells, it is hard to conceive
' how the Lobſter is able to draw the fiſh of their
' large claws out, leaving the ſhells entire and
' attached to the ſhell of their body; in which
' ſtate they are conſtantly found. The fiſhermen
' ſay the Lobſter pines before caſting, till the fiſh
' in its large claw is no thicker than the quill of a
' gooſe, which enables it to draw its parts through
' the joints and narrow paſſage near the trunk. The
' new ſhell is quite membraneous at firſt, but
' hardens by degrees. Lobſters only grow in ſize
' while their ſhells are in their ſoft ſtate. They are
' choſen for the table, by their being heavy in
' proportion to their ſize; and by the hardneſs of
' their

' their shells on their sides, which, when in per-
' fection, will not yield to moderate pressure.
' Barnacles and other small shell-fish adhering to
' them are esteemed certain marks of superior good-
' ness. Cock-Lobsters are in general better than
' the Hens in winter ; they are distinguished by
' the narrowness of their tails, and by their having
' a strong spine upon the center of each of the
' transverse processes beneath the tail, which sup-
' port the four middle plates of their tails. The
' fish of a Lobster's claw is more tender, delicate,
' and easy of digestion than that of the tail. Lob-
' sters are not taken here in pots, as is usual where
' the water is deeper and more still than it is upon
' our coast. Our fishermen use a bag-net fixed to
' an iron hoop, about two feet in diameter, and
' suspended by three lines like a scale. The bait is
' commonly fish-guts tied to the bottom and middle
' of the net. They can take none in the day-time,
' except when the water is thick and opake ; they
' are commonly caught in the night, but even
' then it is not possible to take any when the sea
' has that luminous appearance which is supposed
' to proceed from the *nereis noctiluca*. In summer,
' the Lobsters are found near the shore, and thence
' to about six fathoms depth of water ; in winter,
' they are seldom taken in less than twelve or
' fifteen fathoms. Like other insects, they are
' much more active and alert in warm weather
' than

' than in cold.   In the water they can run nim-
' bly upon their legs or fmall claws, and if alarmed
' can fpring tail-foremoft, to a furprifing diftance,
' as fwift as a bird can fly.   The fifhermen can
' fee them pafs about thirty feet, and by the
' fwiftnefs of their motion, fuppofe they may go
' much farther.   *Athenæus* remarks this circum-
' ftance, and fays, that *the incurvated* Lobfters *will*
' *fpring with the activity of* dolphins.   Their eyes
' are raifed upon moveable bafes, which enables
' them to fee readily every way. When frightened,
' they will fpring from a confiderable diftance to
' their hold, in the rock ; and what is not lefs fur-
' prifing than true, will throw themfelves into
' their hold in that manner, through an entrance
' barely fufficient for their bodies to pafs ; as is
' frequently feen by the people who endeavor to
' take them at *Filey Bridge.* In frofty weather,
' if any happen to be found near the fhore, they
' are quite torpid and benumbed.   A fizeable
' Lobfter is commonly from one pound to two in
' weight.   There was one taken here this fummer
' which weighed above four, and the fifhermen fay
' they have feen fome which were of fix pounds,
' but thefe are very rare.

' I am, Sir, &c.'

I conclude with faying, that the Lobfter was
well known to the ancients, and that it is well de-

3   fcribed

fcribed by *Ariftotle*, under the name of Αϛαχος * ;
that it is found as far as the *Hellefpont*, and is called,
at *Conftantinople*, † *Liczuda*, and *Lichuda*.

Homarus.
22. SPINY.
Cancer. *Lin. Syft.* 1053.
Locufta. la Langoufle. *Rondel. pifc.* 535.

L. with a front broad, armed with two large fpines,
and between them a fmaller, guards to the eyes,
which are prominent; *Antennæ* longer than body
and tail; fpiny at their origin; beneath them two
leffer; claws fhort, fmall, fmooth; fangs fmall,
fingle, hinged; legs flender and fmooth; body and
thorax horrid with fpines; tail longer than that of
the common Lobfter; on each part, above, is a
white fpot, the bottoms are crooked and ferrated;
the tail-fin, partiy membranaceous, partly cruftaceous.

Inhabits our rocky coafts; often taken about
the promontory of *Liŷn*, and *Bardfey* ifle.

The *French* name of this fpecies has been bar-
baroufly tranflated into the *Long-oyfter*.

---

* *Hift. An. lib.* iv. *c.* 2.
† *Belon Hift. Poiffons.* 357.

C. *Lin.*

C. *Lin. Syft.* 1053. *No.* 75.   *Faun. Suec. No.* 2040.
Squilla lata.   *Rondel.* 545.

Arctus.
23. Broad.

L. with two broad ferrated plates before the eyes ;
fhort furcated *antennæ*; body and tail flat and
broad.

Size of the fpiny Lobfter.

Found by Doctor *Borlafe* on *Careg Killas*, in
Mounts-Bay. Is common to the four quarters of
the world.

Cancer Norvegicus.   *Lin. Syft.* 1053.
Sundfiord.   *Pontop. Norway.* ii. 175. *tab. p.* 177.

Norvegicus.
24. Nor-
way.

L. with a long fpiny fnout ; thorax flightly fpiny ;
body marked with three ridges ; claws very long,
angular, and (along the angles) fpiny ; *antennæ*
long ; legs flender, clawed ; tail long ; elegantly
marked with fmooth and fhort-haired fpaces,
placed alternately.

Common length, from tip of the claws to the
end of the tail near nine inches.

Leo.   *Rondel.* 542.

Bamffius.
25. Long-
clawed.

L. with a fmooth thorax, with three fharp flender
fpines in front ; claws fix inches and a half long,
flender and rough ; fangs ftrait ; legs weak, briftly ;

Vol. IV.   C   *antennæ*

antennæ flender, two inches and a half long; tail and body about five inches.

Taken near *Bamff*. Communicated to me by the Reverend Mr. *Cordiner*, and engraven from his beautiful drawing.

Strigofus.
26. PLATED.

Cancer. *Lin. Syft.* 1052.

L. with a pyramidal fpiny fnout; thorax elegantly plated; each plate marked near its junction with fhort *ftriæ*; claws much longer than the body, thick, echinated, and tuberculated; the upper fang trifid; only three legs, fpiny on their fides; tail broad.

The largeft of this fpecies is about fix inches long.

Inhabits the coafts of *Anglefea*; under ftones and *fuci*. Very active. If taken, flaps its tail againft the body with much violence and noife.

Aftacus.
27. CRAW-
FISH.

Cancer. *Lin. Syft.* 1051.

L. with a projecting fnout flightly ferrated on the fides; a fmooth thorax; back fmooth, with two fmall fpines on each fide; claws large, befet with fmall tubercles; two firft pair of legs clawed; the two next fubulated; tail confifts of five joints; the caudal fins rounded.

5

Inhabits

Inhabits many of the rivers of *England* ; lodged in holes which they form in the clayey banks. *Cardan* fays that this fpecies is a fign of the good-nefs of water ; for in the beft water, they are boiled into the reddeft color *.

Squilla Crangon. *Rondel.* 547.

Serratus.
28. Prawn.

L. with a long ferrated fnout bending upwards ; three pair of very long filiform feelers ; claws fmall, furnifhed with two fangs ; fmooth thorax ; five joints to the tail ; middle caudal fin fubu-lated ; two outmoft flat and rounded.

Frequent in feveral fhores, amidft loofe ftones ; fometimes found at fea, and taken on the furface over thirty fathoms depth of water ; cinereous when frefh ; of a fine red when boiled.

Cancer Squilla. *Lin. Syft.* 1051. *Faun. Suec. No.* 2037.
Squilla Batava. *Seb. Muf.* iii. *p.* 55. *tab.* xxi. *fig.* 9. 10.
Squilla fufca. *Bafter* ii. 30. *tab.* iii. *fig.* 5.
Squilla Gibba. *Rondel.* 549.

Squilla.
29. White.

L. with a fnout like the prawn, but deeper and thinner ; and feelers longer in proportion to the bulk ; the fub-caudal fins rather larger ; is at full growth not above half the fize of the former.

* Quoted by *Plot. Hift. Staffordf.* 185.

C 2

Inhabits

Inhabits the coasts of *Kent*; is sold in *London* under the name of *the white shrimp*, as it assumes that color when boiled.

Crangon.	Cancer Crangon. *Lin. Syst.* 1052.
30. Shrimp.	Squilla marina Batava. *Baster.* ii. 27. *tab.* iii. *fig.* I. 11.
	Reesel insect. iii. *tab.* lxiii.

L. with long slender feelers, and between them two thin projecting *laminæ*; claws with a single-hooked moveable fang; three pair of legs; seven joints in the tail; the middle caudal fin subulated; the four others rounded and fringed; a spine on the exterior side of each of the outmost.

Inhabits the sandy shores of *Britain*, in vast quantities. The most delicious of the genus.

Linearis.	Cancer. *Lin. Syst.* 1056.
31. Linear.	Lesser garnel or shrimp. *Marten's Spitzberg.* 115. *tab.* P.
	fig. 1.

L. with long slender claws, placed very near the head, with a slender body, and six legs on each side; is about half an inch long.

Found in the sand, on the shore of *Flintshire*; is very frequent in *Spitzbergen*.

Cancer.

Cancer. *Lin. Syst.* 1056.   *Atomos,*
Mirum animalculum in corallinis, &c. *Baster.* i. 43. *tab.* iv. 32. Atom.
  fig. 11.

L. with a slender body; filiform *antennæ*; three
pair of legs near the head; behind which are two
pair of oval *vesiculæ*; beyond, are three pair of legs,
and a slender tail between the last pair.

Very minute. The help of the microscope
often necessary for its inspection.

C. *Lin. Syst.* 1055. *No.* 81.   *Pulex.*
    33. Flea.

L. with five pair of legs, and two pair of claws
imperfect; with twelve joints in the body.

Very common in fountains and rivulets; swims
swiftly in an incurvated posture on its back; em-
braces and protects its young between the legs;
does not leap.

L. *Lin. Syst.* 1055. *No.* 82.   *Locusta.*
Rosel Insect. iii. *tab.* 62.   34. Locust.

L. with four *antennæ*; two pair of imperfect claws;
the first joint ovated; body consists of fourteen
joints, in which it differs from the former.

Abounds in summer-time on the shores, beneath
stones and *algæ*; leaps about with vast agility.

C 3   Cancer.

Salinus.
35. SALT.

Cancer. *Lin. Syst.* 1056.

L. with jointed body; hands without claws; *an-tennæ* shorter than the body; ten pair of legs; tail filiform, subulated; very minute.

Discovered by Doctor *Maty* in the salt pans at *Limington.*

Stagnalis.
36. POND.

Cancer. *Lin. Syst.* 1056.

L. with jointed body; hands without claws; a bifid tail.

Inhabits the crannies of rocks, in fresh waters; suspected by *Linnæus* to be the *larva* of an *Ephemera.*

The two last never fell under my notice.

Mantis.
37. MANTIS.

C. *Lin. Syst.* 1054. *No.* 76.

L. with short *antennæ*; short *thorax*, and two pinnated substances on each side; three pair of claws with hairy ends; the body long, divided by eight segments: two fins on each side of the tail; tail conoid, with spines on the margin.

From the PORTLAND cabinet.
Weymouth.

Cancer

Cancer. *Lin. Syft.* 1049.   *Bernardus.*
                                                        38.Hermit.

C. with rough claws; the right claw is the longer; the legs fubulated, and ferrated along the upper ridge; the tail naked and tender, and furnifhed with a hook, by which it fecures itfelf in its lodging.

This fpecies is parafitic, and inhabits the empty cavities of turbinated fhells, changing its habitation according to its increafe of growth, from the fmall *nerite*, to the large *whelk*. Nature denies it the ftrong covering behind, which it has beftowed on others of this clafs, and therefore directs it to take refuge in the deferted cafes of other animals.

Ariftotle defcribes it very exactly under the name of Καρκινιον *. By the moderns it is called the *foldier*, from the idea of its dwelling in a tent; or the *hermit*, from retiring into a cell.

* *Hift. An. lib.* iv. *c.* 4. *lib.* v. *c.* 15.

C 4

TABLE XVIII.

MARINE INSECTS.

I. Oniseus Pfora.

II.   Linearis.

III.   Marinus. *Pallas Spicil.* fafc. ix. tab. iv. f. 6.

IV.   Oceanicus.

V.   Entomon.

VI.   Oeftrum.

VII. Phalangium Balænæ.

N° III. of TABLE XXV.

Scolopendra Marina.

CLASS VI.

VERMES,

WORMS.

Div. I. INTESTINE.

    II. SOFT.

    III. TESTACEOUS.

C L A S S VI.

V E R M E S.

W O R M S.

MIHI CONTUENTI SESE PERSUASIT RERUM NATURA,
NIHIL INCREDIBILE EXISTIMARE DE EA.

Plinii lib. xi. c. 3.

SLOW, foft, expanding, tenacious of life, fometimes capable of being new formed from a
part; the enliveners of wet places; without head
or feet; hermaphroditical; to be diftinguifhed by
their feelers.

Not improperly called by the ancients, *imperfect
animals*; being deftitute of head, ears, nofe, and
feet, and for the moft part of eyes; moft different
from infects; from which LINNÆUS has long fince
removed thefe works of Nature.

They may be divided into INTESTINE, SOFT,
TESTACEOUS, LITHOPHYTES, and ZOOPHYTES.

The

The Intestine (heretofore ftyled *the earthly*) perforate all things by help of the great fimplicity of their form. The Gordius pierces the clay, that the water may percolate; the Lumbricus, the common foil, leaft it fhould want moifture; the Myxine, dead bodies, in order that they may fall innoxioufly to pieces; the Teredo, wood, to promote its decay. In like manner, Pholades, and fome forts of muffels penetrate even rocks, to effect their diffolution.

The Mollusca, or Soft, are naked, furnifhed with arms; for the moft part wander through the vaft tract of ocean; by their phofphoreous quality illuminate the dark abyfs, reflecting lights to the heavens; thus what is below correfponds with the lights above.

Thefe Mollusca often become the inhabitants of teftaceous calcareous covers, which they carry about with them, and often they themfelves penetrate calcareous bodies; like infects, are multiplied into infinite variety: and exhibit, both in form and colors, fplendid examples of the excelling powers of the all-mighty Artificer. Nor are they without their ufes; feveral fpecies afford a delicious and nourifhing nutriment. The healing art calls in the fnail in confumptive cafes; and the fhells calcined are of known efficacy in ftubborn acidities. Shells are the great manure of lands in many parts of thefe kingdoms. The pearls of *Great Britain* have been celebrated from the time of *Cæfar*.

CLASS

C L A S S   VI.

V E R M E S.

W O R M S.

Div. I.   I N T E S T I N E.

I. With a filiform body, of equal thickneſs;   *GORDIUS.*
   ſmooth.   HAIR-WORM.

Gordius.   *Lin. Syſt.* 1075.   *Faun. Suec. No.* 2068.   *Aquaticus.*
Vitulus aquaticus.   *Geſner aq.*   1. WATER.

G. **O**F a pale color, with the ends black.
      Inhabits boggy places, and clay at the
bottom of water.

G.   *Lin. Syſt.* 1075.   *Faun. Suec. No.* 2069.   *Argillaceous.*
                                                   2. CLAY.

G. of an uniform yellow color.

                                          G. *Lin.*

Marinus.   G. *Lin. Syst.* 1075.
3. MARINE.

G. filiform, twifted fpirally and lying flat. *Tab.* xx. *fig.* 3.

Common in the inteftines of the herring and other fea-fifh. *Ariftotle* * remarks that the *Ballerus* and *Tillo* are infefted in the dog-days with a worm that torments them fo much, that they rife to the top of the water, where the heat deftroys them. *Bleaks* are obferved to rife at certain feafons to the furface, and tumble about for a confiderable fpace, in feeming agonies. I fufpect them to be affected in the fame manner with thofe *Ariftotelian* fifh.

ASCARIS.   II. Slender filiform body, attenuated at each end.

Vermicularis.   Afcaris. *Lin. Syft.* 1076.
4. VERMI-
CULAR.

Asc. With faint annular *rugæ*; thicker at one end than the other; mouth tranfverfe.

Inhabits, according to *Linnæus*, boggy places, and under the roots of decayed plants; found in the *rectum* of children and horfes; often obferved in the dung of the laft; emaciates children greatly; is fometimes vomited up.

* *Hift. An. lib.* viii. *c.* 20.

Afcaris.

Afcaris.   *Lin. Syft.* 1076.   *Lumbricoides.*
                                                     5. Common.

Asc. with a flender body, fubulated at each end;
but the tail triangular; grows to the length of
nine inches; viviparous; and produces vaft numbers.

Inhabits the human inteftines.

III. Slender annulated body, furnifhed with a   *LUMBRICUS.*
        lateral pore.   DEW-WORM.

Lumbricus.   *Lin. Syft.* 1076.   *Faun. Suec. No.* 2073.   *Terreftris.*
Raii infect. 1.   6. Dew.

L. with a hundred and forty rings; head taper;
mouth, at the end, round; fore part of the worm
cylindric, the reft depreffed; at about one third of
its length is a prominent annulated belt; on each
fide of the belly a row of minute fpines, diftin-
guifhable only by the touch; affiftant in motion.
Tab. xix. *fig.* 6.

A variety only of the former; excepting in fize,   *Minor.*
refembling it.   *Raii infect.* 2.   A. Lesser.

    Inhabits the common foil, and by perforating,
renders it apt to receive the rain; devours the

    Vol. IV.   D   *cotyledons*

cotyledons of plants, or part of the feed that vege-
tates ; comes out at night to copulate ; is the food
of moles, hedge-hogs, birds, &c. In *English*,
the *Dew* or *Lobworm*. *Tab.* xix. *fig.* 6. A.

Inteſtinalis.
B. INTESTI-
NAL.

Inhabits the leſſer inteſtines of the human
ſpecies, chiefly of children ; does not differ in the
left from the former kinds.

Marinus.
7. LUG.

L. marinus. *Lin. Syſt.* 1077. *Faun. Suec. No.* 2074. *Belon.*
aq. 444.

L. with round mouth, and circular body annulated with greater and leſſer rings ; the firſt prominent ; on each of them are two tufts of ſhort
briſtles placed oppoſite ; the tail-part is ſmooth ;
elegant ramifications are obſerved to iſſue from
among the tufts in the living worm ; is ſoft and
full of blood.

Inhabits ſandy ſhores, burying itſelf deep ; but
its place diſtinguiſhable by a little riſing, with an
aperture on the ſurface ; of great uſe as a bait
for fiſh. *Tab.* xix. *fig.* 7.

IV. Flattiſh

IV. Flattifh body ; a pore at the extremity, and on *FASCIOLA.* the belly.   FLUKE.

Fafciola. *Lin. Syft.* 1077. *Faun. Suec. No.* 2075. *Amæn. Acad.* *Hepatica.* 8. LIVER,
Ræfel. app. tab. xxxii. *f.* 5. *Borlafe Nat. Hift. Cornwall, tab.* xx. *fig.* 10.

F. with an ovated body, a little fharper on the fore part; in the centre is a white fpot, with a line of the fame color paffing towards each extremity.

Infefts the livers of fheep and hares.

Fafciola. *Lin. Syft.* 1078. *Faun. Suec. No.* 2076. *Inteftinalis,*
Lin. Syft. ed. vi. 70. *tab.* vi. *f.* 1.   9. INTES-
                                                              TINE,

F. with a long flender body, if extended ; when contracted, of a fub-oval form.

Inhabits the inteftines of frefh-water fifh ; dif-covered in *breams* and *fticklebacks.*

D 2   V. A

IPUNCULUS.
ГUBE-WORM.

V. A flender lengthened body.

Mouth, at the very end; attenuated, cy-lindric.

Aperture on the fide of the body.

Nudus.
10. NAKED,

Sipunculus. *Lin. Syft.* 1078.
Vermis macrorhynchopterus. *Rondel. Zooph.* 110. *Gefner.*
aq. 1026.
Syrinx. *Bohedfch. marin.* 93. *tab.* vii. *fig.* 6. 7.

T. With a cylindric extended mouth, laciniated round the inner edges; body rounded, taper, at the end globofe; about eight inches long; aper-ture at the fide, a little below the mouth. *Tab.* xx. *fig.* 10.

Inhabits the fea.

HIRUDO.
LEECH.

VI. Body oblong; moves by dilating the head and tail, and raifing the body into an arched form.

Medicinalis.
11. MEDI-CINAL.

H. *Lin. Syft.* 1079. *Faun. Suec. No.* 2079. *Raii insect.* 3. *Gefner pifc.* 425.

L. With a brown body, marked with fix yellow lines.

Inhabits

Inhabits ftanding waters.   The beft of phlebo-
tomifts, efpecially in *hæmorrhoids*.   The practice
is as old as the time of *Pliny*, who gives it the
apt name of *hirudo fanguifuga*.   Leeches were ufed
inftead of cupping-glaffes for perfons of plethoric
habits, and thofe who were troubled with the gout
in the feet.   He afferts, that if they left their head
in the wound, as fometimes happened, it was in-
curable ; and informs us, that *Meffalinus*, a perfon
of confular dignity, loft his life by fuch an accident *.

H. *Lin. Syft.   Faun. Suec. No.* 2078.   *Sanguifuga.*
Hirudo maximè apud nos vulgaris.   *Raii infect.* 3.   12. HORSE.

L. with a deprefled body ; in the bottom of the
mouth are certain great fharp tubercles or whitifh
caruncles.   The flendereft part is about the mouth ;
the thickeft towards the tail ; the tail itfelf very
flender ; the belly of a yellowifh green ; the back
dufky.

Inhabits ftanding waters.

Leeches are good barometers, when preferved
in glaffes, and predict bad weather by their great
reftleffnefs and change of place.

* *Lib.* xxxii. *c.* 10.

D 3   H. *Lin.*

Geometra.
13. GEOME-
TRICAL.

H. *Lin. Syst.* 1080. *Faun. Suec. No.* 2083.
Ræsel. App. tab. xxxii. *f.* 1. 4.

L. with a filiform body; greenish, spotted with white; both ends dilatable, and equally tenacious.

Inhabits the same places; moves as if measuring like a compass, whence the name; found on trout and other fish, after the spawning season. *Tab.* xx. *fig.* 13.

Muricata.
14. TUBER-
CULATED.

H. *Lin. Syst.* 1080. *Faun. Suec. No.* 2080. *Mus. Ad. Fr.* i. 93.
Hirudo marina. *Rondel. aquat.*
Hirudo piscium. *Baster,* i. 82. *tab.* x. *f.* 2.

L. with a taper body; rounded at the greater extremity, and furnished with two small horns; strongly annulated, and tuberculated upon the rings; the tail dilated.

Inhabits the sea; adheres strongly to fish, and leaves a black mark on the spot. *Tab.* xx. *fig.* 14.

VII. Slender

VII. Slender body, carinated beneath.
      Mouth at the extremity, cirrated.
      The two jaws pinnated.
      An adipofe or raylefs fin round the tail, and
         under the belly.

MYXINE.
HAG.

M.   *Lin. Syft.* 1080. Putaohl. *Faun. Suec. No.* 2086.
Muf. *Ad. Fr.* i. 91. *tab.* viii. *f.* 4.
Lampetra cæca.   *Wil. Iæb.* 107.   *Raii pifc.* 36.

Glutinofa.
15. Glutinous.

This fpecies is amply defcribed in the definition ; is about eight inches long.

Inhabits the ocean ; enters the mouths of fifh, when on the hooks of lines that remain a tide under water, and totally devours the whole, except fkin and bones. The *Scarborough* fifhermen often take it in the *robbed fifh*, on drawing up their lines. They call it the *hag*. *Linnæus* attributes to it the property of turning water into glue. *Tab.* xx. *fig.* 15.

D 4   Div. II.

Div. II. M O L L U S C A.   S O F T.

Animals of a fimple form, (naked) without a
Shell; furnifhed with members.

LIMAX.   VIII. Oblong body; attenuated towards the tail.
SLUG.   Above, is a flefhy buckler, formed convexly;
                         fiat beneath.
                         A lateral hole on the right fide, for its geni-
                         tals, and difcharge of excrements.

Ater.   L. *Lin. Syft.* 1081. *Faun. Suec. No.* 2088. *Lift. Angl.* 131.
16. BLACK.   *Gefner. aq.* 254.

SL. wholly black.

Rufus.   L. *Lin. Syft.* 1081. *Faun. Suec. No.* 2089.
17. BROWN.   *Lift. Angl. App.* 6. *tab.* ii. *fig.* 1.

SL. of a brownifh color.

L. *Lin.*

L. *Lin. Syst.* 1081. *Faun. Suec. No.* 2090. *List. Angl.* *Maximus.*
 App. 6. *tab.* ii. *fig.* 2.   18. GREAT.
List. Angl. 127.

SL. with a cinereous ground ; the head reticulated
with black ; on the back three pale lines and four
dusky; the last spotted with black.

 These vary ; at times, part is of an amber color.
The largest of the genus, five inches long.

L. *Lin. Syst.* 1082.   *Agrestis.*
Limax cinereus parvus immaculatus. *List. Angl.* 130. 19. FEILD,

SL. small, and of an uniform cinereous color ;
are very common in gardens, and destructive to
plants.

 These have sometimes been swallowed by persons
in a consumptive habit, who thought them of service.

L. *Lin. Syst.* 1082. *Faun. Suec. No.* 2092. *Flavus.*
              20.YELLOW.

SL. of an amber color, marked with white.

IX. Body

LAPLYSIA.   IX. Body covered with membranes reflected.
                 A shield-like membrane on the back.
                 A lateral pore on the right side.
                 The vent on the extremity of the back.
                 Four feelers, resembling ears.

Depilans.   Lepus marinus. *Plinii, lib.* ix. *c.* 48. *Rondel. pisc.* 520.
21. DEPILA-   Lernæa.   *Bohadsch.* 3. *tab.* i. *fig.*
TORY.   Laplysia.   *Lin. Syst.* 1082.

Described in the character.   The specimen engraven shews its size.   Those of *Italy* grow to
the length of eight inches.   *Pliny* calls it *offa
informis,* and placing it among the venomous
marine animals, says, that even the touch is in-
fectious. The smell is extremely nauseous. *Tab.* xxi.
fig. 21.

      Taken off *Anglesea.*

                                            X. Body

X. Body oblong, flat beneath; creeping. *DORIS.*
Mouth placed below.
Vent behind; surrounded with a fringe.
Two feelers, retractile.

Doris. *Lin. Syst.* 1083. *Bohadsch. tab.* v. *fig.* 4. 5. *Argo.*
22. LEMON.

D. with an oval body, convex, marked with numerous punctures; of a lemon color; the vent beset with elegant ramifications.
Inhabits different parts of our seas; called, about *Brighthelmstone*, the *sea-lemon*. *Tab.* xxii. *fig.* 22.

Doris. *Lin. Syst.* 1083. *Verrucosa.*
23. WARTY.

D. of an ovated form, convex, tuberculated.
Tab. xxi. *fig.* 23.
Inhabits the sea, near *Aberdeen*.

D. with the front abrupt; body has the appear- *Electrina.*
ance of a snail; bilamellated; size of the figure; 24. AMBER.
amber colored.
Taken off *Anglesea*. *Tab.* xxiv. *fig.* 24.

2 XI. Body

APHRODITA. XI. Body oval; numbers of faſciculi, ſerving the uſes of feet, on each ſide.

Mouth cylindric, retractile, placed at the extremity.

Two ſetaceous feelers.

Aculeata.
25. Aculeated.

Aph. *Lin. Syſt.* 1084. *Faun. Suec. No.* 2099. *Baſter,* ii. 62.
tab. vi. *fig.* 12.
Muſ. Ad. Fr. i. 93.
Eruca marina. *Seb. Muſ.* i. *tab.* xc. 1. 111. *tab.* iv. *f.* 7. 8.
Sea mouſe. *Dale's Harwich.* 394. *Boate's Nat. Hiſt. Ireland,* 172.

Aph. with the back cloathed with ſhort brown fur ; the ſides, with rich pavonaceous green hairs, mixed with ſharp ſpines ; vent covered with two ſcales ; belly covered with a naked ſkin ; mouth placed beneath ; each foot conſiſts of a *faſciculus* of five or ſix ſtrong ſpines ; on each ſide about thirty-ſix ; grows to the length of between four and five inches. *Tab.* xxiii. *fig.* 25.

Inhabits all our ſeas ; often found in the belly of the cod-fiſh.

Squammata.
26. Scaled.

Aph. *Lin. Syſt.* 1084. *Baſter,* ii. 66. *tab.* vi. *fig.* 5.

Aph. with the back covered with two rows of large ſcales, deciduous ; about an inch long.

Taken

Taken off *Anglesea*.
Tab. xxiii. *fig.* 26.

APH. with two rows of scales on the back, placed *Pedunculata.* alternately; the mouth cylindric, projecting; an 27. PEDUN-CULATED. inch long.

Taken off *Brighthelmstone.   Tab.* xxiv. *fig.* 27.

APH. oblong; fusiform; annulated; smooth, ex- *Annulata.* cepting a row of minute spines, one on each ring, 28. ANNU-LATED. running along the back; feet small; size two inches and a quarter; of a pale yellow color.

Tab. xxiv. *fig.* 28.

APH. Lepidota. *Pallas. Miscel. Zool.* 209. *tab.* viii. *fig.* 1. *Minuta.* 2. vii. 15.   29. LITTLE.

APH. with small scales; slender; not an inch long.

Taken off *Anglesea.   Tab.* xxiv. *fig.* 29.

XII. Oblong

NEREIS. XII. Oblong flender body.

Feet formed like a pencil of rays, and numerous on each fide.

Mouth at the extremity, unguiculated.

Feathered feelers above the mouth.

Noctiluca. N. Segmentis xxiii. corpore vix confpicuo. *Lin. Syft.* 1085.
30. NOCTI- Noctiluca marina. *Amœn. Acad.*
LUCOUS. *Bafter,* i. *tab.* iv. *fig.* 3.

Thefe are the animals that illuminate the fea, like glow-worms, but with brighter fplendor. I have at night, in rowing, feen the whole element as if on fire round me ; every oar fpangled with them ; and the water burnt with more than ordinary brightnefs. I have taken up fome of the water in a bucket, feen them for a fhort fpace illuminate it ; but when I came to fearch for them, their extreme fmallnefs eluded my examination.

Lacuftris. Nereis. *Lin. Syft.* 1085.
31. BOG. *Rœfel. infect. Polyp. tab.* lxxix.

N. with a linear jointed body, with a filiform foot iffuing from each ; the whole animal of the fize of a fhort briftle of a hog ; an object of the microfcope.

Inhabits wet places.

Nereis.

Nereis. *Lin. Syst.* 1086. *Faun. Suec. No.* 2095.

Cærulea.
32. Blue.

N. fmooth; depreffed; with 184 fegments of a bluifh-green color, femi-pellucid; a longitudinal *fulcus* runs along the belly, about four inches long.

Inhabits the deeps. Two figures are given, *fig.* 1. on its belly, 2. on its back, fhewing the *fulcus.*

N. with a very flender depreffed body; two black fpots on the front; attenuated at the end when it draws in its forceps; a blood-red longitudinal line along the middle of the back; the fegments very numerous; about four inches long.

Rufa.
33. Red.

Taken off *Anglefea. Tab.* xxv. *fig.* 33.

Nereis. *Pallaf. Mifc. p.* 131. *tab.* ix. *fig.* 17.

Conchilega.
34. Shell.

N. with a flat body, attenuated towards the tail; pellucid; about thirteen feet on each fide; about the mouth a feries of very fine filaments.

Inhabits the Sabella *Tubiformis.* No. 163. of this work.

XIII. Body

ASCIDIA.   XIII. Body fixed to a fhell, rock, &c.

Two apertures, one on the fummit.

The other lower, forming a fheath.

Ruftica?   Afc. *Lin. Syft.* 1087.
35. RUSTIC.

Asc. with fcabrous extremities; one end bending
upwards; middle part fmooth; lower flat; of a
brown color.

Taken off *Scarborough.* Animals of this genus
have the faculty of fquirting out the water they
take in. *Tab.* xxiii. *fig.* 35.

ACTINIA.   XIV. Body oblong, round, affixing itfelf to fome
other fubftance.

The top dilatable, furrounded within with
numberlefs *tentacula.*

Mouth the only aperture; furnifhed with
crooked teeth.

Sulcata.   HYDRA tentaculis denudatis, numerofiffimis, corpore longi-
36. SUL-   tudinaliter fulcato. *Gaertner, Ph. Tr.* 1761. *p.* 75. *tab.* i. *b.*
CATED.   *fg.* 1. A. B.

Ac. with a body marked with trifurcated fulci;
and fummit furrounded with long flender *tentacula,*

9   from

from 120 to 200 in number; color of the body pale chefnut; of the tentacula a fea-green, varied with purple.

Inhabits the rocks of the *Cornifh* and *Anglefea* feas.

Hydra calyciflora, tentaculis retractilibus variegatis corpore verrucofo. *Ibid. fig.* 2. A. B. C.   *Pedunculata.* 37. Stalky.

Ac. with a long cylindric ftalk, expanding at top, and tuberculated. The *tentacula* difpofed in feveral ranges, fhort, and when open, form a radiated angular circumference, like a beautiful flower, with a fmooth polygonal difc; the color of the ftalk, a fine red; of the *tentacula* varied with feveral colors. This fpecies is retractile.

Inhabits *Cornwall*.

Hydra difciflora, tentaculis retractilibus fubdiaphanis; corpore cylindrico, miliaribus glandulis longitudinalitèr ftriato. *Ibid. fig.* 4. A. B.   *Verrucofa.* 38. Stubded.

Ac. with a long cylindric ftalk; marked with elegant fmall tubercles, difpofed in ftrait lines from top to bottom; the circumference of the mouth ftriated, furrounded with fhort petals, like thofe of the fun-flower; and thofe again with white *tentacula*, barred with brown. When drawn in, it affumes the form of a bell; and the lines of tu-

Vol. IV.   E   bercles

bercles converge to the central of the summit.
Body of a pale red.

Inhabits *Cornwall.*

Hemiſpherica. Hydra diſciflora, tentaculis retraƈilibus, extimo diſci mar-
39.Button.　　gine tuberculato. *Ibid. fig.* 5. A. B.

Ac. with a ſmooth ſhort thick ſtalk ; the edge
of the diſc ſurrounded with a ſingle row of tubercles ; the *tentacula* numerous and ſlender. Color
a dull crimſon. Retraƈtile, and flings itſelf in that
ſtate into the form of a conoid button.

Inhabits moſt of our rocky ſhores.

Pentapetala. Actinia dianthus. Ellis. *Ph. Tr.* 1767. *p.* 436. *tab.* xix.
40.Cinque-　　*f.* 8.
foil.

Ac. with a circular contraƈted mouth ; the diſc
divided into five lobes, covered with ſeveral ſeries
of ſhort ſubulated *tentacula.* Stalk ſhort and thick.
When contraƈted, aſſumes the form of a long white
fig.

Inhabits the rocks near *Haſtings.* Sussex.

XV. Body

XV. Body not affixed; naked; gibbous.   *HOLOTHURI*
    Many *tentacula* at one extremity, furround-
      ing the mouth.

Hol. *Lin. Syft.* 1091.   *Pentaktes.*
Hydra corolliflora tentaculis retraĉtilibus frondofis.   *Gaertner.*   41. FIVE-
   Ph. Tr. 1761. *p.* 75. *tab.* i. *b. fig.* 3. A. B.   ROWED.

H. with an incurvated cylindric body, marked with
longitudinal rows of *papillæ*; out of the centre of
each iffue, at will, flender feelers like the horns
of fnails; the upper extremity retraĉtile; when
exerted, affumes a cordated form, furrounded at
the apex with eight tentacula, elegantly ramified,
of a yellow and filver color.

    Found on the fhore between *Penfance* and *New-
land*. Suppofed to inhabit the deep.

    The figure engraven to illuftrate this genus was
dredged up near *Weymouth*. *Tab.* xxvi. *fig.* 41.

    Ariftotle and *Pliny* make ufe of the words
Ολοθούρια and *Holothuria* *; but I fhould imagine,
from the context, that they intend thofe marine
bodies, which modern naturalifts ftyle *Zoophyta*,
perhaps *Alcyonia*: for both of the former make
them analogous with plants. Yet *Ariftotle* hints that
they have life; a difcovery affumed in later times.

    * *Ariftot. Hift. An. lib.* i. *c.* 1. *de Part. An. lib.* iv. *c.* 5.
Plinii Hift. Nat. lib. ix. *c.* 47.

            E 2   XVI. Body

LERNEA.   XVI. Body oblong ; roundish ; which affixes itself
to other animals by its tentacula.

A thorax heart-shaped.

Two, or three tentacula in form of arms.

Salmonea.   L. *Lin. Syst.* 1093.   *Faun. Suec. No.* 2102.
42. SALMON.

L. with an ovated body, cordated thorax, and
two linear arms approaching nearly to each other.

Inhabits the gills of salmon. Observed in great
numbers on the first arrival of that fish out of the
sea ; but after being a little time in fresh waters,
drops off and dies. The salmon is reckoned in
highest season when these *vermes* are found in them.
Called by the fishermen, *salmon-lice.*

SEPIA.   XVII. Eight arms placed round the mouth, with
CUTTLE.   small concave discs on their insides.

Often two long *tentacula.*

Mouth, formed like a horny beak.

Eyes, placed beneath the *tentacula.*

Body fleshy, a sheath for the breast.

A tube at the base of the last.

Loligo,

Loligo, five Calamarus. *Matthiol. in Diofcorid.* 327.
Loligo magna. *Rondel.* 506.
Le Cafferon. *Belon. aquat.* 342.
Sepia. *Lin. Syft.* 1096. *No.* 4. *Seb. Muf.* iii. *tab.* iv. *fig.* 1, 2.
Faun. Suec. No. 2107. *Borlafe Cornwall. tab.* xx. *fig.* 27.

Loligo.
43. Great.

S. with fhort arms and long *tentacula*; the lower
part of the body rhomboid and pinnated, the upper
thick and cylindric.

   Inhabit all our feas; are gregarious; fwift in
their motions; take their prey by means of their
arms; and embracing it, bring it to their central
mouth. Adhere to the rocks, when they wifh to
be quiefcent, by means of the concave difcs that are
placed along their arms. *Tab.* xxvii. *fig.* 43.

Le Pourpre. *Belon. aquat.* 336.
Polypi prima fpecies. *Rondel.* 513.
Sepia. *Lin. Syft.* 1045. *No.* 1. *Seb. Muf.* iii. *tab.* ii. *fig.* 1.

Octopodia.
44. Eight-
armed.

S. with a fhort round body, without fins or tentacula; with only eight arms; connected at their
bottom by a membrane. This is the *Polypus* of
Pliny, which he diftinguifhes from the *Loligo* and
Sepia, by the want of *tentacula*.

   Inhabits our feas. In hot climates thefe are
found of an enormous fize. A friend of mine,
long refident among the *Indian* ifles, and a dili-
gent obferver of nature, informed me that the
natives affirm, that fome have been feen two fa-

E 3

thoms

thoms broad over their centre, and each arm nine fathoms long. When the *Indians* navigate their little boats, they go in dread of them; and leaft thefe animals fhould fling their arms over, and fink them, they never fail without an ax to cut them off. *Tab.* xxviii. *fig.* 44.

Media.   **S.** *Lin. Syft.* 1093.
45. Middle. Loligo Parva. *Rondel,* 508. *Seb. Muf.* iii. *tab.* iv. *fig.* 5.

S. with a long, flender, cylindric body; tail finned, pointed, and carinated on each fide; two long ten-tacula; the body almoft tranfparent; green, but convertible into a dirty brown, confirming the re-mark of *Pliny**, that they change their color thro' fear, adapting it, *Chameléon* like, to that of the place they are in. The eyes are large and fmaragdine. *Tab.* xxix. *fig.* 45.

Sepiola.   **S.** *Lin. Syft.* 1096.
46. Small. Sepiola. *Rondel.* 519.

S. with a fhort body, rounded at the bottom; a round fin on each fide; two *tentacula*.
Taken off *Flintfhire*. *Tab.* xxix. *fig.* 46.

* *Lib.* ix. *c.* 29.

La

La Seiche. *Belon. aquat.* 338. *Matthiol.* in *Diofcorid.* 326. *Officinalis.*
　　Sepia. *Rondel.* 498.　　　　　　　　　　　　47. OFFICI‑
Seb. Muf. iii. *tab.* iii. *fig.* 1, 2. S. Officinalis. *Lin.* NAL.
Syft. 1095. *Faun. Suec. No.* 2706. *Amæn. Acad.*

S. with an ovated body; fins along the whole of
the fides, and almoft meeting at the bottom;
two long tentacula; the body contains the bone,
the *cuttle-bone* of the fhops, which was formerly
ufed as an abforbent.

The bones are frequently flung on all our
fhores; the animal very rarely.

This (in common with the other fpecies) emits,
when frighted or purfued, the black liquor which
the antients fuppofed darkened the circumambient
wave, and concealed it from the enemy.'

Σηπία αυτε δολοφροσυνησι, &c.

Th' endanger'd *Cuttle* thus evades his fears,
And native hoards of fluid fafety bears.
A pitchy ink peculiar glands fupply,
Whofe fhades the fharpeft beam of light defy.
Purfu'd he bids the fable fountains flow,
And wrapt in clouds eludes th' impending foe.
The fifh retreats unfeen, while felf-born night,
With pious fhade befriends her parent's flight*.

* *Jones*'s Tranflation of *Oppian's Halieut. lib.* iii.

E 4　　　　　　　The

The antients fometimes made ufe of it inftead of ink. *Perfius* mentions the fpecies in his defcription of the noble ftudent.

Jam liber, et bicolor pofitis membrana capillis,
Inque manus chartæ, nodofaque venit arundo.
Tum querimur, craffus calamo quòd pendeat
      humor;
Nigra quòd infufa vanefcat S<small>EPIA</small> Lympha *.

At length, his book he fpreads; his pen he takes:
His papers here, in learned order lays;
And there, his parchment's fmoother fide difplays.
But oh! what croffes wait on ftudious men,
The C<small>UTTLE</small>'s juice hangs clotted at our pen.
In all my life fuch ftuff I never knew,
So gummy thick—Dilute it, it will do.
Nay, now 'tis water!   D<small>RYDEN</small>.

This animal was efteemed a delicacy by the antients; and is eaten even at prefent by the *Italians.* *Rondeletius* gives us two receipts for the dreffing †, which may be continued to this day. *Athenæus* ‡ alfo leaves us the method of making an antique Cuttle-fifh faufage; and we learn from *Ariftotle* ‖,

---

* *Sat.* iii.   † *De Pifc.* 510.   ‡ *Lib.* vii. *p.* 326.
‖ *Lib.* viii. *c.* 30. *Hift. An.*

that

that thofe animals are in higheft feafon, when
pregnant.

XVIII. Body gelatinous, orbicular, convex above; *MEDUSA.*
        flat or concave beneath.
        Mouth beneath, in the middle.
        Tentacula placed below.

Borlase's *Cornwall*, p. 256. *tab.* xxv. *fig.* 7, 8.   *Fufca.*
                                                                          48. Brown.

M. with a brown circle in the middle ; fixteen rays
of the fame color pointing from the circumference
towards the centre.   On the circumference a range
of oval tubercles, and crooked fangs placed alternately.   Four ragged *tentacula* extend little farther
than the body.

Borlase's *Cornwall*, p. 257, *tab.* xxv. *fig.* 9, 10.   *Purpura.*
                                                                          49. Purple.

M. with a light-purple crofs in the centre ; between
each bar of the crofs, is a horfe-fhoe-fhaped mark
of deep purple ; from the circumference diverge
certain rays of pale purple.   Four thick *tentacula,*
fhort, not extending farther than the body.

                                BORLASE's

Tuberculata. BORLASE's *Cornwall,* p. 257. *tab.* xxv. *fig.* 11, 12.
50. TUBER-
CLED.

M. with fifteen rays pointing to and meeting at
a fmall fpot in the centre. Round the edges are
fmall oval tubera; four plain *tentacula* extending
far beyond the body.

Undulata. BORLASE's *Cornwall,* p. 257. *tab.* xxv. *fig.* 15.
51. WAVED.

M. with undulated edges, with fangs on the projecting parts; four orifices beneath; between which
rifes a ftem, divided into eight large ragged *tentacula.*

Lunulata. BORLASE's *Cornwall,* p. 258. *tab.* xxv. *fig.* 16, 17.
52. LUNU-
LATED.

M. with the circumference tuberculated on the
edges; in the center of the lower part are four
conic appendages forming a crofs; feveral others,
like ferrated leaves, furround it. Eight *tentacula,*
not exceeding the edges of the body; eight femilunar apertures, one between each *tentaculum.*

Simplex. BORLASE's *Cornwall,* p. 257. *tab.* xxv. *fig.* 13, 14.
53. ARM-
LESS.

M. with a plain circumference; four apertures beneath; no *tentacula.*

Thefe

Thefe animals inhabit all our feas; are gregarious; often feen floating with the tide in vaft
numbers; feed on infects, fmall fifh, &c. which
they catch with their clafpers or arms. Many fpecies, on being handled, affect with a nettle-like
burning, and excite a rednefs. The antients, and
fome of the moderns, add fomething more *. They
were known to the *Greeks* and *Romans* †, by the
names of Πνευμα Θαλλασσιος, and *Pulmo marinus*,
Sea-Lungs. They attributed medicinal virtues
to them. *Diofcorides* ‡ informs us, that if rubbed
frefh on the difeafed part, they cured the gout in
the feet, and kibed heels. *Ælian* ‖ fays, that they
were depilatory, and if macerated in vinegar,
would take away the beard. Their *phofphorous*
quality is well known; nor was it overlooked by
the antients. *Pliny* notes, that if rubbed with a
ftick it will appear to burn, and the wood to
fhine all over §. The fame elegant naturalift remarks, that when they fink to the bottom of the
fea, they portend a continuance of bad weather.
I muft not omit, that *Ariftotle*, and *Athenæus* after

* Pruritum in pudendis, et uredinem in manibus et oculis
movent, atque acrimonia fua, venerem fopitam, vel extinctam
excitant. *Rondel.* 532. In feveral languages they are called
by an obfcene name.

† *Arift. Hift. An. lib.* v. c. 15. *Diofcorides* notis *Mattbiol.*
341. *Plinii, lib.* ix. c. 47.

‡ P. 341. ‖ *De Animal. lib.* xiii. c. 27.

§ *Lib.* xviii. c. 35.

him,

him, give to fome fpecies the apt name of Κνιδη,
or the *nettle*, from their ftinging quality *.

The antients divided their Κνιδη into two claffes,
thofe that adhered to rocks, the *Actinia* of *Linnæus* ; and thofe that wandered through the whole
element. The laft are called by later writers *Urticæ
Solutæ* ; by *Linnæus*, *Medufæ* ; by the common
people *Sea Gellies* and *Sea Blubbers*.

I do not find that the moderns make any ufe
of them. They are left, the prey of bafking
fharks, perhaps of other marine animals.

ASTERIAS.
SEA-STAR.

XIX. Depreffed body ; covered with a coriaceous
coat ; furnifhed with five or more rays,
and numerous retractile *tentacula.*
Mouth in the center.

* F I V E - R A Y E D.

Glacialis.
54. COM-
MON.

Ast. *Lin. Syft.* 1099. *Faun. Suec. No.* 2113.
Stella coriacea acutangula lutea vulgaris LLUIDII. *Linckii,*
p. 31. *tab.* xxxvi. *No.* 61.

Ast. with five rays depreffed ; broad at the bafe ;
fub-angular, hirfute, yellow ; on the back, a round
ftriated opercule.

* *Arift. Hift. An. lib.* v. *c.* 16. *Athenæus, lib.* iii. *p.* 90.

Thefe

These are found sometimes defective, or with only four rays.   See *Linckius, tab.* xxxv. *fig.* 60.

Common in all our seas ; feed on oysters, and are very destructive to the beds.

Stella pentapetalos cancellata anomalos.
Linckii, p. 32. *tab.* xiv. *No.* 23. and *tab.* vii. *No.* 9.

Clathrata.
55.Cancel-
lated.

Ast. with five short thick rays ; hirsute beneath ; cancellated above.

Found with the former ; more rare. *Tab.* xxx. *fig.* 1.

Pentadactylosaster oculatus. *Linckii, p.* 31. *tab.* xxxvi. *No.* 62.

Oculata.
56.Dotted.

Ast. with five smooth rays, dotted or punctured ; of a fine purple color.

Anglesea. Tab. xxx. *fig.* 56.

Astropecten Irregularis.   *Linckii, p.* 27. *tab.* vi. *fig.* 13.

Irregularis.
57.Rimmed.

Ast. with five smooth rays ; the sides surrounded with a regular scaly rim ; on the mouth, a plate in form of a cinquefoil ; of a reddish hue.

Stella

Hiſpida.
58. Hispid.

Stella coriacea acutangula hiſpida. *Linckii, p.* 31. *tab.* ix. *No.* 19.

Ast. with five rays, broad, angulated at top; rough, with ſhort briſtles; brown.
Angleſea. Tab. xxx. *fig. 58.*

Gibboſa.
59. Gib-
bous.

Pentaceros gibbus et plicatus, altera parte concavus. *Linckii, p.* 25. *tab.* iii. *No.* 20.
Borlase's *Cornwall, p.* 260. *tab.* xxv. *fig.* 25, 26.

Ast. with very ſhort broad rays ſlightly projecting; a pentangular ſpecies, much elevated, ſmall, covered with a rough ſkin; brown; the mouth in the midſt of a pentagon.

Placenta.
59.A. Flat.

Stella quinquefida palmipes. *Linckii, p.* 29. *tab.* i. *fig.* 2.
Pontoppidan's Norway, part. ii. 179.

Ast. with five very broad and membranaceous rays, extremely thin and flat.
Tab. xxxi. *fig.* 59. A.
Weymouth. From the Portland cabinet.

Spinoſa.
60. Spiny.

Pentadactyloſaſter ſpinoſus regularis. *Linckii, tab.* iv. *No.* 7.
Borlase's *Cornwall, p.* 259. *tab.* xxv. *fig.* 18.

Ast. with five rays of almoſt equal thickneſs, beſet with numerous ſpines.

3   ** FIVE-

** FIVE-RAYED, with flender or ferpenti-
form rays.

Hirfuta, feu ftella grallatoria vel macrofceles LUIDII. *Linckii,*   *Minuta.*
p. 50.   61. MI-
NUTE.

Ast. with a round body, and five very flender and
long hirfute rays.
Found by Mr. *Lluyd* near *Tenbigh.*

Stella lacertofa.   *Linckii, p. 47. tab.* ii. *No.* 4.   *Lacertofa.*
62. LIZARD.

Ast. with five fmooth flender rays, fcaled, jointed,
white. *Linckius* calls this *Lacertofa,* from the likenefs of the rays to a Lizard's tail.
Anglefea.   Tab. xxxii. *fig.* 62.

Ast. with a pentagonal indented body, fmooth   *Sphærulata.*
above the aperture; below five-pointed; between   63. BEADED.
the bafe of each ray a fmall globular bead; the
rays flender, jointed, taper; hirfute on their
fides.
Anglefea.   Tab. xxxii. *fig.* 63.

BORLASE'S

Pentaphylla.
64. CINQUE-
FOIL.

BORLASE's *Cornwall, b. 260. tab.* xxv. *fig. 24.*

AST. with the body regularly cinquefoil; rays very flender; hirfute on the fides, teffulated above and below with green, fometimes with fky-blue.
Cornwall.

Varia.
65. PIED.

BORLASE's *Cornwall, p. 259. tab.* xxv. *fig. 21.*

AST. with a circular body, with ten radiated ftreaks; the ends of a lozenge form; the rays hirfute, annulated with red.
Cornwall.

Aculeata.
66. RADI-
ATED.

BORLASE's *Cornwall, p. 259. tab.* xxv. *fig. 19.*

AST. with a round body, with ftreaks from its centre alternately broad and narrow; the rays flender, hirfute.
Cornwall.

Haftata.
67. JAVELIN.

BORLASE's *Cornwall, p. 259. tab.* xxv. *fig. 22.*

AST. with a pentagonal body indented; of a deep brownifh-red hue, marked with ten ochraceous ftreaks;

ftreaks ; five of the ftreaks flender, with javelin-
fhaped extremities ; rays hirfute, jointed.
Cornwall.

Borlase's *Cornwall, p.* 259. *tab.* xxv. *fig.* 20.   *Fiffa.*
                                                                        68. Indent-
                                                                        ed.

Ast. with a circular body, with five equidiftant
dents, penetrating deep into the fides ; five light-
colored ftreaks darting from the centre ; rays flender, hirfute.
Cornwall.

Borlase's *Cornwall, p.* 260. *tab.* xxv. *fig.* 23.   *Nigra.*
                                                                        69. Black.

Ast. with a pentagonal body, black, with five
radiating ftreaks of white ; rays hirfute olivaceous,
teffulated with deeper fhades.
Cornwall.

***   With more than F I V E   R A Y S.

Stella decacnemos rofacea, feu decempeda *Cornubienfium.* *Bifida.*
Linckii, *p.* 55. *tab.* xxxvii. *fig.* 66.   70. Bifid.

Ast. with ten flender rays, befet with tendrils on
their fides ; the mouth furrounded with fhort filiform rays.
Cornwall.
Vol. IV.   F   Stella

Decacnemos.
71. TEN-
RAYED.

Stella decacnemos barbata, feu fimbriata, *Barrelier.* *Linckii,*
    p. 55. *tab.* xxxvii. *fig.* 64.

AST. with ten very flender rays, with numbers of
long beards on the fides; the body fmall, furrounded beneath with ten fmall filiform rays.

Inhabits the weftern coafts of *Scotland.* *Tab.* xxxiii.
fig. 71.

Helianthe-
moides ?
72. TWELVE-
RAYED.

Stella dodecactis Helianthemo fimilis. *Linckii,* *p.* 42. *tab.*
    xvii. *fig.* 28.

AST. with twelve broad rays finely reticulated,
and roughened with fafciculated long papillæ on
the upper part; hirfute beneath; red.

Thefe vary into thirteen, fuch as the *Trifcaide-*
cactis of *Linckius.* *Tab.* xxxiv. *fig.* 54. I have
had one of fourteen rays.

Ariftotle and *Pliny* * called this genus Αϛηρ, and
ftella marina ; fays the firft, from their refemblance
to the pictured form of the ftars of heaven. They
afferted that they were fo exceedingly hot, as inftantly to confume whatfoever they touched.

* *Ariftot. Hift. An. lib.* v. *c.* 15: *Plinii Hift. Nat. lib.* ix.
c. 60.

Afterias

Afterias caput medufæ. *Lin. Syft.* 1101.

Soe-Soele. *Pontop. Norway,* ii. 180.

Ast. with five rays iffuing from an angular body; the rays dividing into innumerable branches, growing flenderer as they receded from the bafe; the moft curious of the genus.

Found, as I have been told, in the north of *Scotland.* The late worthy Doctor *William Borlafe* informed me that it had been taken off *Cornwall.*

XX. Body covered with a futured cruft, often ECHINUS.
    furnifhed with moveable fpines.
    Mouth quinquevalve, placed beneath.

Echinus. *Lin. Syft.* 1102. *Lift. Angl.* 169. *tab.* iii.
Εχινος ωα. *Ariftot. Hift. An. lib.* iv. *c.* v.
Tab. xxxiv. *fig.* 74.

Ech. of a hæmifpherical form, covered with fharp ftrong fpines, above half an inch long; commonly of a violet color, moveable; adherent to fmall tubercles elegantly difpofed in rows. Thefe are their inftruments of motion, by which they change their place.

This fpecies is often taken in dredging, and often lodges in cavities of rocks juft within low-water mark.

F 2   Are

Are eaten by the poor in many parts of *England*, and by the better fort abroad. In old times a favorite difh. They were dreffed with vinegar, honied wine, or mead, parfley and mint; and efteemed to agree with the ftomach *. They are the firft difh in the famous fupper of *Lentulus* †, when he was made *Flamen Martialis*, prieft of *Mars*. By fome of the concomitant difhes, they feem defigned as a whet for the fecond courfe, to the holy perfonages, priefts, and veftals invited on the occafion. Many fpecies of fhell fifh made part of the feaft. The reader will perhaps find fome amufement in learning the tafte of the *Roman* people of fafhion in thefe articles.

Echini, the fpecies here defcribed.

Oftreæ Crudæ, raw oyfters.

Peloridæ ‡, a fort of *Mya*, ftill ufed as a food in fome places. *Vide No.* 15.

Sphondyli, a fort of Bivalve, with ftrong hinges, found in the *Mediterranean* fea. Not the griftly part of oyfters, as Doctor *Arbuthnot* conjectures.

Patina Oftrearum. Perhaps ftewed oyfters.

Pelorides. Balani nigri et albi; two kinds of *Lepades.*

Sphondyli, again.

---

* *Athenæus, lib.* iii. *p.* 91.
† *Macrobius*, as quoted by *Arbuthnot.*
‡ *Rondel. Teftacea, p.* 11.

Glycymerides

Glycymerides *. A fhell. I fufpect to be the fame with the *Mactra Lutraria* of this work, *No.* 44.

Murices, Purpuræ. Turbinated fhells, whofe fpecies I cannot very well determine, there being more than one of each in the *Italian* feas.

Echinus fpatagus. *Lin. Syft.* 1104. *Lift. App. tab.* i. *fig.* 13. *Cordatus.*

75.CORDAT-
ED.

ECH. of a cordated fhape, gibbous at one end, and marked with a deep *fulcus* at the other; covered with flender fpines refembling briftles. Shell moft remarkably fragile.

Length, two inches. *Tab.* xxxiv. *fig.* 75.

Lin. Syft. 1104. *Argenville,* 310. *tab.* xxv. *fig.* K.   *Lacunofus.*
Rumph. Muf. tab. xiv. *fig.* 2.   76. OVAL.

ECH. of an oval depreffed form; on the top of a purple color, marked with a quadrefoil, and the fpaces between tuberculated in waved rows; the lower fide ftudded; and divided by two fmooth fpaces.

Length, four inches. When cloathed, is covered with fhort thickfet briftles mixed with very long ones.

* *Rondel. Teftacea, p.* 13.

F 3   *Weymouth,*

Weymouth, from the Portland cabinet. *Tab.* xxxv. *fig.* 76.

Doctor *Borlafe* gives a figure of an *Echinus*, found in *Mount's Bay*, that refembles in fhape the above; but I cannot, either from defcription or print, determine whether it be the young, or diftinct. Vide *Nat. Hift. Cornwall*, p. 278. *tab.* xxviii. *fig.* 26.

Div. III.

DIV. III.   T E S T A C E A.

VERMES of the foft kind, and fimple make, commonly covered with a calcareous habitation.

DIV. I. MULTIVALVE SHELLS.

I. The animal, or inhabitant of its fhell, the DORIS. CHITON, The fhell plated, confifting of many parts, lying upon each other tranfverfely.

SECT. I. MULTIVALVE SHELLS.

CH. **W**ITH feven valves; thick fet with *Crinitus.* fhort hairs; five-eighths of an inch 1. HAIRY. long.

Of the natural fize. A. 1. magnified.

Inhabits the fea near *Aberdeen.* *Tab.* xxxvi. *fig.* 1,

CH. with eight valves; with a ferrated reflected *Marginatus.* margin, fmooth; fize of the figure. *Tab.* xxxvi. 2. MARGI- NATED. *fig.* 2.

Inhabits the fea near *Scarborough.*

F 4   CH. with

Lævis.
3. Smooth.

Ch. with eight valves; quite fmooth, with a longitudinal mark along the back; a little elevated,
Size of a wood-loufe. *Tab.* xxxvi. *fig.* 3.

Inhabits the fhores of *Loch Broom* in *Weft Rofs-
fhire.*

The inhabitant of this fhell is a fpecies of the
Doris.

The name *Chiton,* taken from χιτων, *Lorica,* a
coat of mail.

LEPAS.
ACORN.

II. Its animal the Triton.
    The fhell multivalve, unequal, fixed by a
    ftem: or feffil.

Balanus.
4. Common.

Lepas. *Lin. Syft.* 1107. *Faun. Suec. No.* 2122.
Common *Englifh* Barnacle. Ellis *Ph. Tr.* 1758. *Tab.* xxxiv.
    fig. 17.

L. of a conoid form, fmooth, and brittle; the lid
or *operculum* fharp pointed.

Found adhering to rocks, oyfters, and fhell-fifh
of various forts. *Tab.* xxxvii. *fig.* 4.

Balanoides.
5. Sulcat-
ed.

L. *Lin. Syft.* 1108. *Faun. Suec. No.* 2123. *Lift. Angl. tab.* v.
    fig. 41.

L. with ftrong fulcated fhells; aperture fmaller in
proportion than the former.

Adheres

Adheres to the fame bodies.   *Tab.* xxxvii. *fig. 5.*

Quere, the figure, **A.** *5.* if not an accidental variety ?

Lepas Cornubienfis. *Ellis Ph. Tr.* 1758. *tab.* xxxiv. *fig.* 16.   *Cornubienfis,* 6. Cornish.
Borlafe Nat. Hift. Cornwall.

L. in form of a limpet, with a dilated bottom, and rather narrow aperture; the fhell fulcated near the lower edges.   *Tab.* xxxvii. *fig.* 6.

L. with the fhells lapping over each other, and   *Striata.* 7. Striated.
obliquely ftriated.

The fea near *Weymouth.*   *Tab.* xxxviii. *fig.* 7.
From the Portland cabinet.

L. *Lin. Syft.* 1108.   *Tintinnabulum.* 8. Bell.

L. with a large deep fhell, rugged on the outfide, of a purple color.

As large as a walnut.

Found frequently adhering to the bottom of fhips, in great clufters. Probably originated in hot climates.

L. *Lin.*

Anatifera.
9. ANATI-
FEROUS.

L. *Lin. Syst.* 1109.   *Faun. Suec. No.* 2120.   *List. Conch.*
tab. 439.

L. confifting of five fhells, depreffed, affixed to
a pedicle, and in clufters. *Tab.* xxxviii. *fig.* 9.

Adheres to fhips bottoms by its pedicles.

The *tentacula* from its animal are feathered ; and
have given our old *English* hiftorians and naturalifts
the idea of a bird. They afcribed the origin of the
Barnacle Goofe to thefe fhells. The account given
by the Sage *Gerard*, is fo curious, that I beg leave
to tranfcribe it.

‘ But what our eyes have feene, and hands have
‘ touched, we fhall declare.   There is a fmall
‘ ifland in *Lancafhire* called the *Pile* of *Foulders*,
‘ wherein are found the broken pieces of old
‘ and bruifed fhips, fome whereof have been caft
‘ thither by fhipwracke, and alfo the trunks and
‘ bodies with the branches of old and rotten trees,
‘ caft up there likewife ; whereon is found a cer-
‘ taine fpume or froth that in time breedeth unto
‘ certaine fhels, in fhape like thofe of the Mufkle,
‘ but fharper pointed, and of a whitifh colour ;
‘ wherein is contained a thing in form like a lace
‘ of filke finely woven as it were together, of a
‘ whitifh colour ; one end whereof is faftened unto
‘ the infide of the fhell, even as the fifh of Oifters
‘ and Mufkles are: the other end is made faft
‘ unto

' unto the belly of a rude maffe or lumpe, which
' in time commeth to the fhape and form of a
' bird: when it is perfectly formed, the fhell
' gapeth open, and the firft thing that appeareth
' is the forefaid lace or ftring; next come the legs
' of the bird hanging out, and as it groweth
' greater it openeth the fhell by degrees, till at
' length it is all come forth, and hangeth onely by
' the bill: in fhort fpace after it commeth to full
' maturitie, and falleth into the fea, where it ga-
' thereth feathers, and groweth to fowle bigger
' than a Mallard and leffer than a Goofe, having
' blacke legs and bill or beake, and feathers blacke
' and white, fpotted in fuch manner as is our *Mag-*
' *Pie*, called in fome places a *Pie-Annet*, which
' the people of *Lancafhire* call by no other name
' than a tree Goofe: which place aforefaid, and
' all thofe parts adjoyning, do fo much abound
' therewith, that one of the beft is bought for three
' pence. For the truth hereof, if any doubt, may
' it pleafe them to repaire unto me, and I fhall
' fatisfie them by the teftimonie of good witneffes.'
Vide Herbal, p. 1587, 1588.

This genus is called by *Linnæus*, Lepas, a name
that is given by the antients to the *Patella*. Shells
of this clafs are called by *Ariftotle*, Βαλανοι *, from
the refemblance fome of them bear to acorns.
We have feen before, in the account of the fupper

* *Hift. An. lib.* v. *c.* 15.

of

of *Lentulus*, that they were admitted to the greateſt tables.

PHOLAS. III. Its animal an ASCIDIA.

Shell bivalve, opening wide at each end, with ſeveral leſſer ſhells at the hinge.

The hinges folded back, united with a cartilage.

An incurvated tooth in the inſide beneath the hinge.

Dactylus. PH. *Lin. Syſt.* 1110. *Faun. Suec. No.* 2124. *Liſt. Angl. App.*
10. DAC- *Tab.* xi. *fig.* 3.
TYLE.

PH. with an oblong ſhell, marked with echinated *ſtriæ*; the tooth broad; the ſpace above the hinge reflected, and cancellated beneath; breadth four inches and a half; length one and a quarter. *Tab.* xxxix. *fig.* 10.

Candidus. PH. *Lin. Syſt.* 1111. *Liſt. Angl. tab.* v. *fig.* 39.
11. WHITE.

PH. with a brittle ſhell, and ſmoother than the former; the tooth very ſlender; breadth an inch and an half; length near an inch. *Tab.* xxxix. *fig.* 11.

PH. *Lin.*

P<small>H</small>. with a ftrong oval fhell; the half next to the hinge waved and ftriated; tooth large and ftrong; breadth three inches and a half; length one and three quarters. *Tab.* xl. *fig.* 12.

This genus takes its name from φωλεω, to lurk in cavities. A fhell of the name of *Pholis* and *Pholas*, is mentioned by *Ariftotle* and *Athenæus*; but I fufpect it to be the *Dactylus* of *Pliny*. A fpecies now called *Datyl*, abounding within the rocks of the *Mediterranean*, is much admired as a food *.

P<small>H</small>. with a fhell thinner than the former; and *Parvus.* the tooth very flender and oblique; in externals 13. L<small>ITTLE</small>. refembling the former, only never found larger than a hazel nut.

I have often taken them out of the cells they had formed in hard clay, below high-water mark, on many of our fhores. They alfo perforate the hardeft oak plank that accidentally is lodged in the water. I have a piece filled with them, which was found near *Penfacola* in *Weft Florida*, and pre-fented to me by that ingenious naturalift the late J<small>OHN</small> E<small>LLIS</small>, Efquire.

* *Pliny, lib.* ix. *c.* 61. *Armftrong's Hift. Minorca,* 173.

I have

I have alſo found them in maſſes of foſſil wood, in the ſhores of *Abergelli* in *Denbighſhire*. The bottom of the cells are round, and appear as if nicely turned with ſome inſtrument.

Tab. xl. *fig.* 13.

Div. II. Bivalve Shells.

MYA.
GAPER.

IV. Its animal an Ascidia.

A bivalve ſhell gaping at one end.

The hinge, for the moſt part, furniſhed with a thick, ſtrong, and broad tooth, not inſerted into the oppoſite valve.

Truncata.   M. Truncata. *Lin. Syſt.* 1112.   *Faun. Suec. No.* 2126.
14. Abrupt.   *Liſt. Angl. tab.* v. *fig.* 36.

M. WITH a broad, upright, blunt tooth, in one ſhell; the cloſed end rounded; the open end truncated, and gaping greatly; the outſide yellow, marked with concentric wrinkles. *Tab.* xli. *fig.* 14.

Lodged under ſlutchy ground, near low-water mark; diſcovered by an aperture in the ſlutch, beneath which it is found in coarſe gravel.

M. with

M. with a brittle half-tranfparent fhell, with a hinge *Declivi:.* flightly prominent ; lefs gaping than the *truncata* ; near the open end floping downwards.   15. Slop-ing.

Frequent about the *Hebrides* ; the fifh eaten by the gentry.

M. Arenaria.   *Lin. Syft.* 1112.   *Faun. Suec. No.* 2127.   *Arenaria.* 16. Sand.

M. with a tooth like the former ; mouth large, rough at the bafe ; the whole fhell of an ovated figure, and much narrower at the gaping end.

Three inches and a half broad ; two inches long in the middle.   *Tab.* xlii.

M. Pictorum.   *Lin. Syft.* 1112.   *Faun. Suec. No.* 219.   *Lift.*   *Pictorum.* *Angl. App. tab.* i. *fig.* 4.   17. Paint-ers.

M. with an oval brittle fhell ; with a fingle longi-tudinal tooth like a lamina in one fhell, and two in the other.   *Tab.* xliii. *fig.* 17.

Breadth a little above two inches ; length one.

Inhabits rivers.

Ufed to put water-colors in ; whence the name. Otters feed on this and the other frefh-water fhells.

Lin.

Margariti-
fera.
18. Pearl.

Lin. Syst. 1112. *Faun. Suec. No.* 2130. *Lift. Angl. App.*
   tab. i. *fig.* 1.
Scheffer Lapland, 145.

M. with a very thick coarfe opake fhell; often much decorticated ; oblong, bending inward on one fide ; or arcuated ; black on the outfide ; ufual breadth from five to fix inches ; length two and a quarter. *Tab.* xliii. *fig.* 18.

Inhabits great rivers, efpecially thofe which water the mountanous parts of *Great Britain.*

This fhell is noted for producing quantities of pearl. There have been regular fifheries for the fake of this pretious article in feveral of our rivers. Sixteen have been found within one fhell. They are the difeafe of the fifh, analogous to the ftone in the human body. On being fqueezed, they will eject the pearl, and often caft it fpontaneoufly in the fand of the ftream.

The *Conway* was noted for them in the days of *Cambden.* A notion alfo prevales, that Sir *Richard Wynne,* of *Gwydir,* chamberlain to *Catharine* queen to *Charles* II. prefented her majefty with a pearl (taken in this river) which is to this day honored with a place in the regal crown. They are called by the *Welfh Cregin Diluw,* or Deluge Shells, as if left there by the flood.

The *Irt* in *Cumberland* was alfo productive of them. The famous circumnavigator, Sir *John Hawkins,*

5

Hawkins *, had a patent for fifhing that river. He had obferved pearls plentiful in the Straits of *Magellan*, and flattered himfelf with being inriched by procuring them within his own ifland.

In the laft century, feveral of great fize were gotten in the rivers of the county of *Tyrone* and *Donegal*, in *Ireland*. One that weighed 36 *carats* was valued at £. 40, but being foul, loft much of its worth. Other fingle pearls were fold for £. 4. 10 *s*. and even for £. 10. The laft was fold a fecond time to Lady *Glenlealy*, who put it into a necklace, and refufed £. 80 for it from the Duchefs of *Ormond* †.

Suetonius reports, that *Cæfar* was induced to undertake his *Britifh* expedition for the fake of our pearls ; and that they were fo large that it was neceffary to ufe the hand to try the weight of a fingle one ‡. I imagine that *Cæfar* only heard this by report ; and that the cryftalline balls in old leafes, called *mineral pearl*, were miftaken for them ‖.

We believe that *Cæfar* was difappointed of his hope : yet we are told that he brought home a buckler made with *Britifh* pearl §, which he dedicated to, and hung up in the temple of *Venus Genetrix*. A proper offering to the Goddefs of Beauty, who fprung from the fea. I cannot omit

---

* *Camden*. ii. 1003.   † *Ph. Tr. Abridg*. ii. 831.

‡ *Sueton. Vit. Jul. Cæf. c.* xliv.

‖ *Woodward's Method of Foffils*, 29. *part* ii.

§ *Plinii, lib.* ix. *c.* 35. *Tacitus Vit. Agricolæ*.

VOL. IV.   G   mentioning,

mentioning, that notwithftanding the claffics honor our pearl with their notice, yet they report them to have been fmall and ill colored; an imputation that in general they are ftill liable to. *Pliny* * fays, that a red fmall kind was found about the *Thracian Bofphorus*, in a fhell called *Mya*, but does not give it any mark to afcertain the fpecies.

Dubia.
19. Dubi-
ous.

M. with a rudiment of a tooth within one fhell; with an oval and large hiatus oppofite to the hinge. Shells brown and brittle.

Shape of a *piftachia* nut.

Length of a horfe-bean.   *Tab.* xliv.

Found near *Weymouth*. From the Portland cabinet.

* *Plinii, lib.* ix. *c.* 35.

V. Its

V. Its animal an Ascidia.

  A bivalve; oblong; open at both ends.

  At the hinge, a fubulated tooth turned back, often double; not inferted in the oppofite fhell.

SOLEN.
RAZOR.

* With the hinge near the end.

Lin. Syft. 1113.   *Faun. Suec. No.* 2131.   *Lift. Angl. tab.* v. *fig.* 37.
Lift. Conch. tab. 409.

Siliqua.
20. Pod.

S. with a ftrait fhell, equally broad, compreffed, with a double tooth at the hinge, receiving another oppofite; and on one fide another tooth fharp pointed, and directed downwards. Color olive, with a conoid mark of an afh color, dividing the fhells diagonally; one part ftriated lengthways, the other tranfverfely. Breadth ufually five or fix inches, fometimes nine.

  Tab. xlv. *fig.* 20.

Lin. Syft. 1113.   *Lift. Conch. tab.* 410.

Vagina.
21. Sheath.

S. with a fhell-nearly cylindrical, one end marginated; the hinge confifting of a fingle tooth in each fhell placed oppofite. Shell yellow, marked

       G 2   much

much like the former; ufually about five or fix inches broad.

Inhabits *Red Wharf*, *Anglefea*.

Enfis.
22. Scyme-
ter.

Lin. Syft. 1114.   *Lift. Angl. App. tab.* ii. *fig.* 9.   *Lift. Conch. tab.* 411.

S. with a fhell bending like a fcymeter, with hinges like thofe of the *Siliqua*; and colored and marked like it. The fhell thin, and rounded at each end. Ufual breadth four or five inches. *Tab.* xlv. *fig.* 22.

Pellucidus.
23. Pellu-
cid.

S. fub-arcuated and fub-oval; with the hinge con-fifting of a fharp double tooth on one fide, receiv-ing a fingle one from the oppofite, with a procefs in each fhell, pointing towards the cartilage of the hinge. Shell fragile, pellucid; about an inch broad. *Tab.* xlvi. *fig.* 23.

Inhabits *Red Wharf*, *Anglefea*.

** With the hinge near the middle.

Legumen.
24. Sub-
oval.

Lin. Syft. 1114.   *Lift. Conch. tab.* 420.

S. with a ftrait fub-oval fhell; with teeth exactly refembling thofe of the laft, furnifhed likewife with fimilar procefses; one end is fomewhat broader than

the

the other. Ufual breadth about two inches and an half. Shell fub-pellucid, radiated from the hinge to the margin.

Tab. xlvi. *fig.* 24.

Inhabits the fame place.

Lin. Syft. 1114. *No.* 37.   *Lift. Conch.* 421.

Cultellus.
25. Kidney.

S. with a kidney-fhaped fhell; with a fingle tooth in both fides of the hinge. The fhell covered with a rough *epidermis*. Breadth near two inches; length feven-eighths of an inch.

Inhabits the fea near *Weymouth.*

Tab. xlvi. *fig.* 25.

This fpecies borders on the *myæ*, and connects the *genera.*

I am not acquainted with the natural hiftory of the two laft. The three firft lurk in the fand near low-water mark, in a perpendicular direction: and when in want of food, elevate one end a little above the furface, and protrude their bodies far out of the fhell. At approach of danger, they dart deep into the fand, fometimes at left two feet. Their place is known by a fmall dimple on the furface. Sometimes they are dug out of the fand with a fhovel; at other times are taken by a bearded dart fuddenly ftruck into them. They

G 3   were

were ufed as a food by the antients. *Athenæus* *
(from *Sophron*) fpeaks of them as great delicacies,
and particularly grateful to widows.

Μακραὶ κόγχαι σῶλενες τᾶτιγα
Γλυκυκρεων κογχύλιον χηρᾶν γυναικῶν λιγχνεῦμα.
Oblongæ conchæ *folenes*, et carne jucundâ
Conchylium, viduarum mulierum cupediæ.

Thefe are often ufed as a food at prefent; and
brought up to table fried in eggs.

TELLINA.　VI. Its animal a TETHYS.
　　　　　　A bivalve, generally floping down on one fide.
　　　　　　Three teeth at the hinge.

* Ovated.

Fragilis.　　*Lin. Syft.* 1117. *No.* 49.
26.FRAGILE.

T. with a very brittle white fhell, truncated at the
narrower, and rounded at the broader end. An inch
broad.
　　Tab. xlvii. *fig.* 26.

* *Lib.* iii. *p.* 86.

　　　　　　　　　　　　　　　　　　T. with

T. with a very thick depreſſed oblong ſhell; white ; *Depreſſa.*
with concentric *ſtriæ.*   27. Depres-
sed.

Tab. xlvii. *fig.* 27.

T. with very thick, broad, and depreſſed ſhells, *Craſſa.*
marked with numerous concentric *ſtriæ.* Breadth,   28. Flat.
an inch and three quarters ; length, an inch and a
quarter.

Has the habit of the Venus *borealis* ; but the
ſides of this are unequal, one being more extended
than the other.

Tab. xlviii. *fig.* 28.

Lin. Syſt. 1117. *No.* 52.   *Planata.*
29. Plain.

T. with a very flat delicate ſhell, marked with
concentric lines of red ; the ſpace about the hinge
brown. Breadth, two-thirds of an inch.

Tab. xlviii. *fig.* 29.

Lin. Syſt. 1117. *No.* 54.   *Radiata.*
30. Rayed.

T. with very convex ſhells of a faint aſh color,
radiated with red ; tinged within with a faint purple. Breadth an inch and an half.

Tab. xlix. *fig.* 30.

G 4   *Lin.*

Incarnata.
31. CARNA-
TION.

Lin. Syst. 1118. *No.* 58. *Faun. Suec. No.* 2133. *Lift. Angl.*
App. tab. 1. *fig.* 8.

T. oblong, depreffed ; originally covered with a thick brown *epidermis*. When naked, of a whitifh color rayed with red, and croffed again with minute concentric *ftriæ*.

Ufual breadth, one inch and three quarters.
Tab. xlvii. *fig.* 31.

Carnaria.
32. FLESH-
COLORED.

Lin. Syst. 1119. *No.* 66. *Lift. Angl. tab.* iv. *fig.* 25.

T. with a ftrong and rounded fhell, generally of a bloom color within and without; externally marked with belts of deeper red.

Breadth about feven-eighths of an inch.
Sometimes found quite white, as *fig.* 32. A.
Tab. xlix. *fig.* 32.

Trifafciata.
33. TRI-
FASCIATED.

Lin. Syst. 1119. *No.* 58.

T. with a very brittle fhell, radiated like the T. *Incarnata*; but leffer.

Rugofa.
34.RUGGED.

T. with oval fhells, marked with rugged concentric *ftriæ*. This has much the habit of the *Mytilus Lithophagus*.

About

About the fize of a filbert.

Dredged up at *Weymouth*. Mifplaced among the Venuses. *Vide tab.* lvii. *fig.* 34.

Borlafe Hift. Cornwall, tab. xxviii. *fig.* 23.

Cornubienfis.
35. Cor-
nish.

T. with oblong oval fhells, deeply ftriated parallel to the margin.

Defcribed by Doctor *Borlafe.*

Lin. Syft. 1120. *No.* 72.   *Faun. Suec. No.* 2138.   *Lift. Angl. App. tab.* i. *fig.* 5.

Cornea.
36. Horny.

T. with round fhells very convex, marked with a tranfverfe furrow; color brown.

Size of a pea.

Inhabits ponds and frefh waters.

Tab. xlix. *fig.* 36.

VII. Bivalve, nearly equilateral, equivalve.
     Its animal a Tethys.

         Two teeth near the beak : a larger (placed
            remote) on each fide; each locking into
            the oppofite.

CARDIUM.
COCKLE.

Lin.

Aculeatum.
37. ACU-
LEATED.

Lin. Syst. 1122. *No.* 78.

C. with high ribs radiating from the hinge to the edges; each rib fulcated in the middle; and near the circumference befet with large and ftrong proceffes, hollowed. One fide of the fhell projects further than the other, and forms an angle. Color yellowifh-brown.

    As large as a fift. The marginal circumference ten inches and a half.

    Found off the *Hebrides* and *Orknies.*

    Tab. l. *fig.* 37.

Echinatum.
38. ECHI-
NATED.

Lin. Syst. 1122. *No.* 79. *Faun. Suec. No.* 2139.
Lift. Angl. tab. v. *fig.* 33. *Conch. tab.* 324.

C. leffer than the former, being little more than fix inches in circumference; the color white; the ribs echinated higher up; has only fixteen ribs, the former twenty-one; the fhape rounder.

    Found dead on many of our fhores.

Ciliare.
39. FRIN-
GED.

Lin. Syst. 1122. *No.* 80.

C. with a very brittle fhell, and delicate; of a pure white; eighteen ribs rifing into thinner fpines.

                       Of

Of the fize of a hazel nut.

Tab. l. *fig.* 39.

Lin. Syft. 1123. *No.* 88.

Lævigatum.
40. Smooth.

C. of a fub-oval fhape, fomewhat depreffed; of a deep brown color, with obfolete longitudinal *ftriæ*; and a few tranfverfal, concealed by a thin *epidermis.*

Circumference fix inches and a half.

Tab. li. *fig.* 40.

Lin. Syft. 1124. *No.* 90.   *Faun. Suec. No.* 2141.   *Lift. Angl.* *Edule.*
tab. v. *fig.* 34.

41. Edible.

C. with twenty-eight depreffed ribs, tranfverfely ftriated; one fide more falient than the other.

Common on all fandy coafts, lodged a little be-neath the fand; their place marked by a depreffed fpot. Delicious and wholefome food.

Tab. l. *fig.* 41.

VIII. Its animal a Tethys.

MACTRA.

Bivalve, unequal fided, equivalve.

Middle tooth complicated; with a little concavity on each fide; the lateral teeth remote, mutually received into each other.

Lin.

Stultorum.
42. SIMPLE-
TON'S.

Lin. Syst. 1126. *No.* 99.

M. with femi-tranfparent fhells, fmooth, gloffy;
white without; purplifh within.
> Size of a hazel nut.
> *Tab.* lii. *fig.* 42.

Solida.
43. STRONG.

Lin. Syst. 1126. *No.* 100. *Faun. Suec. No.* 2140. *Lift. Angl.*
tab. iv. *fig.* 24.

M. with very ftrong fhells; in a live ftate, fmooth,
white, gloffy, and marked with a few tranfverfe
ftriæ. In dead fhells, the *ftriæ* appear like high
ribs. *Vide fig.* 43. A. *Tab.* l.

Lutraria.
44. LARGE.

Lin. Syst. 1126. *No.* 101. *Faun. Suec. No.* 2128. *Lift.*
Angl. tab. iv. *fig.* 19.

M. with an oblong thin fhell; one fide much extended, and gaping; for which reafon *Linnæus*
once placed it among the *Myæ.*
> Breadth five inches; length two and a half.
> Inhabits the fea near the mouth of rivers; and
> even fometimes within the mouth.
> *Tab.* lii. *fig.* 44.

IX. Its

IX. Its animal a Tethys.   *DONAX.*
Bivalve, with the frontal margin very blunt.

Lin. Syft. 1127. *No.* 105.   *Faun. Suec. No.* 2142.   *Lift. Angl.*   *Trunculus.*
    tab. v. *fig.* 35.   45. Yellow.
Conch. tab. 376. *f.* 217.

D. with a gloffy fhell, of a whitifh color tinged
with dirty yellow, and marked lengthways with
many elegant minute *ftriæ*; the infide purple.
    Breadth an inch and a tenth.
    Tab. lv. *fig.* 45.

Lin. Syft. 1127. *No.* 107.   *Denticulata.*
    46. Purple.

C. of a cuneiform fhape ; extremely blunt at one
end, ftriated like the former, ferrated at the edges ;
color within purple ; tranfverfely tinged with the
fame on the outfide.
    Breadth, a little fuperior to the former.

X. Its animal a Tethys.   *VENUS.*
    Hinge with three teeth near to each other ; one
        placed longitudinally, and bent outwards.

Lin.

Mercenaria.   *Lin. Syft.* 1131. *No.* 123.   *Faun. Suec. No.* 2144.   *Lift. Angl.*
47.Commer-   *tab.* iv. *fig.* 22.
cial.   *Conch. tab.* 272.

V. with a ftrong, thick, weighty fhell, covered with a brown epidermis; pure white within; flightly ftriated tranfverfely.

Circumference above eleven inches.

Thefe are called in *North America Clams*; they differ only in having a purple tinge within. *Wampum* or *Indian* money is made of them *.

Tab. liii. *fig.* 47.

Erycina.   *Lin. Syft.* 1131. *No.* 122.   *Lift. Conch. tab.* 284.
48. Sici-
lian.

V. with a very thick fhell, marked with high-ridged ribs tranfverfely; undulated longitudinally.

Fig. 48. A. a worn fhell.

Circumference about five or fix inches.

Tab. liv. *fig.* 48. 48. A.

Exoleta.   *Lin. Syft.* 1134. *No.* 142.
49. Anti-
quated.

V. with orbicular fhells, with numerous tranfverfal *ftriæ*; white, gloffy.

* *Burnaby's Travels, p.* 104. *ed.* 2.

Diameter

Diameter about two inches.

A. Variety of the fame, marked ftrongly with numerous *ftriæ*, and longitudinally with a few fhort yellowifh lines. *Vide Lift. Conch. tab.* 292. 293.

Tab. liv. *fig.* 49. A.   *Tab.* lvi. *fig.* 49.

Lift. Conch. tab. 281.

Rugofa.
50. WRINK-
LED.

V. with thick fhells, marked with rugofe concentric *ftriæ*.

A. Variety, with *ftriæ* lefs elevated, and marked with yellowifh zigzag lines. *Lift. Conch.* 282.

Length, an inch; breadth, an inch and a quarter.

Tab. lvi. *fig.* 50.

V. with thin convex orbiculated fhells, of a white color, tinged with yellow, and marked with thin concentric *ftriæ*; waved at the edges.

Undata.
51. WAVED.

Size of a hazel nut.

Tab. lv. *fig.* 51.

V. with thin convex fhells, with a very deep obtufe *finus*, or bending on the front.

Sinuofa.
51. A. IN-
DENTED.

Size of the figure.

Weymouth. From the PORTLAND cabinet.

Tab. lv. *fig.* 51. A.

Lin.

2

Borealis.
52. Nor-
thern.
     Lin. Syst. 1134. *No.* 143. *List. Angl. tab.* iv. *fig.* 23. *Conch.* *tab.* 253. *fig.* 88.

V. with thin shells, much depreſſed, marked with slender concentric *striæ*.

Length one inch and a half; breadth near two inches.

Litterata.
53. Letter-
ed.
     Lin. Syst. 1135. *No.* 147. *Faun. Suec. No.* 2146. *List. Conch. tab.* 400. *fig.* 239.

V. with thick shells, marked tranſverſely with fre-quent crenulated *striæ*, ſometimes ſmoother; of a whitiſh color, ſtreaked with lines reſembling cha-racters. In *Britiſh* ſpecimens uſually faint; in fo-reign very ſtrong and elegant.

Length an inch and three quarters; breadth two inches and a half.

Tab. lvii. *fig.* 53.

Deflora'a.
54. Fading.
     Lin. Syst. 1133. *No.* 152.

V. with thin oval shells, ſtriated lengthways, ſemi-pellucid; rayed with purple and white, both within and without.

Size near an inch and half in breadth.

Tab. lvii. *fig.* 54.

3

V. with

V. with depreſſed rhomboidal ſhells, marked with *Rhomboides.*
concentric and very neat *ſtriæ*, of a pale brown 55. RHOM-
color variegated. BOID.

Length three quarters of an inch ; breadth an
inch and three quarters.

V. with ovated ſhells, ſtriated elegantly from hinge *Ovata.*
to margin, and ſlightly ſtriated tranſverſely. 56. OVAL.

Size of a horſe-bean.

Tab. lvi. *fig. 56.*

XI. Its animal a TETHYS? *ARCA.*

Shell bivalve equivalve.

Teeth of the hinge numerous, inſerted between
each other.

Lin. Syſt. 1140. *No.* 168. *Borlaſe Nat. Hiſt. Cornw.* *Tortuoſa.*
tab. xxviii. *fig.* 15, 16. 57. DIS-
Liſt. Conch. tab. 368. TORTED.
Mytilus *Matthiol.* apud *Dioſcor. lib.* ii. *c.* 5. *p.* 301.

A. with a rhomboid ſhell, deeply ſtriated from
the apex to the edges.

Inhabits *Cornwall.* Found alſo near *Weymouth.*

Tab. lviii. *fig. 57.*

VOL. IV. H *Lin.*

Glycymeris.   *Lin. Syſt.* 1143. *No.* 181.   *Liſt. Conch. tab.* 247. *fig.* 82.
58. Orbicu-
lar.

A. with thick orbicular ſhells, marked with concentric *ſtriæ*; white zigzagged with ferruginous;
edges crenulated.

Diameter about two inches.

Tab. lviii. *fig.* 58.

Nucleus.   *Lin. Syſt.* 1141. *No.* 184.
59. Sil-
very.

A. with unequally triangular ſhells; ſmooth, pure
white without, ſilvery within; margin finely crenated.

Size of a pea.

Tab. lviii. *fig.* 59.

Barbata.   *Lin. Syſt.* 1140. *No.* 170.
60. Frin-
ged.

A. with oblong ſhells faintly ſtriated; beſet with
Byſſus ſo as to appear bearded.

In *England* of the ſize of a horſe-bean, the
foreign ſpecimens much larger.

XII. Its

XII. Its animal a Tethys.   *PECTEN.*
    Shell bivalve, unequal.   SCALLOP.

The hinge toothless, having a small ovated
    hollow.

*

Lin. Syst. 1144. *No.* 185. *Faun. Suec. No.* 2148. *List. Maximus.*
Angl. tab. v. *fig.* 29.   61. Great.

P. with fourteen rays, very prominent and broad;
striated lengthways above and below; ears equal.

Grows to a large size. *Tab.* lix. *fig.* 61.

Found in beds by themselves; are dredged up,
and pickled and barrelled for sale.

The antients say, that they have the power of
removing themselves from place to place by vast
springs or leaps *. This shell was called by the
Greeks Κτεις, by the *Latins Pecten*, and was used
by both as a food; and when dressed with pepper
and cummins, was taken medicinally †.

The elegant figure of the crouching *Venus*, in
the *Maffei* collection, is placed sitting in a shell
of this kind. The sculptor probably was taught
by the mythology of his time, that the goddess
arose from the sea in a scallop. This perhaps

* *Arist. Hist. An. lib.* iv. *c.* 4.
† *Athenæus, lib.* iii. *p.* 90.

        H 2   may

may have been the *concha venerea* of *Pliny*, fo ftyled from this circumftance.

Another fhell has the fame name, for a different reafon *.

The fcallop is commonly worn by pilgrims on their hat, or the cape of their coat, as a mark that they had croffed the fea in their way to the *Holy Land*, or to fome diftant object of devotion.

Jacobæus.   *Lin. Syft.* 1144. *No.* 186. *Lift. Conch. tab.* 165. *fig.* 2.
62. Lesser.

P. with fifteen broad rays, rounded on the flat fide, and moft finely tranfverfely ftriated; angulated on the convex, and ftriated lengthways; ears nearly equal; c oncave and fmooth on the upper fide.

A rare fpecies in *Great Britain.*
Tab. lx. *fig.* 62.

** Both Shells convex.

Subrufus.   Pecten tenuis, fubrufus, maculofus, circiter 20 ftriis majori-
63. Red.   bus, at lævibus, donatus. *Lift. Angl. p.* 185. *tab.* v. *fig.* 30.

P. with twenty narrow rays, finely ftriated; ears nearly equal, and alfo ftriated.

* See *No.* 82.
6

A fpe-

A fpecies feldom exceeding two inches and a quarter in length; the breadth nearly the fame.

A thin fhell, generally of a fine pale red.

Tab. lx. *fig.* 63.

Lin. Syft. 1146. *No.* 199.   *Lift. Conch. tab.* 178. *fig.* 15.   *Varius.*
64. VARIE-
GATED.

P. with about thirty echinated imbricated rays; fhells almoft equally convex; one ear vaftly larger than the other.

General length two inches and a half; breadth a little lefs.

Color, a fordid red mixed with white.

Often found in oyfter-beds, and dredged up with them.

Tab. lxi. *fig.* 64.

Lin. Syft. 1146. *No.* 200.   *Pufio ?*
Pecten minimus anguftior inequalis ferè et afper, &c.   65. WRITH-
Lift. Angl. p. 186. *tab.* v. *fig.* 31.   ED.

P. with above forty fmall rays; with unequal ears; the furface always irregularly waved or deformed, as if by fome accident; but this appearance regularly maintained.

Length about two inches.

Colors commonly very brilliant reds.

Tab. lxi. *fig.* 65.

H 3   P. with

Obfoletus.
66. WORN.

P. with one large ftriated ear, with fmooth equal fhells; eight obfolete rays; of a dark purple color.

A fmall fpecies three quarters of an inch long. *Tab.* lxi. *fig.* 66.

Lævis.
67. SMOOTH.

P. with unequal ribbed ears; the reft of the fhell entirely fmooth.

Very fmall.

Anglefea.

Glaber.
68. FUR-
ROWED.

Lin. Syft. 1146. *No.* 201.

P. with a very thin fhell; fifteen faint rays; equal ears. The inner fide of the fhells marked with rays, divided by a fingle *fulcus.*

Anglefea. A fcarce fpecies. Small.

OSTREA.
OYSTER.

XIII. Its animal a TETHYS.

Shell bivalve, roughly plated on the outfide.

Edulis.
69. EDIBLE.

Lin. Syft. 1148. *No.* 211. *Faun. Suec. No.* 2149. *Lift. Angl. tab.* iv. *fig.* 26.

O. commonly of an orbicular form, and very rugged. A defcription of fo well-known a fhell

9   **is**

is needlefs.   Varies in fize in different places. This is figured with an *Anomia* on it, *No.* 70. B.

Britain has been noted for oyfters from the time of *Juvenal* *, who fatyrizing an epicure, fays,

Circæis nata forent, an
Lucrinum ad Saxum, *Rutupinove* edita fundo,
Oftrea, callebat primo deprendere morfu.

He, whether *Circe's* rock his oyfters bore,
Or *Lucrine* lake, or diftant *Richborough's* fhore
Knew at firft tafte

The luxurious *Romans* were very fond of this fifh, and had their *layers* or ftews for oyfters, as we have at prefent. *Sergius Orata* † was the firft inventor, as early as the time of *L. Craffus* the orator. He did not make them for the fake of indulging his appetite, but through avarice, and made great profits from them. *Orata* got great credit for his *Lucrine* oyfters ; for, fays *Pliny*, the *Britifh* were not then known.

The antients eat them raw, and fometimes roafted. They had alfo a cuftom of ftewing them with mallows and docks, or with fifh, and efteemed them very nourifhing ‡.

Britain ftill keeps its fuperiority in oyfters over

---

* *Satyr.* iv, *V.* 140.   † *Plin. Nat. Hift. lib.* ix. *c.* 54.
‡ *Athenæus, lib.* iii, *p.* 92.

H 4   other

other countries. Moft of our coafts produce them naturally, and in fuch places they are taken by dredging, and are become an article of commerce, both raw and pickled. The very fhells, calcined, become an ufeful medicine as an abforbent. In common with other fhells, prove an excellent manure.

Stews or *layers* of oyfters are formed in places, which nature never allotted as habitations for them. Thofe near *Colchefter* have been long famous; at prefent there are others, that at left rival the former, near the mouth of the *Thames.* The *oyfters,* or their fpats, are brought to convenient places, where they improve in tafte and fize. It is an error to fuppofe, that the fine green obferved in oyfters taken from artificial beds, is owing to *copperas*; it being notorious how deftructive the fubftance or the folution of it is to all fifh. I cannot give a better account of the caufe, or of the whole treatment of oyfters, than what is preferved in the learned Bifhop *Sprat's* Hiftory of the ROYAL SOCIETY, from p. 307 to 309.

‘ In the month of *May* the oyfters caft their ‘ fpaun, (which the dredgers call their fpats;) it ‘ is like to a drop of candle, and about the big- ‘ nefs of a halfpenny.

‘ The *fpat* cleaves to ftones, old oyfter-fhells, ‘ pieces of wood, and fuch-like things, at the bot- ‘ tom of the fea, which they call *cultch.*

‘ ’Tis

' 'Tis probably conjectured, that the *spat* in
' twenty-four hours begins to have a shell.

' In the month of *May*, the dredgers (by the
' law of the Admiralty court) have liberty to catch
' all manner of oysters, of what size soever.

' When they have taken them, with a knife
' they gently raise the small brood from the *cultch*,
' and then they throw the *cultch* in again, to pre-
' serve the ground for the future, unless they be
' so newly spat, that they cannot be safely severed
' from the *cultch*; in that case they are permitted
' to take the stone or shell, &c. that the *spat* is
' upon, one shell having many times twenty
' *spats.*

' After the month of *May*, it is felony to carry
' away the *cultch*, and punishable to take any
' other oysters, unless it be those of size, (that is
' to say) about the bigness of an half-crown piece,
' or when the two shells being shut, a fair shilling
' will rattle between them.

' The places where these oysters are chiefly
' catcht, are called the *Pont-Burnham, Malden,*
' and *Colne* waters; the latter taking its name
' from the river of *Colne*, which passeth by *Colne-*
' *Chester*, gives the name to that town, and runs
' into a creek of the sea, at a place called the
' *Hythe*, being the suburbs of the town.

' This brood and other oysters they carry to
' creeks of the sea, at *Brickel-Sea, Merfey, Langno,*

Fingrego,

' *Fringrego, Wivenho, Tolefbury*, and *Saltcoafe*, and
' there throw them into the channel, which they
' call their beds or layers, where they grow and
' fatten, and in two or three years the fmalleft
' brood will be oyfters of the fize aforefaid.

' Thofe oyfters which they would have green,
' they put into pits about three feet deep in the
' falt-marfhes, which are overflowed only at fpring-
' tides, to which they have fluces, and let out
' the fault-water until it is about a foot and half
' deep.

' Thefe pits, from fome quality in the foil co-
' operating with the heat of the fun, will become
' green, and communicate their colour to the
' oyfters that are put into them in four or five days,
' though they commonly let them continue there
' fix weeks or two months, in which time they will
' be of a dark green.

' To prove that the fun operates in the greening,
' *Tolefbury* pits will green only in fummer; but
' that the earth hath the greater power, *Brickel-
' Sea* pits green both winter and fummer: and for
' a further proof, a pit within a foot of a green-
' ing-pit will not green; and thofe that did green
' very well, will in time lofe their quality.

' The oyfters, when the tide comes in, lie
' with their hollow fhell downwards, and when it
' goes out, they turn on the other fide; they re-
' move not from their place, unlefs in cold weather,
' to cover themfelves in the Oufe.

' The

' The reafon of the fcarcity of oyfters, and confe-
' quently of their dearnefs, is, becaufe they are of
' late years bought up by the *Dutch.*

' There are great penalties, by the Admiralty
' court, laid upon thofe that fifh out of thofe
' grounds which the court appoints, or that deftroy
' the *cultch,* or that take any oyfters that are not of
' fize, or that do not tread under their feet, or
' throw upon the fhore, a fifh which they call a
' *Five-finger* *, refembling a fpur-rowel, becaufe
' that fifh gets into the oyfters when they gape,
' and fucks them out.

' The reafon why fuch a penalty is fet upon
' any that fhall deftroy the *cultch,* is, becaufe
' they find that if that be taken away, the Oufe
' will increafe, and the mufcles and cockles will
' breed there, and deftroy the oyfters, they having
' not whereon to ftick their *fpat.*

' The oyfters are fick after they have fpat;
' but in *June* and *July* they begin to mend, and
' in *Auguft* they are perfectly well: the male
' oyfter is *black-fick,* having a black fubftance in
' the fin; the female *white-fick,* (as they term it)
' having a milky fubftance in the fin. They are
' falt in the pits, falter in the layers, but falter
' at fea.'

To this I beg leave to join a fort of prefent ftate
of this article, borrowed from the 84th page of

---

* Asterias *glacialis,* the common Sea Star.

the

the Hiftory of *Rochefter*, in 12mo, publifhed in
1776.

    ' Great part of the inhabitants of *Stroud* are
' fupported by the fifheries, of which the oyfter
' is moft confiderable. This is conducted by a
' company of free dredgers, eftablifhed by pre-
' fcription, but fubject to the authority and go-
' vernment of the mayor and citizens of *Rochefter*.
' In 1729 an act of parliament was obtained, for
' the better management of this fifhery, and for
' confirming the jurifdiction of the faid mayor and
' citizens, and free dredgers. The mayor holds
' a court of admiralty every year, to make fuch
' regulations as fhall be neceffary for the well
' conducting this valuable branch of fifhery. Seven
' years apprenticefhip entitles a perfon to the free-
' dom of this company. All perfons catching
' oyfters, not members of the fifhery, are liable
' to a penalty. The company frequently buy
' brood or fpat from other parts, which they lay
' in this river, where they foon grow to maturity.
' Great quantities of thefe oyfters are fent to *Lon-*
' *don*; to *Holland*, *Wefphalia*, and the adjacent
' countries.

ANOMIA.   XIV. Bivalve, inequivalve.
          One valve perforated near the hinge; affixed
          by that perforation to fome other body.

                                           Lin.

Lin. Syst. 1150. *No.* 218.   *Lift. Conch. tab.* 204. *fig.* 38.   *Ephippium.*
70. LARGER.

A. with the habit of an oyfter; the one fide
convex, the other flat; perforated, adherent to
other bodies, often to oyfter-fhells, by a ftrong
tendinous ligature; color of infide perlaceous.

Size near two inches diameter.

Tab. lxii. fhews the exterior fide of the fhell; and
the interior of the upper valve adhering to an
oyfter.

Lin. Syst. 1151. *No.* 221.   *Squammula.*
71. SMALL.

A. with fhells refembling the fcales of fifh;
very delicate and filvery. Much flatted. Perforated. Very fmall.

Adheres to oyfters, crabs, and lobfters, and
fhells.

The foffil fpecies of the *Anomia* genus are uncommonly numerous in this ifland, in our chalk-
pits and limeftone-quarries; but are foreign to the
work in hand. The reader who wifhes to be
acquainted with their appearance, may fatisfy
himfelf, by confulting *Lifter's* Hiftory of Shells,
appendix to the 3d book, tab. 447, &c. and *Hift.
an. Angl.* tab. viii. and ix. *Plot's* Hift. *Oxfordfhire,*
tab. iii. and his Hiftory of *Staffordfhire,* tab. xi.

XV.

MYTILUS. XV. Its animal an Ascidia.
MUSSEL.
      Bivalve; often affixed to fome fubftance by
      a beard.
      Hinge without a tooth, marked by a longi-
      tudinal hollow line.

Rugofus.   *Lin. Syft.* 1156. *No.* 249. *Lift. Angl. tab.* iv. *fig.* 21.
72.RUGGED.

M. with a brittle fhell, very rugged, and in fhape
moft irregular; ufually oblong, and rounded at
the ends.

    Length near an inch. Color whitifh.

    Always found lodged in *limeftone.* The outfide
generally appears honey-combed; but the apertures
are too fmall for the fhell to pafs through, without breaking into the cell they are lodged in.
Multitudes are found in the fame ftone: but
each has a feparate apartment, with a different external fpiracle.

    Tab. lxiii. *fig.* 72.

Edulis.   *Lin. Syft.* 1157. *No.* 253. *Faun. Suec. No.* 2156. *Lift. Angl.*
73. EDIBLE.   *tab.* iv. *fig.* 28.

M. with a ftrong fhell, flightly incurvated on one
fide: angulated on the other. The end near the
                           hinge

hinge pointed; the other rounded. *Tab.* lxiii. *fig.* 73.

When the *epidermis* is taken off, is of a deep blue color.

Abundance of fmall pearls, called *feed-pearls*, were till of late procured from this fpecies of muffel, for medical purpofes; but I believe they are now difufed, fince crabs-claws and the like have been difcovered to be as efficacious, and a much cheaper abforbent.

Found in immenfe beds, both in deep water; and above low-water mark. A rich food, but noxious to many conftitutions. Affect with fwell-ings, blotches, &c. falfely attributed to the pea-crab. The remedy oil, or falt and water.

N E *fraudentur gloriâ fua littora.* I muft in juftice to *Lancafhire* add, that the fineft muffels are thofe called *Hambleton Hookers*, from a village in that county. They are taken out of the fea, and placed in the river *Wier*, within reach of the tide, where they grow very fat and delicious.

M. very crooked on the fide, near the end; then greatly dilated, and covered with a thick rough *epidermis*. Within has a violet tinge.
Incurvatus.
74. CROOK-ED.

Found on the coaft of *Anglefea*, near *Prieft-holme*; ufually an inch and an half long.

Tab. lxiv. *fig.* 74.

M. with

Pellucidus.
75. Pellu-
cid.

M. with a delicate tranfparent fhell, moft elegantly rayed lengthways, with purple and blue; like the former in fhape, but more oval. Commonly fhorter than two inches.

Anglefea. Found fometimes in oyfter-beds; fometimes in trowling over flutchy bottoms.

Tab. lxiii. *fig.* 75.

Umbilicatus.
76. Umbili-
cated.

M. with a ftrong fhell, and the fpace oppofite to the hinge deeply inflected or umbilicated.

The form nearly oval. The length fometimes five inches.

A rare fpecies, and new. Sometimes dredged up off *Prieftholme* ifland, *Anglefea.* Difcovered by the reverend Mr. *Hugh Davies.*

The pea-crab found in this fpecies of a larger fize than ufual.

Tab. lxv. *fig.* 76.

Curtus.
76. A. Short

M. with a fhort, ventricofe, obtufe fhell, of a dirty yellow color.

Size of the figure.

Weymouth. From the Portland cabinet.

Tab. lxiv. *fig.* 76. A.

Lin.

Lin. Syst. 1158. *No.* 256.   *Lift. Conch. tab.* 356. *fig.* 195.   *Modiolus.*
77. Great.

M. with a ftrong fhell, with a blunted upper end ; one fide angulated near the middle ; from thence dilating towards the end, which is rounded.

The greateft of *Britifh* muffels. Length from fix to feven inches.

Lies at great depths. Often feizes the bait of the ground lines, and is taken up with the hooks.

Tab. lxvi. *fig.* 77.

Lin. Syst. 1158. *No.* 257.   *Lift. Angl. App. tab.* i. *fig.* 3.   *Cygneus.*
78. Swan.

M. with a thin brittle fhell, very broad and convex, marked with concentric ftriæ. Attenuated towards one end ; dilated towards the other. Decorticated about the hinge.

Color, dull green.

Length fix inches ; breadth three and a half.

Inhabits frefh waters. Pearls are found in this and the following fpecies.

Tab. lxvii. *fig.* 78.

Lin. Syst. No. 258. *Faun. Suec. No.* 2158.   *Lift. Angl. tab.* i.   *Anatinus.*
fig. 2.   79. Duck.

M. with a fhell lefs convex, and more oblong than the laft. Very brittle, and femi-tranfparent. Space round the hinges like the laft.

Vol. IV.   I 711   Length

Length about five inches; breadth two and a quarter.

Inhabits fresh waters.

Crows feed on these muffels; and also on different shell-fish. It is diverting to observe, that when the shell is too hard for their bills, they will fly with it to a great height, drop the shell on a rock, and pick out the meat, when the shell is fractured by the fall.

Tab. lxviii. *fig*. 79.

PINNA.
NACRE.

XVI. Its animal a Slug.

 Bivalve, fragil, furnished with a beard.

 Gapes at one end. Hinge without a tooth.

Fragilis.
80. Brittle

P. with a very thin femi-pellucid whitish shell, most opake near to the apex. Marked on the furface with longitudinal flender ribs, roughened with concave fcales; and the whole traverfed by innumerable fine *ftriæ*.

In young shells the ribs and fcales are almost obfolete. The valves of lefler tranfverfe diameter.

The largeft about five inches and a half long; and three and a quarter broad in the broadeft part. The figure is of a broader fpecimen than ufual.

Dredged up at *Weymouth*. From the Portland cabinet.

Tab. lix. *fig*. 80.

3

      I faw

I faw fpecimens of fome vaft *Pinnæ*, found among the farther *Hebrides*, in the collection of Doctor *Walker*, at *Moffat*. They were very rugged on the outfide, but I cannot recollect whether they were of the kind found in the *Mediterranean* or *Weft Indies*.

Ingens.
81. GREAT.

Div. III.   Univalve Shells.

With a regular fpire.

XVII. Its animal a SLUG.
  Shell fub-oval, blunt at each end.
  The aperture the length of the fhell, lon-
   gitudinal, linear.   Toothed.

CYPRÆA.
GOWRIE.

Lin. Syft. 1180. *No.* 364. *Lift. Angl. tab.* iii. *fig.* 17. *Conch. tab.* 706, 707. *fig.* 56 and 57.

Pediculus.
82. COM-
MON.

C. with numerous ftriæ, fome bifurcated.   Varies with having three brown fpots on the back.
 Tab. lxx. *fig.* 82.
 This genus is called *Cypræc*, and *Venerea*, from its being peculiarly dedicated to *Venus*; who was faid to have endowed a fhell of this genus with the

I 2   powers

powers of a *Remora*, fo as to impede the courfe of
the fhip which was fent by *Periander*, tyrant of
Corinth, with orders to caftrate the young nobility
of *Corcyra* *.

BULLA.
DIPPER.

XVIII. Its animal a Slug.
Shell fub-oval.
Aperture oblong, fmooth.
One end a little convoluted.

Lignaria.
83. Wood.

Lin. Syft. 1184. *Lift. Conch. tab.* 714. *fig.* 71.

B. of an oval form, and ftriated tranfverfely. Is
narrower towards one end, which is a little umbilicated.   Of a dirty color, like fome woods,
whence the trivial name.   The infide of the fhell
vifible to the very end, through the *columella*.
Length about two inches.   *Tab.* lxx. *fig.* 83.

Ampulla.
84. Ob-
tuse.

Lin Syft. 1183. *No.* 378.

B. with a brittle fhell, more obtufe at the end;
and the inner fide lapping over the *columella*, fo as
to render it invifible.
Poffibly a young fhell of the B. *Ampulla?*
Found near *Weymouth.*

* *Plinii, lib.* ix. *c.* 25. xxxii. *c.* 1.

Lift.

Lift. Conch. tab. 714. *fig.* 70.

Cylindracea.
85. Cylin-
dric.

B. white, cylindric, a little umbilicated at the end.
About twice the fize of a grain of wheat.
Tab. lxx. *fig.* 85.

B. with one end much produced, and fufiform.
The aperture very patulous.
Weymouth. From the Portland cabinet.
Tab. lxx. *fig.* 85. A.

Patula.
85. A. Open.

XIX. Its animal a Slug.
Aperture narrow, without a beak.
Columella pleated.

VOLUTA.
VOLUTE.

Lin. Syft. 1187. *No.* 394. *Lift. Conch. tab.* 835.

V. exactly oval; acuminated at each end; with a
fingle fold near the mouth, or upper part of the
columella. With five fpires. Striated fpirally.
Pale red, with white *fafciæ.*
Anglefea.
Tab. lxxi. *fig.* 86.

Tornatilis.
86. Oval.

V. with a very thin brittle fhell, with two fmall
fpires.

Jonenfis.
87. Jona.

I 3   Inhabits

Inhabits the ifle of *Jona*, or *Y Columb-kil.*
Tab. lxxi. *fig.* 87.

BUCCINUM. WHELK.	**XX.** Its animal a SLUG. Aperture oval, ending in a fhort canal.

Pullus.
88. BROWN.

Lin. Syft. 1201. *No.* 458. *Gualtieri. tab.* 44. *fg.* N. *Lift. Conch. tab.* 971. *fig.* 26.

B. with five fpires ftriated, waved, and tubercu-
lated. Aperture wrinkled; upper part replicated.
Length five-eighths of an inch.
Tab. lxxii. *fig.* 88.

Lapillus.
89. MASSY.

Lin. Syft. 1202. *No.* 467. *Faun. Suec. No.* 2161. *Lift. Angl.
tab.* iii. *fig.* 5, 6. *Lift. Conch. tab.* 965.

B. with about five fpires, often obfolete; infide
of the mouth flightly toothed. A very ftrong
thick fhell, of a whitifh color.

A variety yellow; or fafciated with yellow on a
white ground; or fulcated fpirally, and fometimes
reticulated.

See *figures* 89. *tab.* lxxii.

In many, which I fufpect to be fhells not
arrived at full growth, the lip is thin and cultra-
ted.

Length

Length near an inch and a half.

Inhabits (in vaft abundance) rocks near low-water mark.

This is one of the *Englifh* fhells that produces the purple dye, analogous to the *purpura* of the antients : our fhell has been made ufe of as an object of curiofity.

The antient has been long fince fuperfeded by the introduction of the infect *Coccus Cacti,* or the *Cochineel* beetle. The fhells were of the genus of *Murex,* mentioned by *Linnæus, pp.* 1214, 1215. But one was a fort of *Buccinum. Pliny* defcribes both *. The fineft was the *Tyrian.*

     ' Tyrioque ardebat Murice lana ;'

A ftrong expreffion of *Virgil,* who defcribes the cloth,

     ' Glowing with the Tyrian Murex.'

The fpecies of fhells are found in various parts of the *Mediterranean.* Immenfe heaps of them are to be feen about *Tarentum* † to this day, evincing one place where this precious liquor was extracted.

The procefs of obtaining the *Englifh Purpura* is well defcribed by Mr. *William Cole,* of *Briftol,* in 1684, in the following words ‡.

---

* *Lib.* ix. *c.* 36.   † *Baron Riedefel's Travels. p.* 174.
‡ *Ph. Tr. Abr.* ii. 826.

       I 4   ' The

' The shells being harder than most of other
' kinds, are to be broken with a smart stroke with
' a hammer, on a plate of iron, or firm piece of
' timber, (with their mouths downwards) so as
' not to crush the body of the fish within; the
' broken pieces being pick'd off, there will appear
' a white vein, lying transversely in a little furrow
' or cleft, next to the head of the fish, which
' must be digged out with the stiff point of a
' horse-hair pencil, being made short and taper-
' ing.   The letters, figures, or what else shall be
' made on the linnen, (and perhaps silk too) will
' presently appear of a pleasant light-green color,
' and if placed in the sun, will change into the
' following colours, i. e. if in winter, about noon;
' if in the summer, an hour or two after sun-rising,
' and so much before setting; for in the heat of
' the day, in summer, the colours will come on
' so fast, that the succession of each colour will
' scarcely be distinguished.   Next to the first light-
' green, it will appear of a deep-green, and in
' few minutes change into a sea-green, after which,
' in a few minutes more, it will alter into a
' watchet-blue; from that, in a little time more,
' it will be of a purplish-red; after which, lying
' an hour or two, (supposing the sun still shining)
' it will be of a very deep purple-red, beyond
' which the sun can do no more.
   ' But then the last and most beautiful colour,
' after washing in scalding water and soap, will
                                        ' (the

' (the matter being again put into the fun or
' wind to dry) be of a fair bright crimfon, or near
' to the prince's colour, which afterwards, not-
' withftanding there is no ufe of any ftiptick to
' bind the colour, will continue the fame, if well
' ordered; as I have found in handkerchiefs,
' that have been wafhed more than forty times;
' only it will be fomewhat allayed, from what it
' was, after the firft wafhing. While the cloth
' fo writ upon lies in the fun, it will yield a very
' ftrong and fœtid fmell, as if garlick and *affa-*
' *fœtida* were mixed together.'

Lin. Syft. 1204. *No.* 475. *Faun. Suec. No.* 2163. *Lift. Undatum.*
  Angl. tab. iii. *fig.* 2.   90. WAVED.
Lift. Conch. tab. 962. *fig.* 14.

B. with feven fpires, fpirally ftriated, and deeply
and tranfverfely undulated.

    Length three inches.

    Inhabits deep water.

    Tab. lxxiii. *fig.* 90.

B. Leve tenue ftriatum et undatum. *Lift. Angl. p.* 157. *tab.* iii. *Striatum.*
  fig. 3.   91. STRIAT-
                                                          ED.

B. with eight fpires, with elevated *ftriæ*, undulated
near the apex.

    Length near four inches.

    Tab. lxxiv. *fig.* 91.

                                Lin.

Reticulatum.   *Lin. Syst.* 1204. *No.* 476.   *List. Conch. tab.* 966. *fig.* 21.
92. RETI-
CULATED.

B. with fpires fcarcely raifed, and ftrongly reticulated; of a deep brown color, and of an oblong
oval form. The aperture white, gloffy, and den-
ticulated.

Size of a hazel-nut.

Tab. lxxii. *fig.* 92.

Minutum.   B. with five fpires, ftriated fpirally; ribbed tranf-
93. SMALL.   verfely.

Size, lefs than a pea.

Found alfo in *Norway.* Vide *Act. Nidr.* tom. iv.
tab. 16. *fig.* 25.

Tab. lxxix.

STROMBUS XXI. Its animal a SLUG.
Shell univalve, fpiral.
The opening much dilated, and the lip
expanding, produced into a groove.

Pes Pelecani.   *Lin. Syst.* 1207. *No.* 490.   *Faun. Suec. No.* 2164. *List.*
94. CORVO-   *Conch. tab.* 866. *fig. min.*
RANT'S FOOT

STR. with ten fpires, tuberculated along their
ridges, with the lip expanding and digitated. The
                                                    fpires

spires end in a most exquisite point. Length
about two inches. Extent of the expanse an inch
and a quarter.

Tab. lxxv. *fig.* 94.

XXII. Its animal a Slug.   *MUREX.*
    The aperture oval; the beak narrows into a
        canal or gutter, a little ascending.

Lin. Syst. 1206. *No.* 526. *Gualtieri. tab.* 49. *fig.* H.   *Erinaceus.*
                                                    95.Urchin.

M. with an angular shell, surrounded with tubular ribs; each rib ending with its mouth on the
angle. Consists of six spires on the whole; a most
rugged shell. The aperture exactly oval; the
gutter or canal covered.
    Length near two inches.
    Tab. lxxvi. *fig.* 95.

M. with five or six spires, the body ventricose: *Carinatus.*
the spires rising into angulated ridges. The aper- 96. Angu-
ture semicircular.   lated.
    Length near four inches.
    From the Portland cabinet.
    Tab. lxxvii. *fig.* 96.

Lin.

Antiquus.
97. AN-
TIQUE. *Lin. Syst.* 1222. *No.* 558. *Gualtieri, tab.* 46. E. *Faun. Suec.* *No.* 2165.

M. with eight spires finely striated; the first very ventricose. Color a dark dirty yellow. Length three inches and a half.

Despectus.
98. DESPI-
SED. *Lin. Syst.* 1222. *No.* 559. *Faun. Suec. No.* 2166. *List. Angl. tab,* iii. *fig.* 1.

M. with eight spires. The first large, ventricose, and produced; the others more prominent than those of the preceding. Striated and somewhat rugged. The outside white, the inside glossy and yellow.

Length near five inches.

Inhabits the deep sea. Dredged up in plenty with oysters. Eaten by the poor; but oftener used for baits for cod and ray.

Tab. lxxviii. *fig.* 98.

Corneus.
99. HORNEY. *Lin. Syst.* 1244. *No.* 565. *List. Angl. tab.* iii. *fig.* 4. *Conch. tab.* 913. *fig.* 5.

M. with a narrow oblong shell of eight striated spires. Snout much produced. Color pure white, covered with a brown *epidermis.*

Length

Length near three inches.

Tab. lxxvi. *fig.* 99.

M. with an oblong fhell of fix fpires, neatly rib- *Coftatus.*
bed. Vide *tab.* lxxix. 100. Ribbed.

 Minute.

Anglefea. Inhabits alfo *Norway.* *Act. Nidr.*
tom. iv. *tab.* 16, *fig.* 26.

M. with a narrow oblong fhell, acuminated fpires, *Acuminatus.*
ribbed. Vide *tab.* lxxix. 101. Sharp.

 Minute.

Lin. Syft. 1226. *No.* 578. *Decollatus?*
102. Short-
ened.

A fpecies offered with doubts. Perhaps accidentally mutilated. Let the critical conchyliolo-
gift confult *tab.* lxxix.

 Minute.

  XXIII. Its animal a Slug. *TROCHUS.*
     Shell conic. TOP.
     Aperture fub-triangular.

Lin

Ziziphinus. *Lin. Syſt.* 1231. *No.* 599. *Liſt. Conch. tab.* 616. *No.* 1.
103. Livid. *Liſt. Angl. tab.* iii. *fig.* 14. *Faun. Suec. No.* 2168.

Tr. with a ſharp apex, imperforated bottom ; with
a *ſtria* elevated above the reſt. Each is ſmooth.
The color livid, much ſpotted with deep red.
Tab. lxxx. *fig.* 103.

Conulus. *Lin. Syſt.* 1230. *No.* 598.
104. Co-
nule.

Tr. with an imperforated baſe, and a prominent
line along the ſpires. Scarcely diſtinct from the
laſt.
Tab. lxxx. *fig.* 104.

Exaſperatus. Trochus pyramidalis parvus, ruberrimus faſciis crebris exaſ-
105. Rough. peratus. *Liſt. Conch. tab.* 616. *fig.* 2.

I am unacquainted with this ſpecies ; ſo refer the
reader to *Liſter*, who deſcribes it as above ; and
marks the figure with A. as an *Engliſh* ſhell.

Umbilicaris. *Lin. Syſt.* 1229. *No.* 592. *Liſt. Conch. tab.* 641. *fig.* 31, 32.
106. Umbi- *Liſt. Angl. tab.* iii. *fig.* 15.
lical.

Tr. with a perforated baſe, and of a convex conic
form ; dirty white waved with purple. Varies
much in colors.
A moſt

A moft common fhell on all our fhores.
Tab. lxxx. *fig.* 106.

Lin. *Syft.* 1229. *No.* 590.

Cinerarius.
106*. CINE-
REOUS.

T. with a perforated bafe; fpires a little prominent.
Of a cinereous color, ftriped obliquely.
    Size of a pea.
    Anglefea.

Lin. *Syft.* 1228. *No.* 585.

Magus.
107. TUBER-
CULATED.

TR. with a perforated bafe; fomewhat depreffed:
ftriated; with the ridges of the fpires rifing into
blunt diftinct tubercles. Color white, ziz-zagged
with red.
    When the upper coat is taken off, the next is of
a rich mother-of-pearl color.
    Anglefea.
    Tab. lxxx. *fig.* 107.

Minute, conic, livid.
    A new fpecies, difcovered in the mountains of
Cumberland, by Mr. *Hudfon.*
    Tab. lxxx. *fig.* 108.

Terreftris.
108. LAND.

XXIV. Its

TURBO.
WREATH.

XXIV. Its animal a Slug.
Aperture round.

* Ventricofe.

Littoreus.
109. Perri-
winkle.

Lin. Syft. 1232. *No.* 607. *Lift. Angl. tab.* iii. *fig.* 9. *Fauni Suec. No.* 2169.

T. with five fpires, the firft ventricofe, in younger fubjects ftriated fpirally; in the old fmooth, and of a dufky color.

Tab. lxxxi. *fig.* 109.

Abundant on moft rocks, far above low-water mark. The *Swedifh* peafants believe, that when thefe fhells creep high up the rocks, they indicate a ftorm from the fouth.

They are called *Perriwinkles*; are fold commonly in *London*, and eaten by the poor; as they are in moft parts of the kingdom.

Tumidus.
110. Tumid.

Lift. Angl. tab. ii. *fig.* 5.

T. with five tumid fpires, the firft ventricofe, and all moft elegantly ftriated; of a pale-red color.

A rare fhell. Inhabits woods in *Cambridgefhire*, and fome other counties in *England*.

Tab. lxxxii. *fig.* 110.

10   ** Taper.

** Taper.

Lin. Syſt. 1237. *No.* 631.   *Faun. Suec. No.* 2170. *Liſt. Conch.*   *Clathrus.*
    tab. 588. *fig.* 51.   III. BAR-
                                                            RED.

T. with a taper ſhell of eight ſpires, diſtinguiſhed
by elevated diviſions, running from the aperture to
the apex.

A. A variety? Pellucid; ridges very thin.

Theſe are analogous to that curious and ex-
penſive ſhell the *Wentle-trap.*

T. with about twelve ſpires of a duſky color, finely   *Tuberculata.*
tuberculated.   III*. STUD-
                                                         DED*

From the coaſt of *Northumberland.*

Tab. lxxxii. *fig.* *111.

Lin. Syſt. 1239. *No.* 645.   *Liſt. Angl. tab.* iii. *fig.* 7.   *Duplicatus.*
                                                            112. DOUB-
                                                            LED.

T. with a ſtrong taper ſhell, each ſpire marked
with two prominent *ſtriæ.* Has about twelve
ſpires.

Found by Doctor *Liſter* at *Scarborough,* who
ſays it was five inches long.

Tab. lxxxi. *fig.* 112.

VOL. IV.   K   *Lin.*

Terebra.
113.Auger. | *Lin. Syſt.* 1139. *No.* 645.　*Seb. Muſ.* iii. *tab.* lvi. *fig.* 40.
Liſt. Angl. tab. iii. *fig.* 8.　*Faun. Suec. No.* 2171.

T. with a taper ſhell of twelve ſpires, ſpirally ſtriated.

　Tab. lxxxi. *fig.* 113.

Albus.
114.White.

T. with eight ſpires, ſtriated tranſverſely; white.
　Tab. lxxix.

Lævis.
115.Smooth

T. with eight ſmooth ſpires, nearly obſolete.
　Tab. lxxix.

Both about a third of an inch long. Found on the ſhores of *Angleſea.*

Perverſus.
116. Re-
versed.

Lin. Syſt. 1250. *No.* 650. *Faun. Suec. No.*　*Liſt. Angl. tab.* ii.
fig. 11.

T. with eleven ſpires of a duſky color. The mouth turned a contrary way to moſt others of the genus.

　Length four-tenths of an inch; very taper.
　Found in moſſes, eſpecially among the *Hypna.*
　Tab. lxxxii. *fig.* 116.

Lin.

Lin. Syst. 1249. *No.* 649.   *Lift. Conch. tab.* 41. *fig. maj.*   Bidens.
117. Bi-
dent.

T. at firft fight to be diftinguifhed from others
of this genus by two teeth in the aperture. Agrees
with the laft in the contrary turn of the fpires,
which are twelve in number, and of a dufky
hue.

Tab. lxxxi.

Lin. Syft. 1249. *No.* 651. *Faun. Suec. No.* 2173. *Lift.*   Muscorum.
Angl. tab. ii. *fig.* 6.   *Conch. tab.* 41. *fig. min.*   118. Moss.

T. of an oval fhape, of the fize of a grain of white
muftard. With four fpires, very fhining and
brittle.

Found with the *Perverfus.*

Tab. lxxxii. *fig.* 118.

Buccinum exiguum fafciatum & radiatum.   Fafciatus.
Lift. Conch. tab. 19. *fig.* 14.   119. Fasciated.

T. with fix fpires; white, marbled or fafciated
with black.

Length half an inch.

Very frequent in *Anglefea,* in fandy foils near the
coafts.

Tab. lxxxii. *fig.* 119.

K 2   T. with

Ulvæ.
120. ULVA.

T. with four fpires, the firft ventricofe; of a deep brown color; aperture oval.

Size of a grain of wheat.

Tab. lxxxvi. *fig.* 120.

Inhabits the *Ulva Lactuca* on the fhores of *Flint-fhire.*

HELIX.
SNAIL.

XXV. Its animal a SLUG.

Shell fpiral, fub-pellucid.

Semi-lunar aperture.

* Depreffed.

Lapicida.
121. ROCK.

Lin. Syft. 1241. *No.* 656. *Lift. Angl. tab.* ii. *fig.* 14. *Faun. Suec. No.* 2174.

SN. with five fpires, externally carinated or de-preffed to an edge. Umbilicated; of a deep brown color.

A land fhell. Inhabits clefts of rocks.

Tab. lxxxiii. *fig.* 121.

Albella.
122. GREY.

Lin. Syft. 1242. *No.* 658. *Lift. Angl. tab.* ii. *fig.* 13. *Gualtieri, tab.* iii. *fig.* Q. *Faun. Suec. No.* 2175.

SN. with five fpires rounded on the outfide. Thin, prettily fafciated along the fpires with brown and white. Deeply umbilicated.

6　　　　　　　　　　　Inhabits

Inhabits dry fandy banks.

Tab. lxxxv. *fig.* 122.

Lin. Syft. 1242. *No.* 662. *Lift. Angl. tab.* ii. *fig.* 27. *Planorbis.*
Gualtieri, tab. iv. *fig.* E. E. *Faun. Suec. No.* 2176.   123. FLAT.

Sn. with a very flat brown fhell, flightly carinated
on the outfide ; the aperture oblique.

Inhabits ponds.

Tab. lxxxiii. *fig.* 121.

Lin. Syft. 1243. *No.* 667. *Lift. Angl. tab.* ii. *fig.* 28. *Gualtieri. Vortex.*
    tab. iv. *fig.* G. G.   124. WHIRL.
Lift. Conch. tab. 138. *fig.* 43.   *Faun. Suec. No.* 2178.

Sn. with a very flat thin fhell, and fix fmall fpires.
The outmoft carinated.

Found with the former.

Tab. lxxxiii. *fig.* 124.

Sn. with four fpires ; the exterior very large. *Nana.*
Thick in proportion to its diameter.   Umbili- 125. DWARF,
cated.

Whether a young, or a variety of the following ?

Tab. lxxxiii. *fig.* 125.

K 3   *Lin.*

Cornea:
126. HORNY.　　*Lin. Syſt.* 1243. *No.* 671. *Liſt. Angl. tab.* ii. *fig.* 26. *Gualtieri. tab.* iv. D. D. *Faun. Suec.* 2179.

Sɴ. with four rounded ſpires. Umbilicated; of
a horny appearance.

Found in dull deep rivers, and in ponds. The
largeſt of the *Britiſh* depreſſed ſpecies,

Tab. lxxxiii.

** Ventricoſe.

Rufefcens.
127. MOT-
TLED.　　Cochlea dilutè rufefcens, aut ſubalbida, ſinu ad umbilicum
exiguo, circinato. *Liſt. Angl. tab.* ii. *fig.* 12.

Sɴ. with four ſpires, and minutely umbilicated;
the exterior ſpire ſub-carinated. Of a pale browniſh
red mottled with white.

Inhabits woods.

Tab. lxxxv. *fig.* 127.

Pomatia.
128. EXO-
TIC.　　Pomatia *Diofcor. lib.* ii. *c.* 9. *p.* 305. *Gefner. Aq.* 655.
Lin. Syſt. 1244. *No.* 677. *Liſt. Angl. tab.* ii. *fig.* 1. *Faun.
Suec. No.* 2183.

Sɴ. with five ſpires moſt remarkably ventricoſe.
Slightly umbilicated. Faſciated with a lighter and
deeper brown.

9　　　　　　　　　　　Inhabits

Inhabits the woods of the southern counties of *England*.

A naturalized species, introduced, as is said, by Sir *Kenelm Digby*; whether for medical purposes, or as a food, is uncertain. Tradition says, that to cure his beloved wife of a decay was the object.

They are quite confined to our southern counties. An attempt was made to bring them into *Northamptonshire* *, but they would not live there.

These are used as a food in several parts of *Europe* during *Lent*; and are preserved in an *Escargatoire*, or a large place boarded in, with the floor covered half a foot deep with herbs, in which the snails nestle and fatten †. They were also a favorite dish with the *Romans*, who had their *Cochlearia*, a nursery similar to the above. *Fulvius Hirpinus* ‡ was the first inventor of this luxury, a little before the civil wars between *Cesar* and *Pompey*. The snails were fed with bran, and sodden wine. If we could credit *Varro* ‖, they grew so large, that the shells of some would hold ten quarts! People need not admire the temperance of the supper of the younger *Pliny* §, which consisted of only a lettuce a-piece, *three* snails, two eggs, a barley cake, sweet wine, and snow; in case his

---

* *Morton*, 415.   † *Addison's Travels*, 272.
‡ *Pliny*, lib. x. c. 56.   ‖ *De Re Rustica*, lib. iii. c. 14.
§ Epist. *lib.* i. *Epist.* xv.

K 4   snails

ſnails bore any proportion in ſize to thoſe of *Hir-*
pinus.

Its name is derived not from any thing relating
to an orchard, but from Πῶμα, an *operculum*, it
having a very ſtrong one. This ſeems to be the
ſpecies deſcribed by *Pliny,* lib. viii. c. 39, which
he ſays was ſcarce ; that it covered itſelf with the
opercle, and lodged under ground ; and that
they were at firſt found only about the maritime
Alps, and more lately near *Velitræ.*

Tab. lxxxiv. *fig.* 128.

Hortenſis.
129. GAR-
DEN.

Cochlea vulgaris major puila maculata et faſciata hortenſis.
 Liſt. Angl. tab. ii. *fig.* 2. *Gualtieri, tab.* i. *fig.* C.
Helix lucorum. *Lin. Syſt.* 1247. *No.* 692. *Liſt. Conch.*
 tab. 49. *fig.* 47. The common garden ſnail.

Sɴ. in form like the laſt, but leſſer, and not
umbilicated and clouded, or mottled with browns.

Theſe are often uſed with ſucceſs in conſumptive
caſes.

Tab. lxxxiv. *fig.* 129.

Arbuſtorum.
130. SHRUB.

Lin. Syſt. 1245. *No.* 680. *Liſt. Angl. tab.* ii. *fg.* 4. *Faun.*
 Suec. Nᵒ. 2184.

Sɴ. with a gloſſy ſhell, brown, marked with a
ſingle black ſpiral *faſcia :* the rim of the aperture
reflects a little. Sub-umbilicated. Varies with
deeper and lighter colors.

Inhabits

Inhabits woods.

Tab. lxxxv. *fig.* 130.

Lin. Syst. 1247. *No.* 691. *Gualtieri. tab.* i. *fig.* P. *Lift. Nemoralis.*
Conch. tab. 57. *Lift. Angl. tab.* ii. *fig.* 3.   131. VARIE-
                                                    GATED.

SN. with a glossy shell; very thin and pellucid.
The aperture awry. Varies infinitely: often yellow, or light green, or red fasciated with black or
white, along the spires.   Often quite plain.

Inhabits woods and gardens.

Lin. Syst. 1247. *No.* 690. *Lift. Angl. tab.* ii. *fig.* 18.   *Conch. Vivipara.*
tab. 126. *fig.* 26. *Faun. Suec. No.* 2185.   132. VIVI-
                                                              PAROUS.

SN. with six ventricose spires, umbilicated.   The
aperture almost round.   Color brown, with dusky
spiral *fasciæ.*

Inhabits stagnant waters, and semi-stagnant
rivers.

Tab. lxxxiv. *fig.* 132.

Lin. Syst. 1245. *No.* 681. *Gualtieri. tab.* iii. *fig.* L.   *Zonaria.*
                                                                  133. ZON-
                                                                  ED.

SN. with five spires ; the first very ventricose.
Slightly umbilicated. Fasciated spirally with narrow
stripes of white, dusky, and yellow.

Inhabits dry banks.

Variety ?

Variety? of the former. A fhell of a plain color, with the apex a little more projecting. *Fig.* 133. A.

Pellucida.
134. Pel-
lucid.

Cochlea terreftris umbilicata pellucida flavefcens. *Gualtieri,* *tab.* ii. *fig.* G.

Sn. a very thin pellucid fhell, of a yellowifh-green color. Very brittle. With four fpires, the firft very tumid.

Found by me only once; in *Shropfhire.*

*** Of a taper Form.

Oacna?
135. Eight-
spired.

Lin. Syft. 1248. *No.* 698. *Gualtieri. tab.* 6. *fig.* B. ?

Sn. with eight fpires of a brown color. My fpe-cimen was mutilated.

Inhabits ponds.

Tab. lxxxvi. *fig.* 135.

**** Ovated, imperforated.

Stagralis.
136. Lake.

Lin. Syft. 1249. *No.* 703. *Lift. Angl. tab.* ii. *fig.* 21. *Conch. tab.* 123. *fig.* 21. *Faun. Suec. No.* 2188.

Sn. with fix fpires; the firft very large and ventri-cofe,

coie, and the laft quite pointed.   Very brittle;
Length two inches one eighth.

Inhabits ftill waters;   is,   with others of the
kind, the food of trouts.

In younger fpecimens is a duplicature of the
fhell, from the aperture fpreading along the firft
fpire; as in *fig.* A.   In old fhells it vanifhes.

B. Another, which I fufpect to be alfo a variety: leffer and fomewhat ftronger.   Perhaps the
Helix lineofa of *Linnæus, No.* 706.   *Lift. Angl.*
tab. ii. *No.* 22.

  Tab. lxxxvi. *fig.* 136. A. B.

Lin. Syft. 1249. *No.* 705.   *Lift. Angl. tab.* ii. *fig.* 24.   *Conch.* *Putris.*
 tab. 123. *fig.* 23.   *Faun. Succ. No.* 2189.   137. Mud,

Sn. with the firft fpire vaftly large and tumid.
The two others very fmall.

Inhabits ponds, &c.

  Tab. lxxxvi. *fig.* 137.

Lin. Syft. 1250. *No.* 708.   *Lift. Angl. tab.* ii. *fig.* 23.   *Conch.* *Auricularia,*
 tab. 123. *fig.* 22.   *Faun. Suec. No.* 2192.   138. Ear,

Sn. with a very ventricofe firft fpire, fub-umbili-
cated.   The laft forms a minute apex.   Color yellow.   Very brittle.

Inhabits ponds.

  Tab. lxxxvi. *fig.* 138.

             Lin.

Lævigatum ? *Lin. Syft.* 1250. *No.* 709.
139.SMOOTH-
ED.

Sn. with only two fpires : the firft very ventricofe;
the other very minute, and placed laterally.   Of a
pale-red color.   Pellucid.
    Inhabits ponds.
    Tab. lxxxvi. *fig.* 139.

Tentaculata. *Lin. Syft.* 1249. *No.* 707.   *Lift. Angl. tab.* ii. *fig.* 19.   *Conch.*
140. OLIVE.   *tab.* 132. *fig.* 32. *Faun. Suec. No.* 2191.

Sn. of an oval fub-conic form, with five fpires.
Clouded with brown.
    Inhabits ponds.
    Tab. lxxxvi. *fig.* 140.

NERITA.   XXV. Its animal a SLUG.
NERITE.   Shell gibbous, flattifh at bottom.
                     Aperture femi-orbicular.

Glaucina.   *Lin. Syft.* 1251. *No.* 716.   *Lift. Angl. tab.* iii. *fig.* 10. *Faun.*
141. LIVID.   *Suec. No.* 2197.

N. with five fpires, umbilicated.   Of a livid color.
The fpires marked with fhort brown ftripes; but
it varies in colors.
    Tab. lxxxvii. *fig.* 141.

                                                                Lin.

Lin. Syst. 1253. *No.* 723. *Lift. Angl. tab.* ii. *fig.* 20. *Conch.* *Fluviatilis.*
tab. 141. *fig.* 38. *Faun. Suec. No.* 2194.   142. RIVER.

N. with only two spires. Brittle, dusky, marked
with white spots.
    Not half the size of a pea.
    Inhabits still rivers and standing waters.
    Tab. lxxxviii. *fig.* 142.

Lin. Syst. 1253. *No.* 725. *Lift. Angl. tab.* iii. *fig.* 11, 12, 13. *Littoralis.*
Conch. tab. 607. *fig.* 39, &c. *Faun. Suec. No.* 2195.   143. STRAND

N. with a thick shell, with four spires. Generally
of a fine yellow. Varies greatly into other colors.
    Large as a horse-bean.
    Common at the sea rocks.
    Tab. lxxxvii. *fig.* 143.

XXVI. Its animal a SLUG.   *HALIOTIS.*
        Shell of the shape of a human ear, with a
            row of orifices along the surface.
        The spire near one end turned in.

Lin. Syst. 1256. *Lift. Conch. tab.* 611. *Lift. Angl. tab.* iii. *Tuberculata.*
    fig. 16.   144. TUBER-
                                                            CULATED.

H. with a rough shell, the inside like mother-of-
pearl.

                                        Inhabits

Inhabits the sea near *Guernsey*; also frequently cast up on the southern shores of *Devonshire*. When living adheres to rocks.

This was the λεπας αγρια, the wild limpet, and θαλλάττιον ἕς, the sea ear of *Aristotle* *.

Tab. lxxxviii. *fig.* 144.

Div. IV. Univalve Shells.

Without a regular spire.

PATELLA.   XXVII.   Its animal a Slug.
LIMPET.   Conic shell, without spires.

Vulgata.   *Lin. Syst.* 1258. *No.* 758. *List. Angl. tab. v. fig.* 40. *Faun.*
145. Com-   *Suec. No.* 2199.
mon.

P. with rough prominent *striæ*, and sharply crenated edges. Vertex pretty near the centre. The edges often in old subjects are almost smooth.

Tab. lxxxix. *fig.* 145.

Depressa.   *List. Conch. tab.* 538. *fig. inf.*
146. Flat.

P. much depressed; the vertex approximating nearly to one edge. More oblong than the former.
Tab. lxxxix. *fig.* 146.

* *Hist. de lib. iv. c. 4.*

Lin.

P. with a white acuminated ftriated fhell, the top
turning down like a *Phrygian* bonnet.

 Tab. xc. *fig.* 147.

Patella vertice intorto, &c.   *Gualtieri. tab.* ix. *fig.* 10.   *Intorta.*
148. INCLIN-
ING.

P. with an elevated fhell, flightly ftriated; the
vertex bending, but not hooked.

 Inhabits *Anglefea.*   Found on the fhores

 Tab. xc. *fig.* 148.

Lin. Syst. 1260. *No.* 769.   *Lift. Angl. tab.* ii. *fig.* 32. *Conch.* *Lacustris.*
 tab. 141. *fig.* 39.   *Faun. Suec. No.* 2200.   149. LAKE.

P. with a fhell almoft membranaceous; the top
reclined.

 Inhabits frefh waters.

Lin. Syst. 1260. *No.* 770.   *Lift. Conch. tab.* 543. *fig.* 27.   *Pellucida.*
150. TRANS-
PARENT.

P. with a pellucid fhell, marked longitudinally
with rows of rich blue fpots. The vertex placed
near one edge.

 Inhabits the fea rocks of *Cornwall.*

 Tab. xc. *fig.* 150.

Patella

Lævis.
151.Smooth

Patella lævis fufca.   *Lift. Conch. tab.* 542. *fig.* 26.

P. with a fmooth and glofsy fhell, fomewhat de
prefsed; the apex inclining.
Found on the fhores near *Bamff.*
Tab. xc. *fig.* 151.

Fiffura.
152. Slit.

Lin. Syft. 1261. *No.* 778. *Lift. Conch. tab.* 543. *fig.* 28.

P. with a white fhell, of an elevated form, vertex
inclining; elegantly ftriated and reticulated. Has
a remarkable flit in front.
Inhabits the feas of the Weft of *England.*
Tab. xc. *fig.* 152.

Græca?
153. Stri-
ated,

Lin. Syft. 1262. *No.* 780.   *Lift. Conch. tab.* 527. *fig.* 2.

P. with an oblong fhell, perforated vertex, ftriated
roughly to the edges.
Inhabits the Weft of *England.*
Tab. lxxxix. *fig.* 153.
This genus was called by the *Greeks* λεπας, and
is mentioned by *Ariftotle* and *Athenæus*; who
acquaint us, that it was ufed for the table; and alfo

---

* *Arifot. Hift. An. lib.* iv. *c.* 4. *Athenæus, lib.* iii. *p.* 85.

inform

inform us of its nature of adhering to rocks. *Ari-stophanes* with much humour speaks of an old woman who stuck as close to a young fellow as a *Lepas* to a rock.

Linnæus has adopted the *Latin* name of *Patella*, a sort of dish; and has applied it (as some other modern writers have before) to this genus.

XXVIII. Its animal a Terebella.   *DENTALIUM.*
    A slender tubiform shell.   TOOTH-SHELI

Lin. Syst. 1263. *No.* 786. *List. Conch. tab.* 547. *fig.* 2.   *Faun. Entalis.*
  Suec. No. 2201.   154. Com-
                         mon.

D. with a slender shell, a little bending.   Pervious.
  Length near an inch and a half.
  Inhabits most of our seas.
  Tab. xc. *fig.* 154.

XXIX. Its animal a Terebella.   *SERPULA.*
    Tubular shell adhering to other bodies.

Lin. Syst. 1264. *No.* 794.   *Faun. Suec. No.* 2204.   *Spirorbis.*
                        155. Spiral.

S. with a shell spiral or wreathed, like the *cornu ammonis.*

Vol. IV.   L   Very

Very fmall; adhering to fhells, *cruftacea* and *algæ.*
Tab. xci. *fig.* 155.

Triquetra.
156. Angu-
lar.

Lin. Syft. 1265. *No.* 795. *Faun. Suec. No.* 2206.

S. with a triangular fhell, irregularly twifted.
　Adheres to (in a creeping form) ftones and other
fubftances.

Intricata.
157. Com-
plicated.

Lin. Syft. 1265. *No.* 796.

S. with a flender fhell greatly entwined.
　Adheres to fhells, &c. moft intricately twifted.
　Tab. xci. *fig.* 157.

Contortupli-
cata.
158. Twin-
ed.

Lin. Syft. 1266. *No.* 799. *Lift. Conch. tab.* 29. *fig.* D.
Faun. Suec. No. 2205.

S. with a ftrong, rugged, angulated fhell, entwined.
　Adheres to fhells, &c.
　Tab. xci. *fig.* 158.

Vermicularis.
159. Worm.

Lin. Syft. 1257. *No.* 805. *Ellis Coral. tab.* xxxviii. *fig.* 2.

S. with a flender, incurvated, taper, and rounded
fhell.
　According to Mr. *Ellis,* inhabits all our coafts,

3　　　　　　　　　　　　　　　　　　　Its

Its animal a TEREBELLA.
Shell flender, bending.

TEREDO.
PIERCER.

Lin. Syft. 1267. *No.* 807. *Faun. Suec. No.* 2087.

Navalis.
160. SHIP.

Juftly called by *Linnæus calamitas navium.* Was imported from the *Indies.* Penetrates into the ftouteft oak plank, and effects their deftruction.

XXX. Its animal a NEREIS.
A tubular covering, fabricated with fand and broken fhells, coherent by a gluti-nous cement.

SABELLA.

Lin. Syft. 1268. *No.* 811. *Bafter fubfef.* 1. *p.* 80. *tab.* 9. *fig.* 4.

Rudis.
161. COARSE.

S. with a fingle cafe formed of larger fragments of fhells, with little or no fand.

Found near *Weymouth,* lodged in the fhell of a bivalve. The animal is reprefented magnified in *Tab.* xxvi. marked A. A.

Lin. Syft. 1268. *No.* 812. *Ellis Coral. tab.* xxxvi. *p.* 90.

Alveolata.
162. HONEY-
COMB.

S. with numerous tubes placed parallel ; with the orifices open, forming in the mafs the appearance

L 2   of

of the furface of honey-combs: compofed chiefly of fand, with very minute fragments of fhells. The tubes fometimes above three inches long.

Found on the weftern coafts of *Anglefea*; near *Cricceth, Caernarvonfhire*; and near *Yarmouth*. It covers the rocks for a confiderable fpace near low-water mark.

Tab. xcii. *fig.* 162.

Tubiformis.
163. Tube.
Nereis cylindracea belgica. *Pallas. Mifc. Zool. p.* 211. *tab.* ix. *fig.* 3.

S. with a cafe of a taper ftrait form ; made up of minute particles of fand, moft elegantly put toge-ther.

Its animal defcribed at *No.* 34.

Common on all our fandy fhores.

Tab. xcii. *fig.* 163.

F I N I S.

E R R A T A.

Page 36. —— *For* Sipunculus *read* Siphunculus.

74. —— — Barnacle —— Bernacle.

In the Plates.

Plate XXV. — N° 41. —— Plate XXVI.

LIX. — N° 80. —— LXI.

INDEX.

L 3

Clams,

N. B. The Binders are requeſted to place all the Plates at the End.

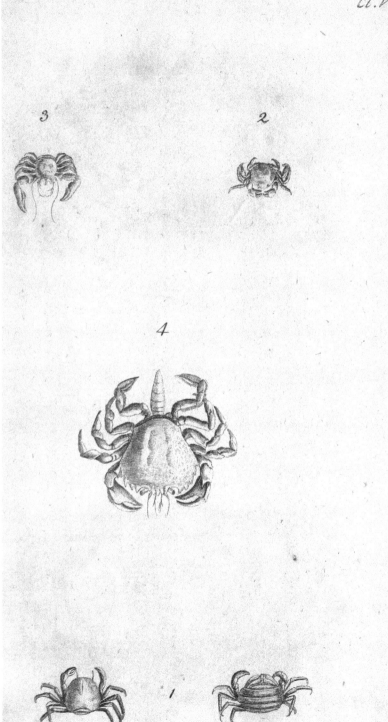

6.

5.

7

A.VI.

8

9.

10.

12.

11.

Pl VII.

Cl. V.

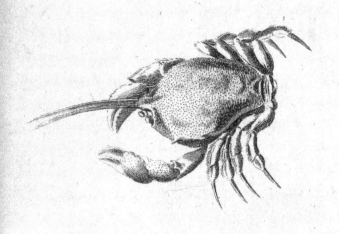

13.

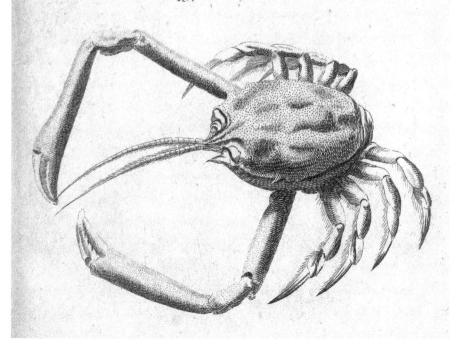

16.

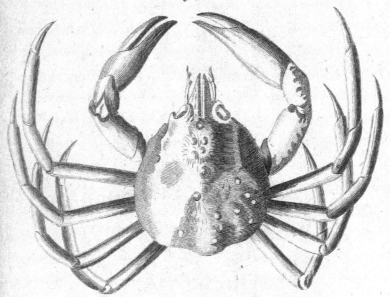

17.

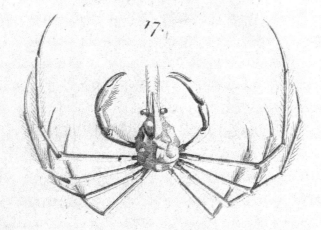

19.

20.

18.

Pl. X.

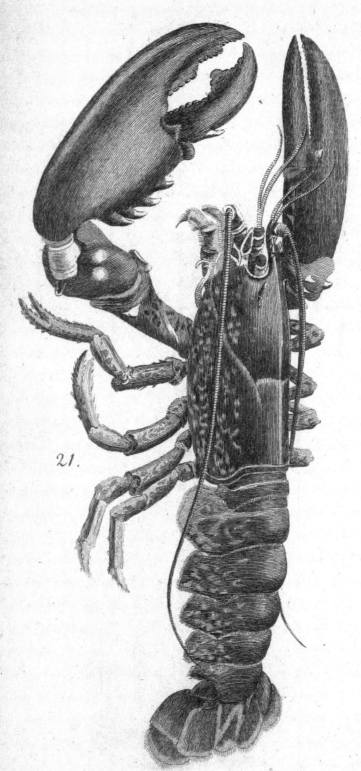

21.

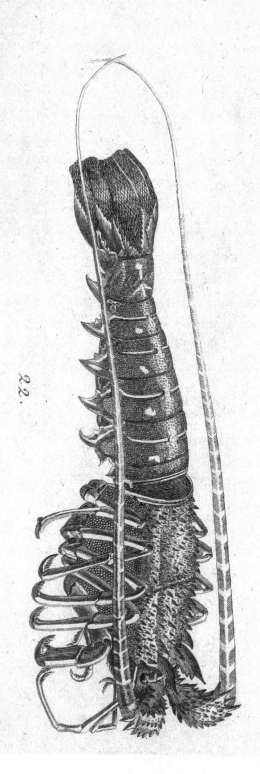

22.

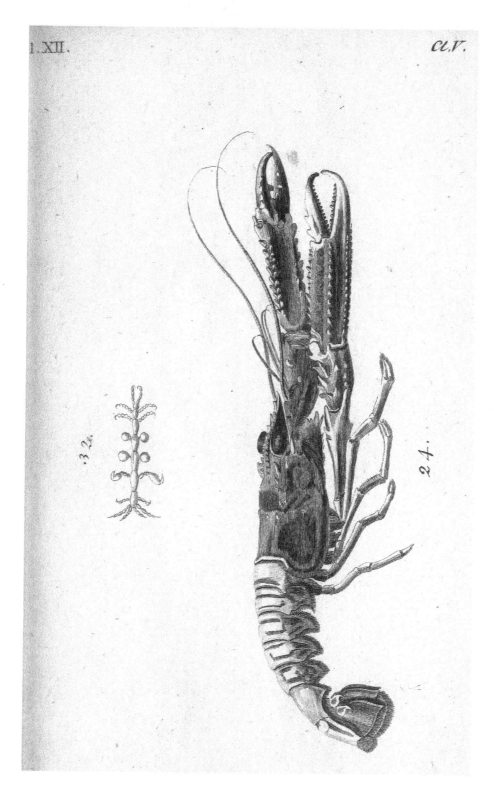

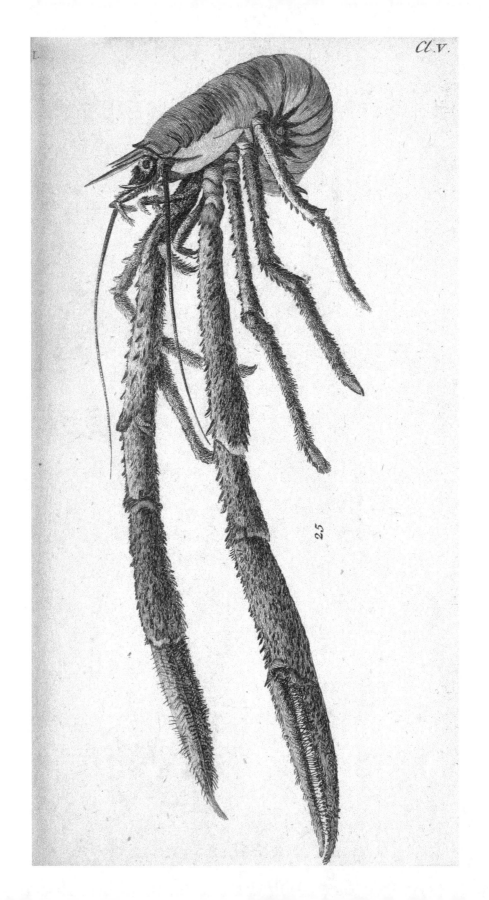

25

26.

27.

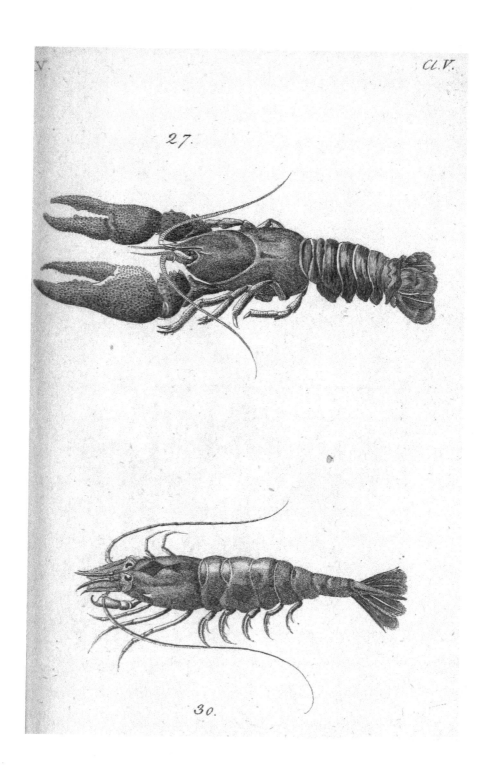

30.

28.

31.

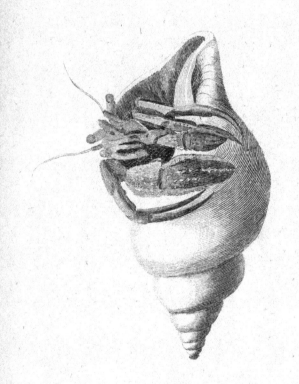

38.

1.

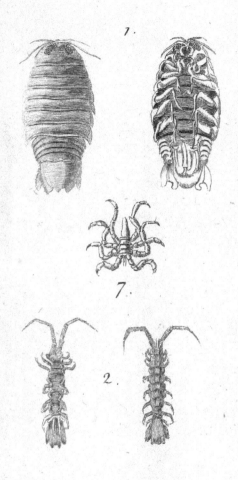

7.

2.

3.

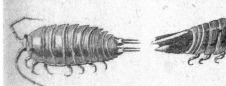

4.   *5.*   *6.*

Pl. XIX.

6

7

6.A.

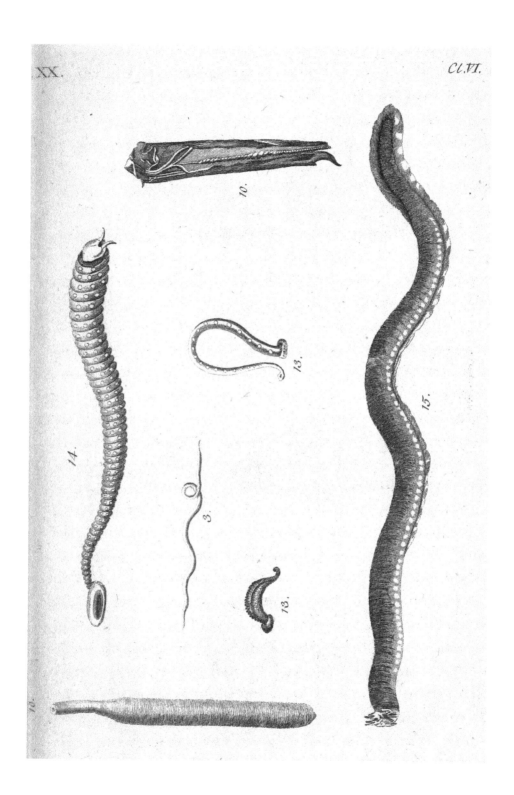

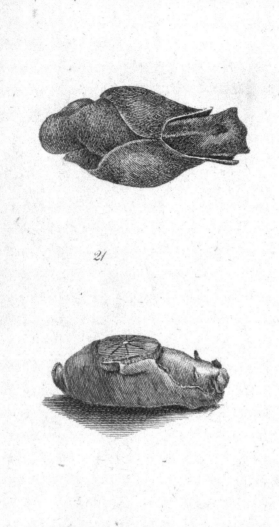

21

23

22.

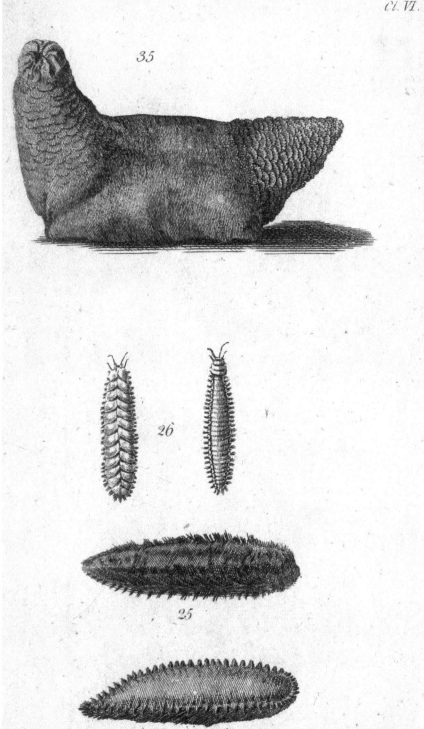

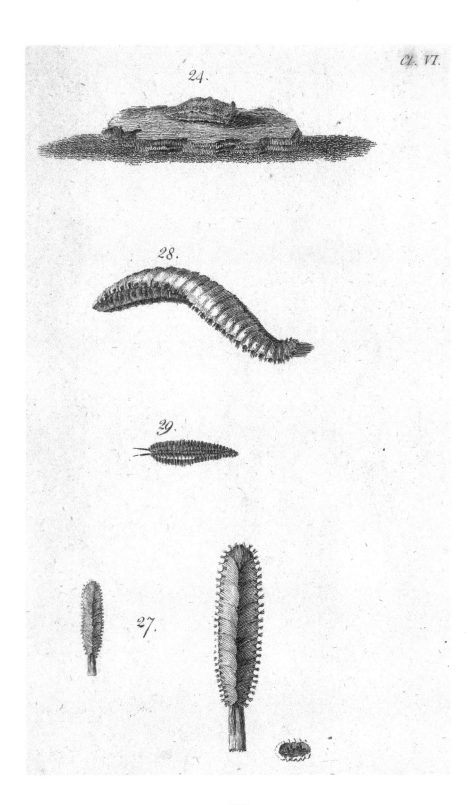

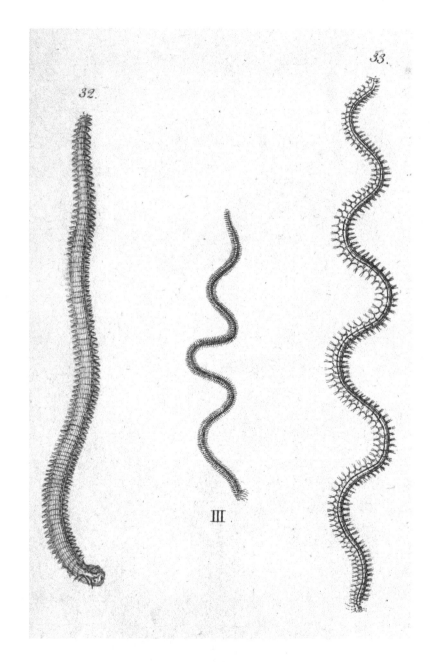

III

41.

43.

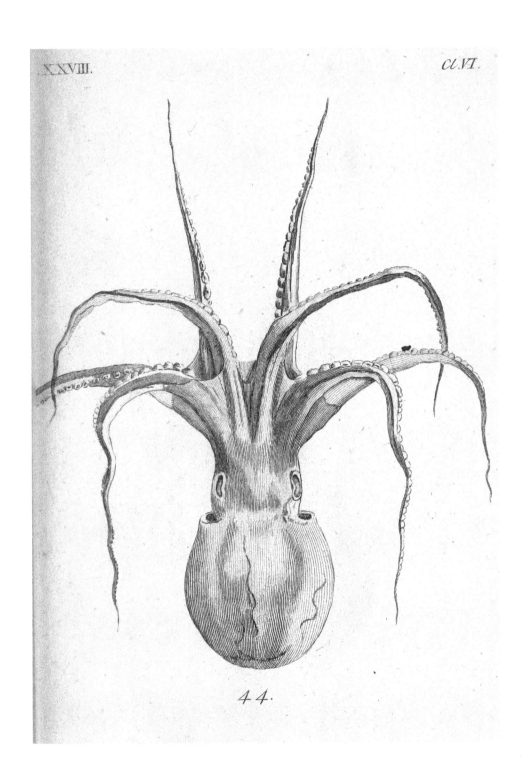

44.

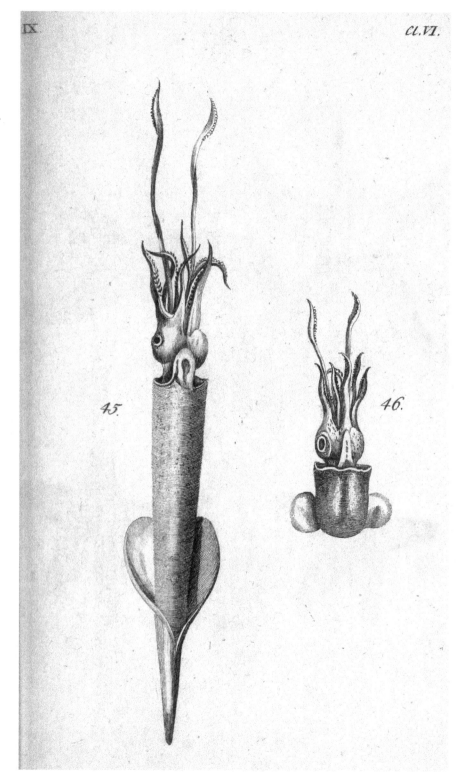

45.

46.

56

58.

56

58.

59. A.

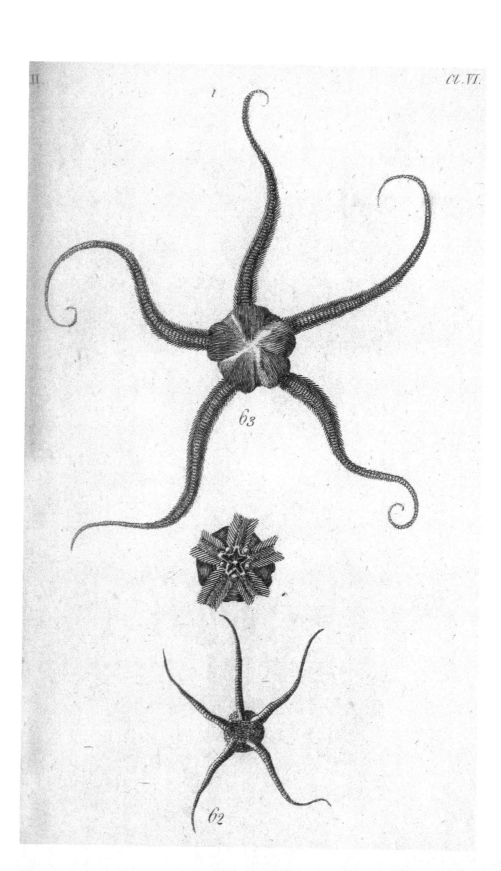

1

63

62

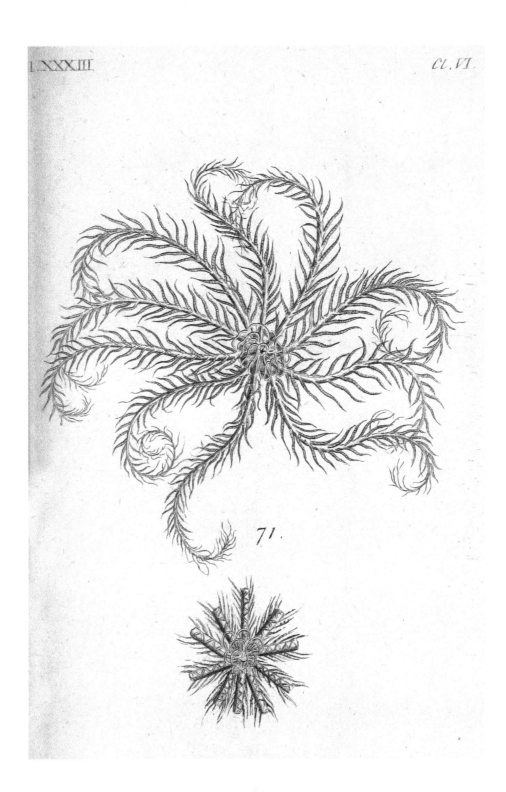

71.

74.

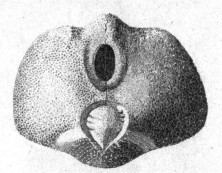

75.

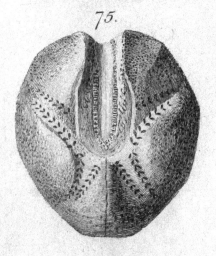

76.

Pl. XXXVI.

A.I.

1.

2.

CHITON.

3

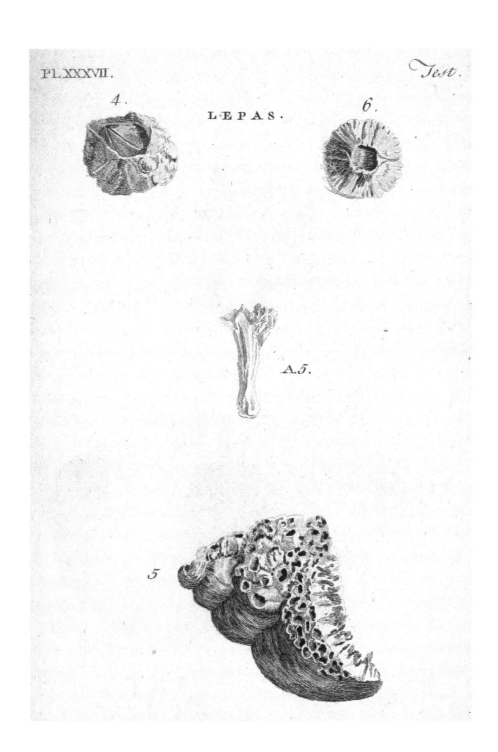

LEPAS.

A.5.

LEPAS.

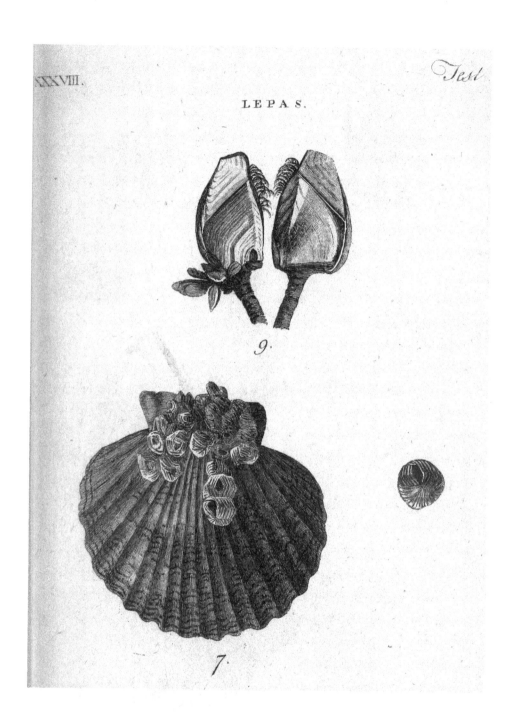

9.

7.

Pl. XXXIX.

PHOLAS

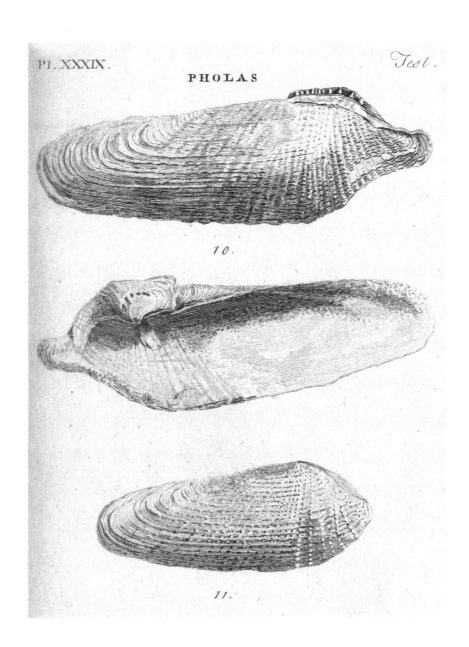

10.

11.

13.

PHOLAS.

12.

MYA

14.

Pl. XLII.

Tab.

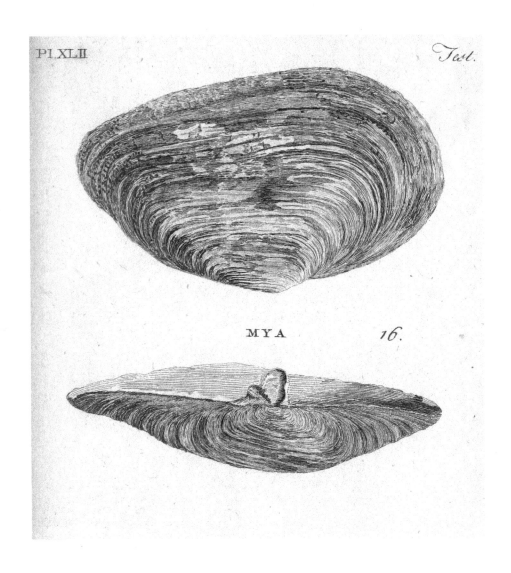

MYA 16.

MYA.

18.

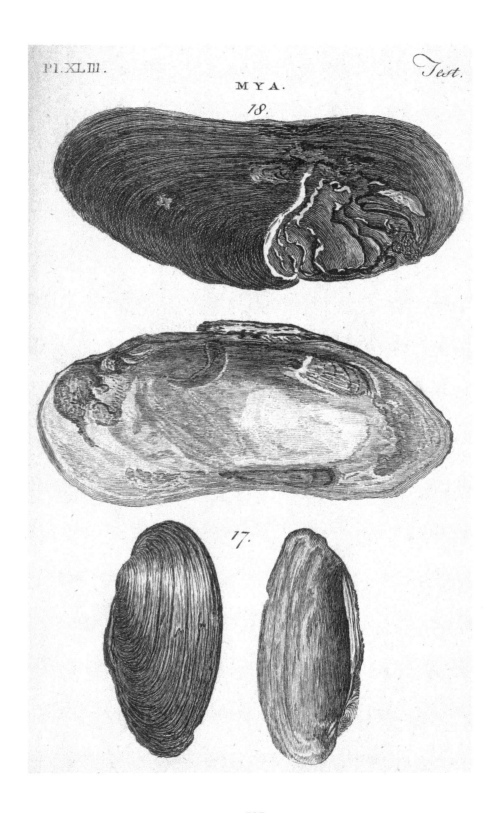

17.

Pl. XLIV.

Test.

MYA.

19.

SOLEN.

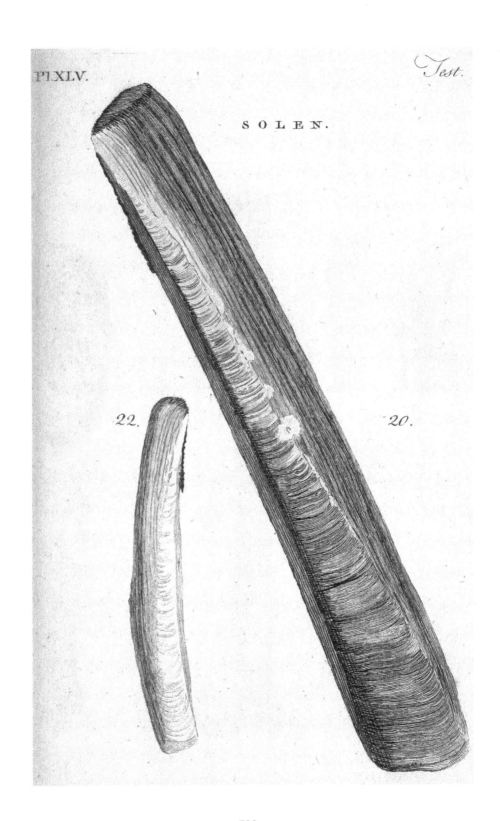

22.

20.

Pl. XLVI . Test

SOLEN.

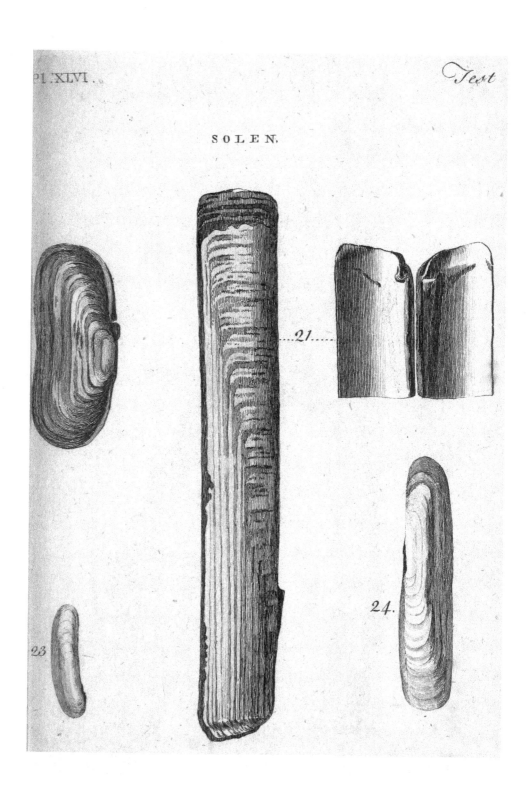

TELLINA.

27

26.

31.

Pl. XLVIII.

TELLINA.

28.

29.

TELLINA.

36.

32.

32.

30.

32.A.

CARDIUM.

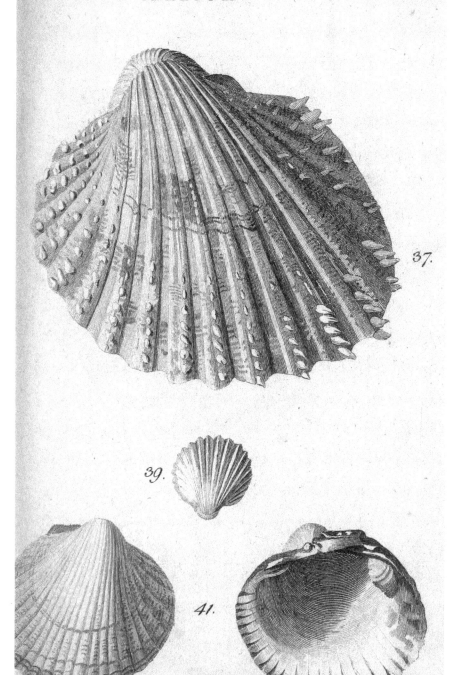

37.

39.

41.

MACTRA.

43.A.

CARDIUM.   40.

P. Mazell sculp.

MACTRA.

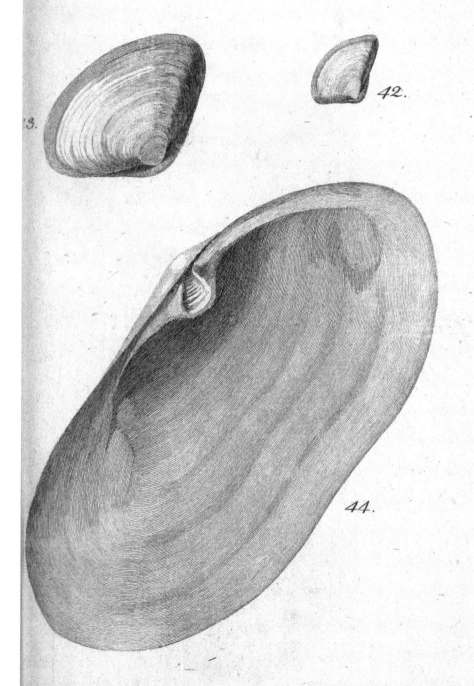

43.

42.

44.

P. Mazell sculp.

VENUS.

47.

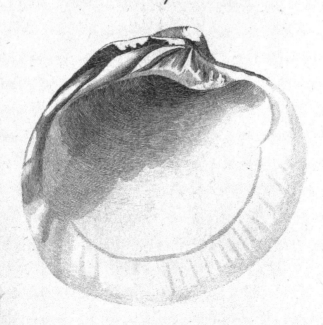

P. Mazell sculp.

V E N U S.

49 A.

48. A.

48.

D O N A X.

45.

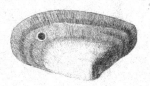

51.

51. A.

46.

VENUS.

56.

49.

50.

P. Mazell sculp.

VENUS.

34.

53.

54.

A R C A.

58.

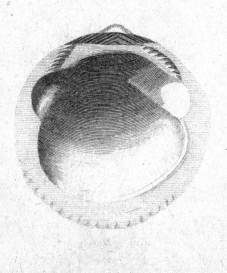

59.

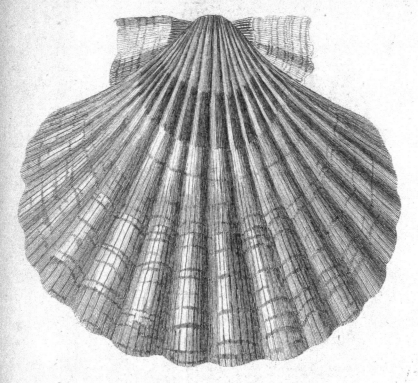

61.

62.

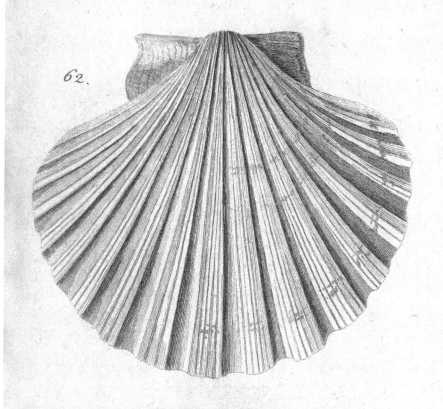

63.

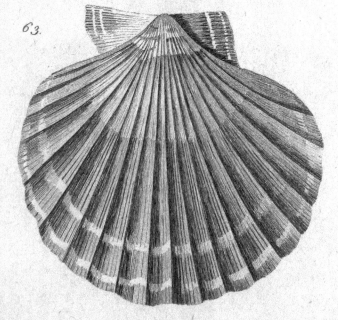

Pl. LXI.

Test.

64.

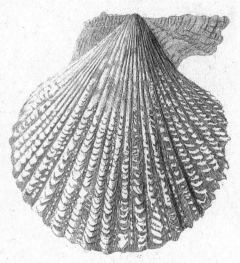

66.

PECTEN.

65.

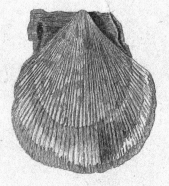

Pl. LXII.

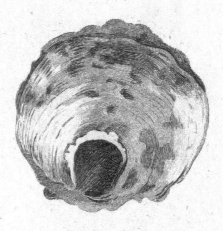

ANOMIA.

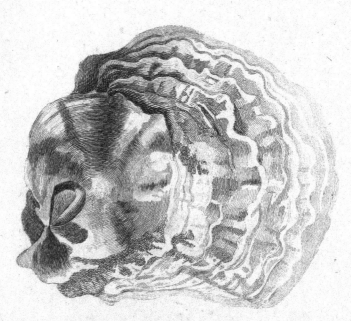

70.

Pl. LXIII.

Test.

MYTILUS.

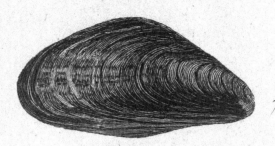

7.3.

72.

75.

MYTILUS.

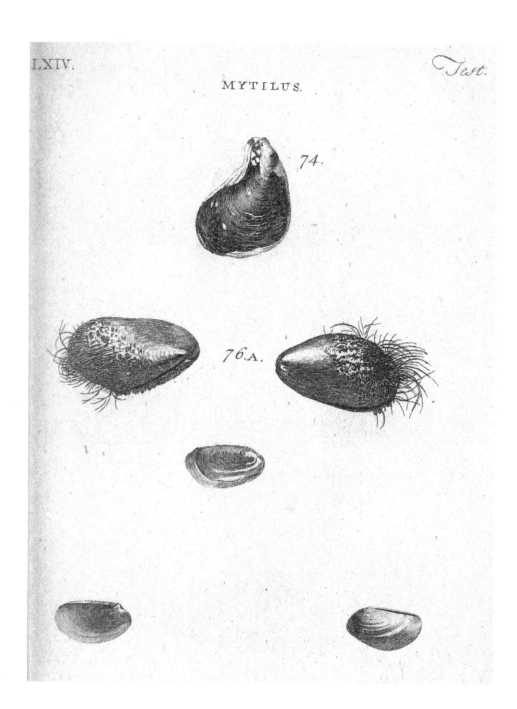

74.

76.A.

Pl LXV.

Test.

MYTILUS.

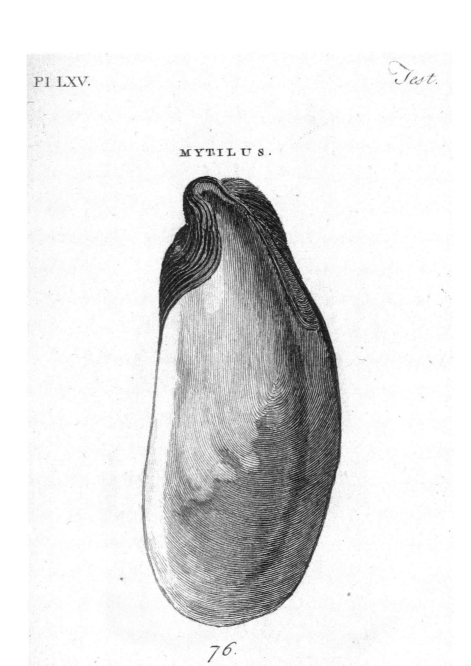

76.

Pl. LXVI.

MYTILUS.

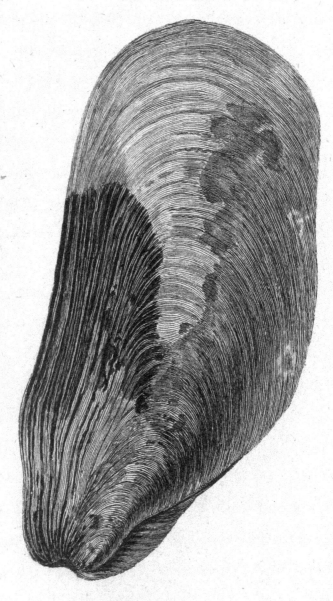

77.

MYTILUS.

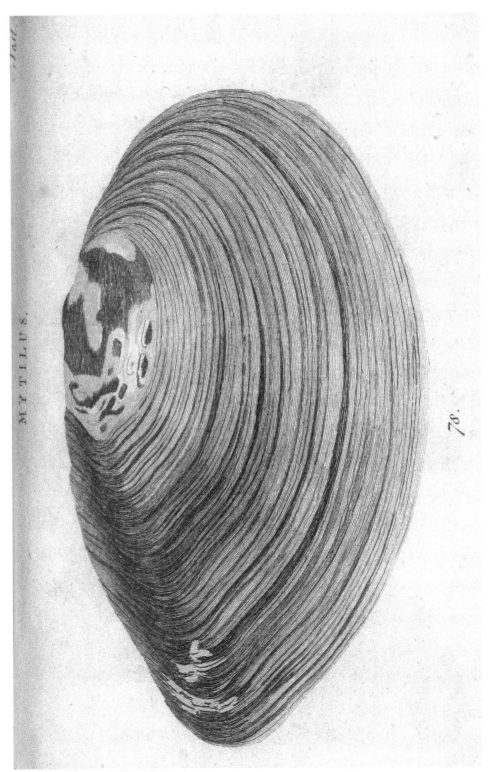

78.

Pl. LXVIII.

MYTILUS

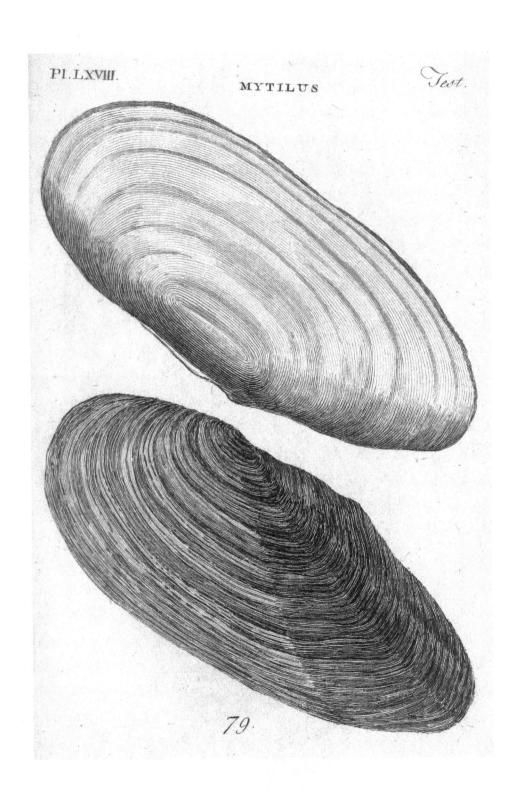

79.

Pl. LXIX

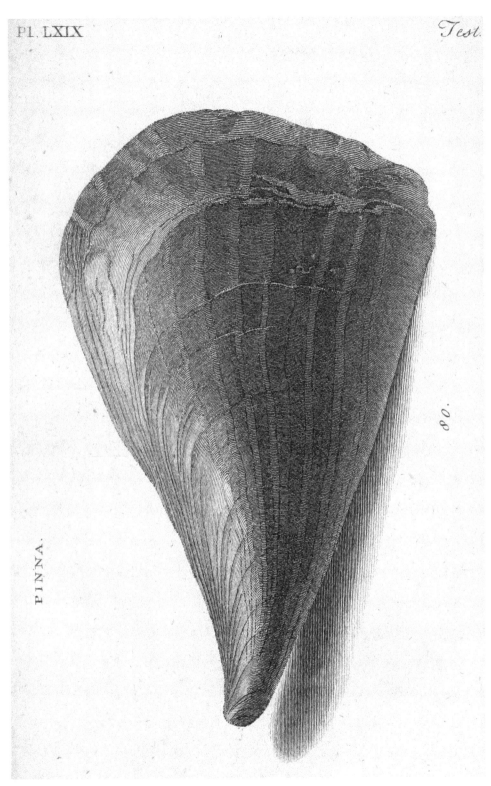

PINNA.

δ0.

Pl. LXX.   Test.

CYPRÆA.

82.

 85. A.

83.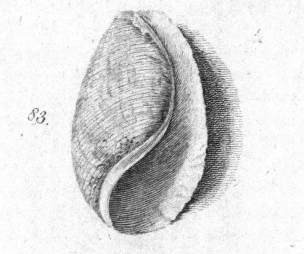

BULLA

 85.

Pl LXXI.

Test.

VOLUTA.

86.

87.

Pl. LXXII.

BUCCINUM.

88.

89.

89.

89.

92.

Pl. LXXIII.

Test.

BUCCINUM

90.

BUCCINUM.

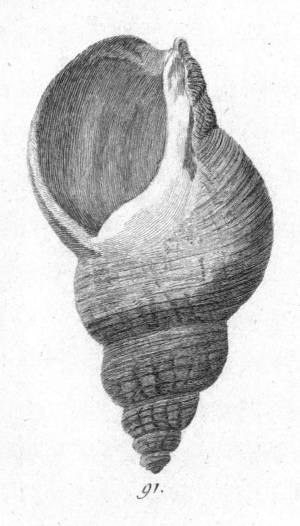

91.

Pl. LXXV.

STROMBUS.

94.

Pl. LXXVI.   *Test.*

99.

M U R E X.

95.

Pl. LXXVII.

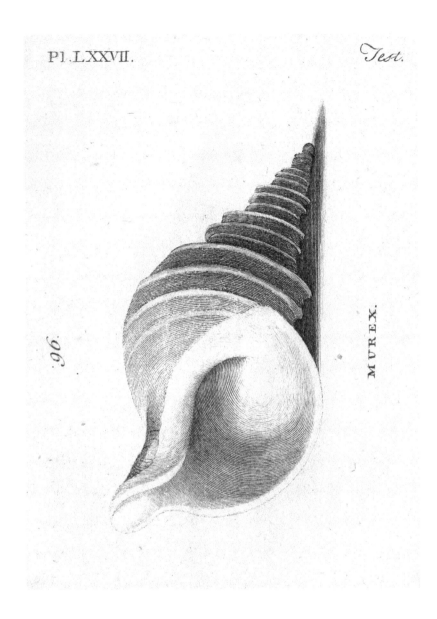

.96.

MUREX.

Pl. LXXVIII.   Test

MUREX.

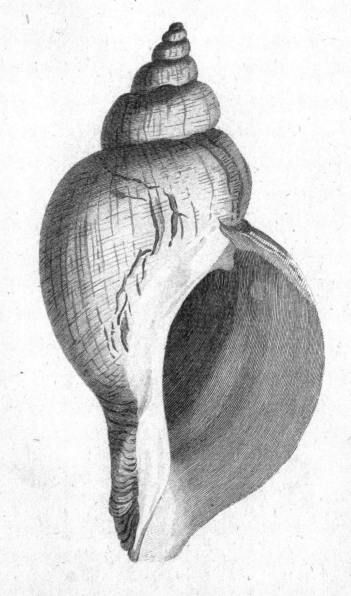

98.

Pl. LXXIX. Test.

App:

Pl. LXXX.

TROCHUS.

103.

106.

106.

104.

108.

107.

108.

HELIX.

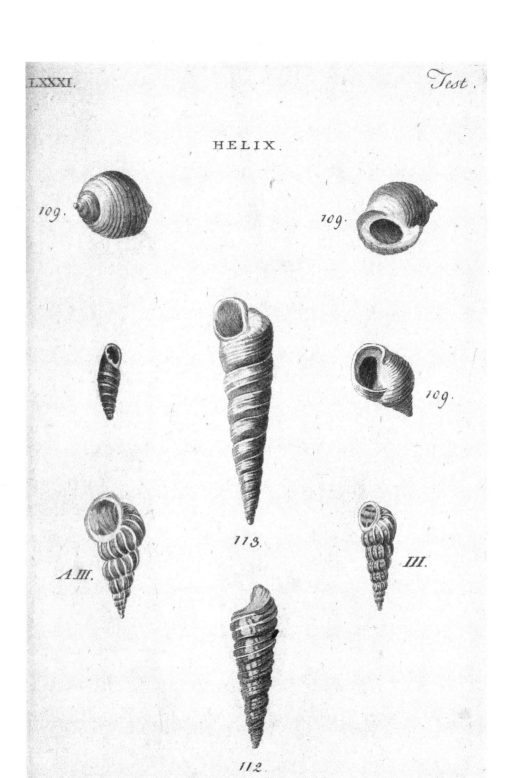

Pl. LXXXII. *Test.*

TURBO.

116.

118.

110.

119.

* III.

Pl. LXXXIII. Test.

HELIX.

 121.

 123.

126.

125.

 125.

124.

HELIX.

132.

128.

129.

HELIX.

133.

130.

133.A.

122.

127.

HELIX.

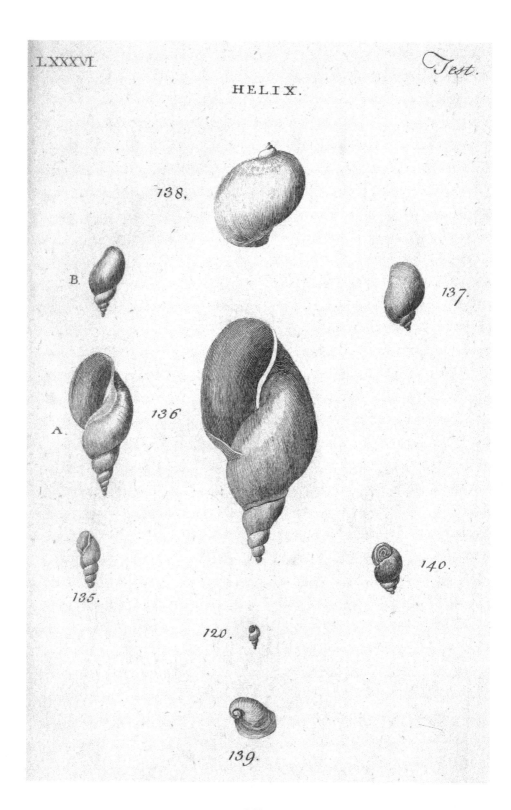

138.

B.

137.

A.   136.

135.

140.

120.

139.

NERITE.

143.

143.

141.
 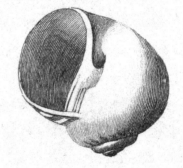

142.

Pl. LXXXVIII.

144.

PATELLA.

145.

146.

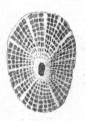

153.

145.

PATELLA.

147.

148.

151.

150.

Pl. XCI. *Test.*

SERPULA.

157.

158.

155.

SABELLA.

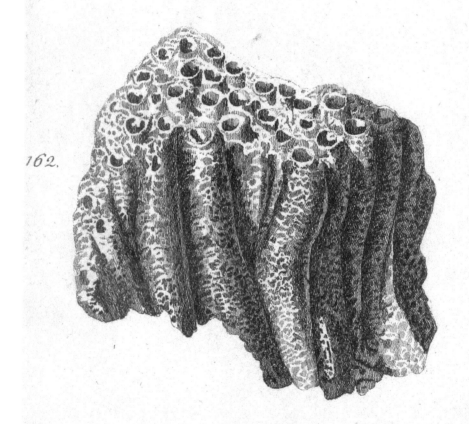

162.

163.

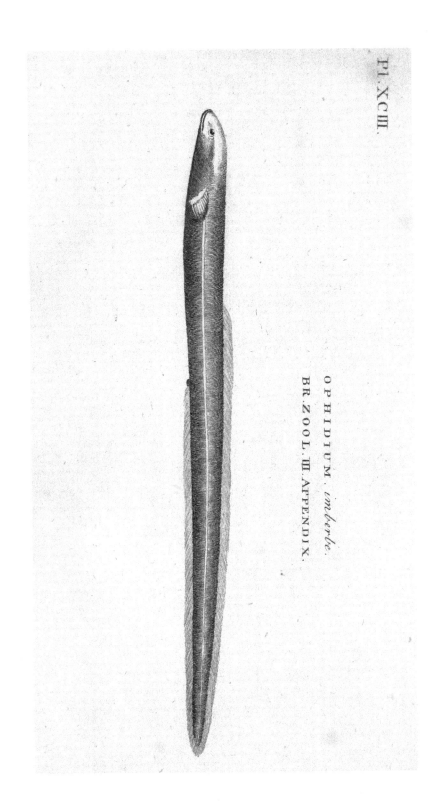

Pl. XCIII.

OPHIDIUM imberbe.
BR. ZOOL. III. APPENDIX.